Sodium-Ion Batteries
Materials and Applications

Edited by

Inamuddin[1,2,3], Rajender Boddula[4], Mohd Imran Ahamed[5] and Abdullah M. Asiri[1,2]

[1]Chemistry Department, Faculty of Science, King Abdulaziz University, Jeddah 21589, Saudi Arabia

[2]Centre of Excellence for Advanced Materials Research, King Abdulaziz University, Jeddah 21589, Saudi Arabia

[3]Department of Applied Chemistry, Faculty of Engineering and Technology, Aligarh Muslim University, Aligarh-202 002, India

[4]CAS Key Laboratory of Nanosystem and Hierarchical Fabrication, National Center for Nanoscience and Technology, Beijing 100190, PR China

[5]Department of Chemistry, Faculty of Science, Aligarh Muslim University, Aligarh-202 002, India

Published by **Materials Research Forum LLC**
Millersville, PA 17551, USA

Published as part of the book series
Materials Research Foundations
Volume 76 (2020)
ISSN 2471-8890 (Print)
ISSN 2471-8904 (Online)

Print ISBN 978-1-64490-082-6
eBook ISBN 978-1-64490-083-3

Distributed worldwide by

Materials Research Forum LLC
105 Springdale Lane
Millersville, PA 17551
USA
https://www.mrforum.com

Manufactured in the United States of America
10 9 8 7 6 5 4 3 2 1

Table of Contents

Preface

Currently, electrochemical energy storage systems play a huge role in human society and are used in all kinds of electronic devices, hybrid electric vehicles, and grid-scale systems, etc. Batteries are one of the most important renewable energy storage systems for real-world applications. In recent years, sodium-ion batteries have stimulated great scientific interest in next-generation power sources due to their safety, and sodium unlimited abundance compared to limited lithium sources. Designing and fabrication of energy storage materials (electrodes and electrolytes) for high-performance sodium-ion battery applications requires meeting sustainable world necessities. Therefore, the values added applications of sodium-ion batteries had drawn attention of research and development specialist of various disciplines including, engineers, chemists, material scientists, and mechanical engineers. The research in the area of sodium-ion batteries has been in progress towards the development of practically viable technologies. The sodium-ion battery-based devices have an incredible future, but still more research and development studies are needed to commercialize them at a large scale.

This book explores the sodium-ion batteries which can replace the lithium-ion battery market. It provides an in-depth overview of the history, theory, and experimental evidence about consumer and industrial applications of sodium-ion batteries. It also focuses on the principles, mechanisms, fundamentals, nanostructured materials, and electrolytes used in sodium-ion batteries. Some of the important materials such as silicon, carbon, conducting polymers, alloys and metals used in the sodium ion batteries are discussed in details. This book is a unique reference guide. This book is well structured essential resource for undergraduate and postgraduate students, faculty, research and development professionals, energy chemists, and industrial experts. It brought together panels of highly-accomplished experts in the field of energy storage technologies and encompasses basic studies and addresses topics of novel issues and challenges with aspects related to sodium-ion battery applications.

Inamuddin[1,2,3], Rajender Boddula[4], Mohd Imran Ahamed[5] and Abdullah M. Asiri[1,2]

[1]Chemistry Department, Faculty of Science, King Abdulaziz University, Jeddah 21589, Saudi Arabia

[2]Centre of Excellence for Advanced Materials Research, King Abdulaziz University, Jeddah 21589, Saudi Arabia

[3]Department of Applied Chemistry, Faculty of Engineering and Technology, Aligarh Muslim University, Aligarh-202 002, India

[4]CAS Key Laboratory of Nanosystem and Hierarchical Fabrication, National Center for Nanoscience and Technology, Beijing 100190, PR China

[5]Department of Chemistry, Faculty of Science, Aligarh Muslim University, Aligarh-202 002, India

Sodium-Ion Batteries: Materials and Applications   Materials Research Forum LLC
Materials Research Foundations 76 (2020) 1-30   https://doi.org/10.21741/9781644900833-1

Chapter 1

NASICON Electrodes for Sodium-Ion Batteries

Rekha Sharma[1], Sapna[1], Kritika S. Sharma[2] and Dinesh Kumar[2*]

[1]Department of Chemistry, Banasthali Vidyapith, Rajasthan 304022, India

[2]School of Chemical Sciences, Central University of Gujarat, Gandhinagar 382030, India

*dinesh.kumar@cug.ac.in

Abstract

The incessantly growing demand for energy storage is attracting the researcher's attention to develop consistent, effective, and ecologically harmless electrochemical systems for energy storage. Sodium-ion batteries (SIBs) are emergent as one of the utmost efficient large-scale systems for energy storage due to the great accessibility of raw sodium resources and their cost-effective assets. Sodium Super Ionic Conductor (NASICON) electrodes-based materials have offered an opportunity precisely for SIBs as a next-generation energy storage device because of their unique properties. NASICON electrodes have been extensively proven to demonstrate enhanced and miscellaneous features for SIBs in terms of high rate competence, flexible battery structures, long cycling life, and high specific capacity, due to their outstanding characteristics such as high charge carrier mobility, excellent mechanical strength, high theoretical ability, high electronic conductivity, and large surface area. This chapter summarizes the important advancement accomplished on NASICON-based electrodes for application in SIBs, including both cathodes and anodes over the past decade. Furthermore, this chapter also focuses on the new challenges and commercial demand for SIBs and some perspectives on the use of NASICON-based electrodes for future SIB applications.

Keywords

NASICON, Electrodes, Energy Storage, SIBs, Cathodes, Anodes

Contents

1.   Introduction

In the fields of information, actuation, and energy conversion, lithium-ion batteries (LIBs) have been initiating extensively spread significance in the last two decades. They demonstrated enormous valuable for various needs, for example, robots, electric vehicles, and portable smart devices because of benefits in terms of volume density (250–650 Wh L^{-1}) and energy capacity (100–300 Wh kg^{-1}) [1-3]. Nowadays, due to the low abundance of lithium, LIBs face severe bottle-necks at a relatively high cost and limited energy densities. Thus, they are entering in the meadows of significant energy storage devices, especially for grid storage [4-6]. Sodium-ion batteries (SIBs) offer numerous applications compared to LIBs, for instance, consistently topographical dispersal, markedly low cost and plenty of elemental sources, henceforth recognized as a grave alternate to LIBs [7-9]. The demand for SIBs in terms of kinetic as well as thermodynamic criteria is certainly predictable to designate competitive with LIB systems [10]. Sometimes, the electrochemical performance of Na-ion batteries is superior to LIBs [11].

On a foremost peek, in comparison to Li, solemn drawbacks predicted due to the different location in the periodic table of the Na. The Na^+ ions desertion and insertion hooked on a large number of host materials are not as good as intended for their Li-analogues, i.e., V_2O_5, $NaFePO_4$, $Na_7V_3(P_2O_7)_4$ and sodium oxides ($NaMO_2$, M = Cu, V, Fe, Ni, Co, and Mn) [12-19]. However, despite these drawbacks, the "extrinsic" benefits of SIBs create them efficient for the use of energy storage devices. At another peek, the seeming "intrinsic" drawbacks must be appraised further: Polarizability of the sodium ion as well as the electro-positivity is higher. As a significance, the Li is used for most systems although the cell potential is not unavoidably higher of their respective Li-analogues. More prominently, the two key advantages of increased ionic radius are: (1) the transference resistance produced through transporting the ions into the electrode from the liquid electrolyte intended for exceedingly polarizing small Li-ion, is characteristically lesser due to the less severe solvation in liquid electrolytes. More prominently, the higher ionic radius is not attainable with Li, which allows solid structures with massive ion conductivities. β-alumina's is a characteristic example which demonstrates that the sodium ion is of flawless polarizability and size. Likewise, sodium ion surpasses Li-ion conductivities having very high conductivities and in the Na-super ionic electrode in the NASICON family. Previously, various metal oxide materials have been synthesized by using chemical precipitation, hydrothermal, and sol-gel method which has been used for the detoxification of water [20-28], but these nanomaterials were not utilized for the conduction of electricity. Hence, some other metal oxides have been synthesized and applied, such as NASICON based materials which were used as conductor materials. As the tools of nanotechnology, because of the limited electronic conductivity, the electrode materials based on NASICON display inadequate chemical diffusion and must be practical to advance performance despite the so-induced varied conductivity [29-31]. The five main categories for present SIBs anodes are: (1) nonmetallic elements (S and P); [32,33] (2) metal oxides/selenides/phosphides/sulphides (SnS_2, $CoSe_2$, Sb_2S_3, WS_2, Sb_2O_4, MoS_2, TiO_2, SnO_2, NiP_3, and $CoSe$); [34-43] (3) NASICON based materials ($NaTi_2(PO_4)_3$); [44,45] (4) carbon-based materials (nitrogen-doped carbon nanofibers, porous, hard carbon); [46,47] and last but not least (5) alloys and metals (Ge, Sb, Sn-Ge, Na, Sn, and Sn-Ge-Sb); [48-53].

2. Machinery of SIBs

Characteristic SIBs are composed of four components which utilize as LIBs based on the rocking chair principle, [54] and i.e., non-aqueous salt containing electrolyte, a negative electrode, a positive electrode, and separator. Comprising $NaClO_4$ (1M) ethylene carbonate/propylene carbonate as electrolyte used for the formation of the typical SIB.

The external circuit passes the electrons flow from the positive electrode towards the negative electrode throughout the charging process, and the Na^+ ions leave characteristic cathode for example layered $NaCoO_2$ or $Na_3V_2(PO_4)_3$ and are fused in the anode while the process is reversed throughout discharging. For the reduction of the overall battery cost up to \$70 per 100 m, battery grade, the aluminum foil can be utilized commercially for cathode and anode as a current collector [55]. In this chapter, we emphasize on NASICON electrode materials having physical steadiness and high ionic conductivity as the major advantages have previously been stated. The list ought to be completed by thermal stability and chemical flexibility. Providing pronounced Na^+ diffusion channels, the NASICON framework construction is built by polyhedral XO_4 groups and corner-shared MO_6 ($M'O_6$) [56]. The cathodes materials in SIBs are made of various materials such as $Na_3V_2(PO_4)_2F_3$, $Na_2FeTi(PO_4)_3$, $Na_{1.5}VPO_4F_{0.5}$, $Fe_2(MoO_4)_3$, and $Na_3V_2(PO_4)_3$ [57-59]. Besides these materials, some other NASICON materials, for example, sodium salt of zirconium phosphate ($NaZr_2(PO_4)_3$), Manganese-titanium mixed salts ($Na_3MnTi(PO_4)_3$), sodium salt of titanium phosphate ($NaTi_2(PO_4)_3$), and sodium salt of vanadium phosphate ($NaV_2(PO_4)_3$) having low redox potential is pondered by means of anode material for SIBs and can be used as Ti^{3+}/Ti^{4+} conversion based $NaTi_2(PO_4)_3$ materials displays a reversible capacity of 120 mAh g^{-1} and a discharge voltage at 2.1 V [44]. The redox conversion of V^{3+}/V^{4+} in $Na_3V_2(PO_4)_3$ cathode material with high capacity (117 mAh g^{-1}) demonstrates a discharge current of 3.4 V, whereas the V^{2+}/V^{3+} redox conversion, chief as an anode material having a capacity (50 mAh g^{-1}) offers a lower energy of 1.6 V [60,61].

2.1   Storing the progression of NASICON materials

Primarily in the 1980s and 1990s, the ionic conductivity for NASICON solid electrolytes has been structurally analyzed and methodically measured. Goodenough, Kefalas, and Hong [62,63] worked on the material science of NASICON materials. Kreuer et al. [64] provided a complete description upon mechanisms of ion conductivity as well as kinetic stability, thermodynamic stability, and phase thermodynamics regarding sodium of the solid electrolyte. The absence of phosphate required for the thermodynamic stability against Na, which previously principals to a considerable subordinate conductivity [65]. If passivation layers are fashioned and lower temperature is pondered, this demonstrates the use of NASICON's as solid electrolyte. However, in the NASICON electrode for the ion transport formulated the origin, the ion transport subjects elucidated in the electrolyte literature also [66,67]. The conversation of storage kinetics is not exclusively transported controlled and non-trivial. This is because of the super-position of diffusion and migration and the heterogeneity of the situation as well. The storing progression is a chemical diffusion problem when given sufficiently fast access of electrons (current collecting phase) and ions (liquid electrolyte) to the electroactive particles, wherever Equation 1 demonstrates the chemical diffusion coefficient [68]

$$D^\delta \propto \frac{\sigma_{ion}*\sigma_{eon}}{\sigma}\left(\frac{\chi_{eon}}{C_{eon}}+\frac{\chi_{ion}}{C_{ion}}\right) \tag{1}$$

The concentration of the ionic and electronic transporters given as C_{ion} and C_{eon}. The χ-factors are typically critical at the process temperature and define internal trapping reactions. If we assume $\sigma_{ion} \gg \sigma_{eon}$ and $C_{ion} \gg C_{eon}$ for our NASICON materials.

$$D^\delta \propto \chi_{eon} D_{eon} \tag{2}$$

The D^δ is strong-minded through the trapping reactions (χ factors) and electronic mobility (proportional to D_{eon}). The conduction of ions and electrons to the particles is next to the share of the transport kinetics. The conductivity of ions of NASICON materials assistances in ion filtration due to the permeable liquid electrolytes and morphologies, which is basically the transport of electrons that is essential.

Table 1 summarizes the distinctive chemical dispersal constants values of NASICON materials. Here, following abbreviation are utilized i.e. propylene carbonate (PC), dimethylsulfoxide (DMSO); fluoroethylene carbonate (FEC); bisphenol A ethoxylate dimethacrylate (BEMA); ethylene carbonate (EC); poly(ethylene glycol) methyl ether methacrylate (PEGMA); Room temperature (RT); Effective diffusion coefficient (EDC); diethyl carbonate (DEC); dimethyl carbonate (DMC).

Regrettably, literature values are essential effective values and affected by phase transformations, interfaces, or issues of size are frequently not measured bulk values consistently. The D^δ values have been dependable at 150 °C to 1.6×10^{-9} cm^2 S^{-1} on single crystalline LiFePO$_4$ [86]. Finally, it was concluded that NASICON based materials offer above standards of most other electrode phases around 10^{-10} cm^2 S^{-1}. The NASICON materials are poor towards electric conductivity because of electronic wiring by the accumulation of current collecting phases, for example, by deliberate carbon coating, which is the example of carbon and additional conductive coating. In the electroactive particles, the transport kinetics is apprehension because of the relaxation time occurred due to two adjusting screws, the typical solutions are shown in Figure 1a, b [68].

$$\tau^\delta \propto L^2/D^\delta \tag{3}$$

Where D is the chemical diffusion coefficient, τ is a biochemical dispersion duration, and L is the length of dispersion. The D^{δ} cannot be significantly stimulated.

Doping utmost assistances through increasing its influence on χ factor and D_{eon} by the increase in σ_{ion} or σ_{eon}, in disrupting, the particles transference is shown in Equation (2) [87]. The particle size is the utmost operative parameter which arrives quadratically. Therefore, Figure 1c shows that nano-structuring is a delicate tool. For a one µm particle size a value leads to a minute storage time at 10^{-10} cm^2 S^{-1}.

Table 1. The distinctive values of chemical diffusion coefficients.

Materials	Test environment	EDC (cm^2 S^{-1})	Ref.
FePO$_4$; LiFePO$_4$	LiPF$_6$ (1 M) at 20 °C in DMC/EC	1.8 X 10^{-14}; 2.2 X 10^{-16}	[69]
Li$_3$V$_2$(PO$_4$)$_3$	LiPF$_6$ (1 M) at 20 °C in DMC/EC (1:1, w/w)	$\approx 10^{-11} - 10^{-9}$	[70]
P$_2$-Na$_{2/3}$[Ni$_{1/3}$Mn$_{2/3}$]O$_2$	LiPF$_6$ in DEC/EC; NaPF$_6$ in DEC/EC at 25 °C	1 X $10^{-9} \approx 1$ X 10^{-10} (Na); 3 X $10^{-9} \approx 2$ X 10^{-11} (Li)	[71]
NaSn$_2$(PO$_4$)$_3$ (NASICON)	25 °C	6.03 X 10^{-12}	[72]
Na$_3$V$_2$(PO$_4$)$_3$@C-B-doped (NASICON)	NaClO$_4$ (0.8 M) at RT in (1:1; v/v) of DEC/EC	4.0 X $10^{-14} \approx 2.48$ X 10^{-13}	[73]
NaTi$_2$(PO$_4$)$_3$	NaCF$_3$SO$_3$ (1 M) at RT in DMSO	1 X $10^{-9} \approx 10^{-12}$	[74]
LiNi$_{0.5}$Mn$_{0.5}$O$_2$; NaNi$_{0.5}$Mn$_{0.5}$O$_2$	LiPF$_6$ (0.5 M) in PC/FEC (98:2; w/w); NaPF$_6$ (0.5 M) at 25 °C in PC with FEC (2 wt.%)	7.7 X 10^{-12}; 6 X 10^{-13}	[75]
Na$_{3.1}$Zr$_{1.95}$Mg$_{0.05}$Si$_2$PO$_{12}$	RT; (Na$_{3.1}$Zr$_{1.95}$Mg$_{0.05}$Si$_2$PO$_{12}$)	5.24 X 10^{-8}	[76]
Carbon fiber	LiPF$_6$ (1 M) at 20 °C in DMC/EC	$\approx 10^{-10}$	[77]
NaTi$_2$(PO$_4$)$_3$ (NASICON)	NaClO$_4$ (1 M) at RT in DEC/EC (1:1; v/v) with FEC (3 wt.%)	2.0 X 10^{-13}	[78]
Na$_{0.44}$MnO$_2$	NaClO$_4$ at 25 °C in (1:2; v/v) DMC/EC	5.75 X $10^{-16} \approx 2.14$ X 10^{-14}	[79]
Graphene (Bilayer)	2 $\approx$ 4% 2-hydroxy-2-methylpropiophenone and LiTFSI at RT in BEMA:PEGMA (7:3)	$10^{-6} \approx 7$ X 10^{-5}	[80]
P$_2$-Na$_{0.62}$Ti$_{0.37}$Cr$_{0.63}$O$_2$	NaClO$_4$ in PC with FEC (2 %) at 25 °C	1 X $10^{-12} \approx 2$ X 10^{-13}	[81]
Graphite	LiPF$_6$ (1 M) at 20 °C in DMC/EC	1.12 X10^{-10}; 1.35 X 10^{-10} (55 °C)	[82]
NaTi$_2$(PO$_4$)$_3$@C (NASICON)	NaClO$_4$ at RT in PC/EC	1.79 ≈ 2.32 X 10^{-10}	[83]
Na$_3$V$_2$(PO$_4$)$_3$-C (NASICON)	25 °C; NaPF$_6$ in PC	2 X $10^{-15} \approx 6$ X 10^{-13}	[84]
Li$_{0.9}$FePO$_4$; Na$_{0.9}$FePO$_4$	LiPF$_6$ in DEC/EC; NaClO$_4$ in DMC/EC at 25 °C	8.63 X 10^{-17}; 6.77 X 10^{-16}	[85]

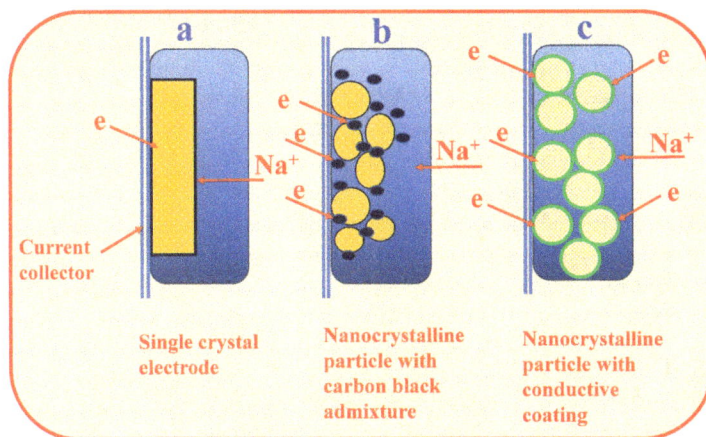

Figure 1. Graphic representation of several electrode morphologies: (a) Insufficient storage kinetics of single crystal electrode (b) electronic transport via nanocrystalline particles with carbon black admixtures permitting for electrolyte access (c) Coated nanoparticles for optimized storage by conductive medium.

2.2 Cathode materials based on NASICON type

2.2.1 NASICON-type nanoparticles of $Fe_2(MoO_4)_3$ wrapped with graphene

Sheng et al. [88] reported nanoparticles of $Fe_2(MoO_4)_3$ (NASICON-type) wrapped through graphene for SIBs as an ultra-high rate cathode. To synthesize SIBs, nanoparticles of $Fe_2(MoO_4)_3$ based on Naþ superionic conductor (NASICON) type is an utmost cathode material with a flat discharge plateau and capacious ion diffusion tunnels. Though, because of the long Naþ diffusion path and poor electron conductivity, the lethargic electrochemical kinetics bounds its additional advancement. The $Fe_2(MoO_4)_3$ nanoparticle composite wrapped with graphene was synthesized by the microemulsion method followed by annealing. After 100 cycles at 10 °C, the $Fe_2(MoO_4)_3$ nanoparticle wrapped with graphene displays virtuous high-rate cycling stability up to 76% capacity retention and an ultra-high rate capability up to 64.1 mAh g^{-1} at 100 °C. The improved electrochemical concerts are ascribed to the inimitable composite structure having high electron conductivity and reduced ion diffusion distance. It was concluded that the nanoparticle of $Fe_2(MoO_4)_3$ wrapped with graphene consumes the excessive latent meant for higher degree SIBs. The $Fe_2(MoO_4)_3$ nanoparticle composite wrapped with graphene displays a significant capacity of 98.4 mAh g^{-1} as a cathode material for SIBs with a flat voltage plateau [88].

2.2.2 NASICON-type materials based on $Na_3V_2(PO_4)_3$

Novikova et al. [89] synthesized NASICON-type materials having the compositions $Na_{3+x}V_{2-x}Ni_x(PO_4)_3$, $Na_3V_{2-x}Cr_x(PO_4)_3$, $Na_3V_{2-x}Fe_x(PO_4)_3$, and $Na_3V_{2-x}Al_x(PO_4)_3$ where x = 0.05, 0.03, 0.1, and 0. The outcomes determine that the doped material (5% Fe: $Na_3V_{1.9}Fe_{0.1}(PO_4)_3$) showed maximum electrical conductivity amongst the samples studied, having ~33 kJ/mol of activation energy at high-temperature conduction. Besides, for low-temperature conduction, the activation energy is decreased from 84 ± 2 to 54 ± 1 kJ/mol. At the elevated charge rate, the more porous material retains its high level of the theoretical capacity [89].

Zheng et al. [90] also described $Na_3V_2(PO_4)_3$ (NVP) type NASICON-based electrode material for SIBs. The remarkably high ion conductivity and distinct structural stability of the synthesized sodium salt of vanadium phosphate $\{(Na_3V_2(PO_4)_3$ (NVP)$\}$ interpreted it as an utmost auspicious conductor used for sodium storage. Therefore, via coating the vigorous materials thru downsizing the NVP particles, assimilation the NVP particle by ion doping strategy and many carbon materials, and with a conductive carbon layer, various actions have been adapted to upsurge the intrinsic electrical conductivity and surface of NVP. Additionally, to gain an improved consideration in the temperature range of −30 °C~225 °C on the sodium storage in NVP, four separate crystal structures, specifically β-NVP, γ-NVP, β′-NVP, and α-NVP. Furthermore, the authors gave an outline of current methods to augment the intrinsic electrical conductivity and surface electrical conductivity of NVP [90].

2.2.3 NASICON-type materials based on $Na_3V_2(PO_4)_2F_3$ and $Na_3V_2(PO_4)_3$

For SIBs, the NASICON materials, i.e., $Na_3V_2(PO_4)_2F_3$ and $Na_3V_2(PO_4)_3$ have resumed designate for electrochemical examination as proficient material for cathode by Song et al. [91]. The $Na_3V_2(PO_4)_2F_3$ has been synthesized by the insertion of fluorine through substituting limited PO_4^{3-} groups in $Na_3V_2(PO_4)_3$. Significantly, to change the configuration of the ion including the quantity of varied Na sites, ions occupation, and for ions extraction, the species leads to different modifications, fluorine is proficient at partaking in structural development. Because of its strong ionicity, the $(PO_4)_3$ polyanion's inductive effects can be enhanced by changing electronic cloud density underneath the consequences of molded F-V bond upon compositional atoms. These effects could restrain the redox pair energetics of transition metal ions from generating moderately high operating potentials. Predominantly, to upsurge achieved voltages through varying atomic surroundings, the replacement of fluorine intended for negative-charge anion or polyanion can be operative to advance the electrochemical possessions. Additionally, one clear couple of redox reaction peaks are accessible in CV (cyclic voltammetry) curves of $Na_3V_2(PO_4)_3$ while in CV curves of $Na_3V_2(PO_4)_2F_3$ two couples have come out [91].

2.2.4 NASICON-type materials of porous $Na_3V_2(PO_4)_3$ and $NaTi_2(PO_4)_3$

Zhang et al. [92] reported that due to their viable price benefits, SIBs grasp the high potential for grid-level energy storage devices. However, to empower high power and extended lifetime, contests persist in the advancement of appropriate conductor composites. Zhang et al. [92] synthesized the porous sodium salt of titanium phosphate $(NaTi_2(PO_4)_3)$ and the sodium salt of vanadium phosphate $(Na_3V_2(PO_4)_3)$ compounds intended to high rate cycling stability in cooperation with their electrochemical concerts. Owing to the upsurge thermal stability of the PO_4^{3-} based anode and cathode materials, the $Na_3V_2(PO_4)_3//NaTi_2(PO_4)_3$ cells display simple production and wellbeing with 94 % of capacity retention over 5000 cycles (ultralong cycle life), 80% efficiency of energy, and 85 mAh g^{-1} of power at 2.4 A g^{-1} based on cathode.

The $NaTi_2(PO_4)_3$ (NTP) and $Na_3V_2(PO_4)_3$ (NVP) composites have been produced by a scalable sol-gel method having a porous structure. Furthermore, these cells showed good concert retaining aluminum consistent with the current collector for the anode, though some enhancements must be required to attain improved power capability and energy efficiency of these kinds of current collector material [92].

2.2.5 A negative electrode of $Mg_{0.5}Ti_2(PO_4)_3$ based NASICON materials

The particles of $Mg_{0.5}Ti_2(PO_4)_3$ was synthesized by sol-gel synthesis process, which is a polyanion material coated with carbon and utilize for SIBs as the negative electrode materials. The material showed a precise capacity in the voltage gap of Na^+/Na^0 vs 0.01-3.0 V around 268.6 mAh g^{-1}. The material showed advanced rate capability in the NASICON based materials, at a current density of 5 A g^{-1} having a specific capacity of 94.4 mAh g^{-1} because of the fast diffusion of Na^+. Additionally, after 300 cycles, the retention capacity was achieved at 99.1 %, signifying outstanding cycle stability.

On comparing with LIBs, $Mg_{0.5}Ti_2(PO_4)_3$ showed 629.2 mAh g^{-1} of capacity ascribed to formation/decomposition of the SEI film in the voltage range of 0.01-3.0 V vs Li^+/Li^0 and the interfacial Li^+ storage. After the first release, the pristine particles of $Mg_{0.5}Ti_2(PO_4)_3$ became encapsulated into Li_3PO_4 in addition to solid electrolyte interphases (SEI) matrix and decayed to metallic Mg and Ti nanocrystals. The surface area of the material closely related to the capability of interfacial Li^+ storage, the specific capacity increases as an increase in the surface area. The sub-micron sized $Mg_{0.5}Ti_2$ showed a 600 mAh g^{-1} capacity of Li^+ $(PO_4)_3$ electrode, which was comparable to the conversion reaction, for instance, NiO [93] and MoS_2 [94] and was considerably higher than other materials based

Materials Research Forum LLC
https://doi.org/10.21741/9781644900833-1

on the intercalation reaction such as TiO_2 [95] and graphite [96]. Through complementary the mass ratio of the negative and positive conductors and regulating the working voltage range, the full battery concert could be improved. To conclude, the $Na_{3.5}Mg_{0.5}Ti_2(PO_4)_3$ electrochemical disintegration was not perceived of the SIB cell because of the large electrode polarization. The manufacturing of composite materials with carbon additives and synthesis of nano $Mg_{0.5}Ti_2(PO_4)_3$ are some strategies utilized for the disintegration of $Na_{3.5}Mg_{0.5}Ti_2(PO_4)_3$. In these types of strategies, a large Na^+ capacity could be obtained thru taking benefit of the interfacial Na^+ storage mechanism [97].

2.2.6 Numerous other NASICON cathode materials

In addition to $Na_3V_2(PO_4)_2F_3$ (NVPF) and NVP there are various extra fascinating materials like $Fe_2(MoO_4)_3$, $Na_3Fe_2(PO_4)_3$, $Na_2TiFe(PO_4)_3$, $Na_{1.5}VOPO_4F_{0.5}$, $NaVPO_4F$, $NaNbFe(PO_4)_3$, and $Na_3(VO_{1-x}PO_4)_2F_{1+2x}$ also utilized as NASICON materials for SIBs [57,59,98-105]. Inadequately, low sodium lodging capability showed by $Na_3Fe_2(PO_4)_3$, which distributed only 45 mAh g^{-1} of the lodging capability [106]. Similar problems are faced by phosphates of a transition metal, for example, Ni, Mn, Cu etc. [58]. The compound $Fe_2(MoO_4)_3$ which is of comprising of tetrahedral MoO_4 and octahedral FeO_6 consistent over corner distributing oxygen atoms is a monoclinic (P21/c) crystal structure considered as more electro-active NASICON compounds. Because of structural and good thermal stability, this compound proved as a capable cathode material for both Li as well as Na ions [59,107-111]. An open tunnel is designed by enabling a robust structure having three MoO_4 tetrahedra and two FeO_6 octahedra connected [103]. Two discharge plateaus have been obtained on addition of two Na-ions at 2.55 V and 2.62 V as a typical charge/discharge. Fu et al. [59] assumed the reduction of particle size to advance the electrochemical concert of $Fe_2(MoO_4)_3$. Furthermore, to augment the electrochemical concerts coating of graphene was also demonstrated [108,112]. Moreover, some other NASICON compounds, for example, $Na_2TiM(PO_4)_3$ are similarly Na-active composites where, M = Cr, Fe having dual transition metals consistent to Cr^{2+}/Cr^{3+}, Fe^{2+}/Fe^{3+}, and Ti^{3+}/Ti^{4+} redox processes. The redox procedures for M = Cr, Fe, two desodiation steps are comprised, the first related to Cr^{2+}/Cr^{3+} and Fe^{2+}/Fe^{3+} value variations, and the second corresponds to Ti^{3+}/Ti^{4+}. These NASICON materials need further optimization to attain higher ionic conductivity and virtuous thermal steadiness to accomplish improved electrochemical performances. For the deliberated NASICON cathode materials, all appropriate figures and parameters of advantages are comprised in Table 2.

Table 2. Appropriate figures and parameters of advantages of NASICON cathode materials.

Materials	Capacities (mAhg^{-1})	Morphology	Size (nm)	Capacity retention (%)	Cycle number	Ref.
$Fe_2(MoO_4)_3$	94, at 1 C	Thin film	200	80.8	100	[59]
$Na_3V_2(PO_4)_3$/CS	94, at 1C	Nanowires (core-shell)	20–50	74	50	[113]
$Na_3V_2(PO_4)_3$@C@rGO	115, at 1 C; 86, at 100 C	3D porous composites	50	64	10,000	[114]
$Na_3V_2(PO_4)_3$@C	104.3, at 0.5 C	Nanoparticles	40	99.6	50	[115]
$Na_3V_2(PO_4)_3$@C-$B_{0.38}$	87.1, at 0.2 C	Nanoparticles	100–800	81	50	[116]
$Na_3V_2(PO_4)_3$-F/C	103, at 0.1 C	Nanofibers	10	99.3	50	[117]
$Na_3V_2(PO_4)_3$@CNT-G	109, at 30 C; 82, at 100 C	3D foam	50–100	96	2000	[60]
$Na_3V_2(PO_4)_3$@AC	100.6, at 5 C	Nanoparticles	5	96.4	200	[118]
$Na_3V_2(PO_4)_2F_3$@C	126, at 0.1 C	Nanorods	200	92	50	[119]
$Na_3V_2(PO_4)_3$@C/G	86.5, at 40 C	3D porous composites	100	80	1500	[120]
$Na_3V_2(PO_4)_3$@C–N-CNT	93.8, at 0.2 C	NPs with CNT	100	92.2	400	[121]
C@$Na_3V_2(PO_4)_3$@pC	103, at 10 C	Double-shell nanospheres	20–40	80.5	1000	[122]
$Na_3V_2(PO_4)_2F_3$@PC	111, at 1.82 C	Porous nanoparticles	<1000	98.2	90	[123]
$Na_3V_2(PO_4)_3$@C	110, at 10 C	3D nanofibers	20–80	95.9	1000	[124]
$Na_3V_2(PO_4)_3$@C-N_{142}	101.9, at 0.2 C	Porous particles	200	95.5	50	[125]
$Na_3V_2(PO_4)_2F_3$/PC	111.5, at 0.091 C	Nanoparticles	<500	93	100	[126]
$Na_3V_2(PO_4)_3$@C@CMK-3	115, at 1 C; 109, at 5 C	3D CMK-3	3	68	2000	[127]
$Na_3V_2(PO_4)_3$@MCNT	146.5, at 0.1 C	MCNT NCs	100	94.3	50	[128]
$Na_3V_2(PO_4)_2F_3$@AC	116, at ≈0.1 C	Nanoparticles	40–200	97.6	50	[129]

2.3 Anode materials based on NASICON-type

The usage of sodium metal is challenging due to safety motives to the extent that SIBs anodic materials are worried. Throughout the Na charge/discharge process, dendrites could be formed due to higher chemical reactivity than lithium metal. Amongst four kinds of storage mechanisms, one has to differentiate for an electrode, i.e. interfacial storage and conversion reaction. This was comprehended in relevant materials, for example, Na-alloys, single phase absorption, ionic in addition to covalently bonded composites, phase-transition, which is analogous to LIB [130-136].

In comparison to Li-based batteries, multi-phase reactions are more probable due to the typically subordinate solubility of Na. NASICON type materials, for example, sodium salt of vanadium phosphate ($Na_3V_2(PO_4)_3$), sodium salt of zirconium phosphate {($NaZr_2(PO_4)_3$) (NZP)}, and sodium salt of titanium phosphate {($NaTi_2(PO_4)_3$) (NTP)} distinct towards transitional metal sulfides/oxide including conversion reactions with high rate capabilities, are deliberated as utmost capable anodes for SIB.

Additionally, as anode materials for SIBs, the rGO-CNT modified sodium salt of vanadium phosphate ($Na_3V_2(PO_4)_3$) compounds have been applied by Zhu et al. [137] having capacity retention ratio of 77% at 10 °C. Here, mainly the property, preparation, and concert of NZP, NTP has been discussed and considered as negative electrodes in NASICON type materials.

2.3.1 $NaTi_2(PO_4)_3$ (NTP) type anode materials

NTP is a triclinic NASICON variant (P1) which is frequently used SIB anode material [137-139]. NTP shaped on Ti^{3+}/Ti^{2+}, and Ti^{4+}/Ti^{3+} valence variations divulge two discharge plateaus between 0 V and 3 V at 2.1 V and 0.4 V. The strategy above using conductive agents is suitable as NTP, NVP particles labor under inadequate electrical conductivity, for example, carbon shell, graphene, and CMK-3. Similarly, particle size reduction is true. The 3D graphene networks entrenching NTP nanoparticles (NTP@GN) has been synthesized using a hydrothermal method. Also, allowing fast charge transport and transfer, the porous NTP nanoparticles having particle size around ≈100 nm is uniformly dispersed inside the graphene nanonetwork, which is preserved throughout cycling. After 200 cycles, the NTP@GN composites delivered a high capacity and high retention ratio (93%) at 109 mAh g^{-1} of capacity. Additionally, a homogeneously spread CMK-3 (≈5 nm) matrix in CMK-3 channels has been utilized by Zhang et al. [140]. At 0.2 C, about 100 mAh g^{-1} of the attained NTP@CMK-3 composites distributed good cycling concert with an alterable capacity between 1.5 V and 3 V. The carbon-coated NVP (NTP@C) composite has been synthesized where titanium isopropoxide is used as titanium source and citric acid as a carbon source [83]. The NTP nanoparticles have virtuous crystallinity having two charge/discharge plateaus and a 7 nm carbon layer exhibited by the ensuing NTP@C composites.

2.3.2 $NaZr_2(PO_4)_3$ (NZP) type anode materials

The sodium salt of zirconium phosphate ($NaZr_2(PO_4)_3$) has been produced and electrochemically examined as a fascinating NASICON-type anode material by Jiao et al.

[141]. The octahedral ZrO_6 and tetrahedral PO_4 structures are composed to synthesize $NaZr_2(PO_4)_3$ in which ZrO_6 behave as the host for Na^+ ions. However, over PO_4-ZrO_6 polyhedral tunnels, a rapid gesture of the Na^+-ions permitted by the open structure. The XRD patterns were utilized to confirm the disparity of the Na content. Throughout charge/discharge processes, the diverse stage conditions of mono and trisodium salt of zirconium phosphate {$(NaZr_2(PO_4)_3$ and $Na_3Zr_2(PO_4)_3)$} has been signified. The two lattice spacings of 6.33 Å and 4.40 Å demonstrated using HRTEM, corresponded to the hkl values of (012) and (110) planes. Through the charge/discharge process of NZP the $Na_{1-x}Zr_2(PO_4)_3$ converted into $Na_{3-y}Zr_2(PO_4)_3$ using redox reactions where ($0 < x < 1, 0 < y < 1$). Due to stable coulombic efficiency [141], with steady cycling and rapid kinetics concert, the NZP compounds considered as anode materials for SIB.

2.3.3 Numerous other NASICON anode materials

For activation in full batteries, amorphous Sn, Sb, P, or carbon needed pre-sodiation because these are non-Na materials [142]. This causes pronounced irreversible capacity losses and devours much of sodium to produce steady layers of SEI. In addition to NZP ($NaZr_2(PO_4)_3$) and NTP ($NaTi_2(PO_4)_3$) compounds, some other anode composites, such as, mono- and trisodium salt of vanadium phosphate ($Na_3V_2(PO_4)_3$), mixed salt of manganese and titanium phosphate ($Na_3MnTi(PO_4)_3$), sodium salt of tin phosphate ($NaSn_2(PO_4)_3$), and sodium salt of titanium phosphate ($Na_3Ti_2(PO_4)_3$) should be mentioned consistent with NASICON based materials. For instance, the mono- sodium salt of vanadium phosphate ($NaV_2(PO_4)_3$) materials in the non-aqueous electrolyte as anodes are of more importance [143-146]. Furthermore, Jian et al. [147] demonstrated $Na_3V_2(PO_4)_3$ with deep sodiation processes at 0.3 V and 1.6 V thru synthesis of tetra- and penta- sodium salt of vanadium phosphate {$Na_4V_2(PO_4)_3$ and $Na_5V_2(PO_4)_3$}, respectively. At 11.7 mAh g^{-1} (0.1 C), $Na_3V_2(PO_4)_3$ exhibited a higher reversible measurement of 149 mAh g^{-1} and virtuous rate measurements [147]. The anode materials mentioned above could be practical to accumulate full cells originated on NASICON-type electrodes [148]. The sodium salt of vanadium and titanium phosphate, respectively {$Na_3V_2(PO_4)_3$, $NaTi_2(PO_4)_3$} and are accumulated utilizing Na_2SO_4 (1 M) aqueous electrolyte solution to form a full cell and the voltage-profiles distributing an over-potential of 1.3 V. In aqueous electrolyte, $Na_3MnTi(PO_4)_3$ considered as both material i.e. cathode and anode which is also another characteristic example of NASICON full cell [149]. Deprived of seeming polarization throughout cycles, around 1.4 V, a rescindable competence of 57.9 mAh g^{-1} has been used for characterization of the charge/discharge processes centred at 0.5 C. Various other NASICON compositions have been prepared with Ti, Ni, Mn etc. For the deliberated anode materials, all appropriate figures of virtues and parameters are listed in Table 3. Masquelier's group recently reported NVP cell in which a NASICON solid electrolyte was utilized to detach the cathode and anode [149]. However, for reasonable performance, the elevated temperature was necessary, demonstrates the difficulty of all-solid-state batteries.

Table 3. The most anode and cathode materials for NIBs their morphology, size and capacity retention.

Materials	Size (nm)	Capacity retention (%)	Cycle number	Capacities (mAh g^{-1})	Ref.
$Na_3V_2(PO_4)_3$/C composites	>1000	91	50	146 (at 11.7 mA g^{-1})	[147]
$NaTi_2(PO_4)_3$@CMK-3 composites	≈3	73	200	101	[140]
$NaTi_2(PO_4)_3$@graphene	100–200	95.7	100	104.4 (at 2 C; aqueous)	[45]
$Na_3MnTi(PO_4)_3$/C composites	≈30	98	100 (1 C)	57.9 (at 0.5 C; 29.4 mA g^{-1})	[149]
$NaTi_2(PO_4)_3$@graphene	30–40	95.5	1000 (10 C)	128.6 (at 0.1 C)	[150]
$NaTi_2(PO_4)_3$@graphene nanosheets	100	93	200	109	[44]
$Na_3Zr_2(PO_4)_3$	400–500	88	100	150 (at 20 mA g^{-1})	[141]
$NaTi_2(PO_4)_3$@C composites	≈7	68	10000 (at 20 C)	220 (cycled 0 V to 3 V)	[83]
$NaSn_2(PO_4)_3$/C composites	≈100	87	120	320 (at 50 mA g^{-1})	[72]

2.4 Commercial prospects of NIB technologies

Diverse NIB chemistries are existence established as by LIBs; materials for electrolytes, cathodes, and anodes are all existence examined minutely. About specific materials, this section précises the effort in emerging fuel-cell technologies. Four groups may be characterized for NIBs through the most cathode and anode materials. For cathodes: polyanionic compounds, layered O_3, Prussian blue analogs, and layered P2; For anodes: carbonaceous, phosphoric, alloy, and metal sulfide/oxide comprised in Table 3. The layered oxide cathode cells in comparison to the polyanion systems, displayed advanced reversible capacities. When extra sodium inserted into the system, the pre-sodiated anode cells exhibited higher reversible specific capacities. Since 2014, numerous vital NIB commercial improvements have been displayed thru numerous corporations worldwide. The O_3-type $NaNi_{0.3}Fe_{0.4}Mn_{0.3}O_2$ and hard carbon anode were demonstrated, which was a prototype pouch cell of capacity around 650 mAh g^{-1} [60]. The SHARP labs of America demonstrated a hard carbon battery vs a 3 V Prussian white cathode in 2015. This showed that the rate of competence was restricted via a 30% first-cycle loss and the hard carbon anode [61]. Faradion established a solid carbon anode and a cathode cell based on Ni containing layered oxide {126 watt-hours (Wh kg^{-1})} with 300 cycles. In 2016, a solid carbon anode of sodium nickelate oxide doped with tin demonstrated by SHARP labs of Europe. These materials exhibited uppermost reported volumetric energy density in NIB up to now. The values of these high energy densities with a 3.4 and a 4.2 Ah pouch cell were found up to 211 and 250 Wh L^{-1}.

Conclusions

NASICON-type composite materials have remarkable characteristics of excellent thermal stability, high ionic conductance, and steady structural framework, which can apply in various areas, like electrodes for gas sensors, membrane for lithium-oxygen batteries, Na-ion batteries, solid electrolytes, fuel cells and Li-ion batteries. Previously, to attain elevated performance SIBs, immense development has been completed in the growth of electrode materials based on NASICON-type. This chapter focused on design principles to optimize the functions of electrode materials based on NASICON-type and offered advantages and disadvantages of these materials. Electrochemical properties and synthesis are thoroughly interlinked because morphology is imperative. For efficient and reliable anode and cathode, the NASICON system presents an exceedingly constructive and multipurpose platform. Currently, to defeat all the discussed materials is within the range of vision display precise short-comings but a variety of potential. Owing to lower potential gaps the energy density of NIBs demonstrates no benefits to LIBs, in so far as the elevated standard functioning potentials of electrode materials based on NASICON-type as an anode, which may be a tentative obstruct for useful relevance. Though, using other anode materials, this discomfited situation can be improved with properties of subordinate discharge potential. Because of the comparatively lower energy density and higher ionic conductivity than other energy-oriented materials, the NASICON-based vigorous materials contrast to another sodium/lithium-based cathode are concerned with additional power materials. In various labs and research, the solid batteries made of materials based on NASICON type are under development, these consents highly prevailing Na-electrolytes due to the higher ionic conductance of materials based on NASICON-type.

Acknowledgment

The authors gratefully acknowledge the support from the Ministry of Human Resource Development Department of Higher Education, Government of India under the scheme of Establishment of Centre of Excellence for Training and Research in Frontier Areas of Science and Technology (FAST), for providing the necessary financial support to perform this study vide letter No, F. No. 5–5/201 4–TS.Vll. Dinesh Kumar is also thankful DST, New Delhi for financial support to this work (sanctioned vide project Sanction Order F. No. DST/TM/WTI/WIC/2K17/124(C).

References

[1] W. Sun, Y. Wang, Graphene-based nanocomposite anodes for lithium-ion batteries, Nanoscale 6 (2014) 11528-11552. https://doi.org/10.1039/C4NR02999B

[2] V. Etacheri, R. Marom, R. Elazari, G. Salitra, D. Aurbach, Challenges in the development of advanced Li-ion batteries: A review, Energy Environ. Sci. 4 (2011) 3243-3262. https://doi.org/10.1039/c1ee01598b

[3] B. Scrosati, J. Garche, Lithium batteries: Status, prospects and future, J. Power Sources 195 (2010) 2419-2430. https://doi.org/10.1016/j.jpowsour.2009.11.048

[4] H. Pan, Y.S. Hu, L. Chen, Room-temperature stationary sodium-ion batteries for large-scale electric energy storage, Energy Environ. Sci. 6 (2013) 2338-2360. https://doi.org/10.1039/c3ee40847g

[5] J.B. Goodenough, K.S. Park, The Li-ion rechargeable battery: A perspective, J. Am. Chem. Soc. 135 (2013) 1167-1176. https://doi.org/10.1021/ja3091438

[6] S. Chen, L. Shen, P.A. Van Aken, J. Maier, Y. Yu, Dual-functionalized double carbon shells coated silicon nanoparticles for high performance lithium-ion batteries, Adv. Mater. 29 (2017) 1605650. https://doi.org/10.1002/adma.201605650

[7] L. Fu, K. Tang, K. Song, P.A. Van Aken, Y. Yu, J. Maier, Nitrogen doped porous carbon fibres as anode materials for sodium ion batteries with excellent rate performance, Nanoscale 6 (2014) 1384-1389. https://doi.org/10.1039/C3NR05374A

[8] M.D. Slater, D. Kim, E. Lee, C.S. Johnson, Sodium-ion batteries, Adv. Funct. Mater. 23 (2013) 947-958. https://doi.org/10.1002/adfm.201200691

[9] Y. Wen, K. He, Y. Zhu, F. Han, Y. Xu, I. Matsuda, Y. Ishii, J. Cumings, C. Wang, Expanded graphite as superior anode for sodium-ion batteries, Nat. Commun. 5 (2014) 4033. https://doi.org/10.1038/ncomms5033

[10] S.P. Ong, V.L. Chevrier, G. Hautier, A. Jain, C. Moore, S. Kim, X. Ma, G. Ceder, Voltage, stability and diffusion barrier differences between sodium-ion and lithium-

ion intercalation materials, Energy Environ. Sci. 4 (2011) 3680-3688. https://doi.org/10.1039/c1ee01782a

[11] C. Wu, P. Kopold, P.A. Van Aken, J. Maier, Y. Yu, MOF-derived hollow Co_9S_8 nanoparticles embedded in graphitic carbon nanocages with superior Li-ion storage, Small 12 (2016) 2354-2364. https://doi.org/10.1002/smll.201503821

[12] A. Ponrouch, E. Marchante, M. Courty, J.M. Tarascon, M.R. Palacin, In search of an optimized electrolyte for Na-ion batteries, Energy Environ. Sci. 5 (2012) 8572-8583. https://doi.org/10.1039/c2ee22258b

[13] D. Su, G. Wang, Single-crystalline bilayered V_2O_5 nanobelts for high-capacity sodium-ion batteries, ACS Nano 7 (2013) 11218-11226. https://doi.org/10.1021/nn405014d

[14] X. Wang, G. Liu, T. Iwao, M. Okubo, A. Yamada, Role of ligand-to-metal charge transfer in O_3-type $NaFeO_2$–$NaNiO_2$ solid solution for enhanced electrochemical properties, J. Phys. Chem. C, 118 (2014) 2970-2976. https://doi.org/10.1021/jp411382r

[15] C. Deng, S. Zhang, B. Zhao, First exploration of ultrafine $Na_7V_3(P_2O_7)_4$ as a high-potential cathode material for sodium-ion battery, Energy Storage Mater. 4 (2016) 71-78. https://doi.org/10.1016/j.ensm.2016.03.001

[16] H. Kabbour, D. Coillot, M. Colmont, C. Masquelier, O. Mentré, α-$Na_3M_2(PO_4)_3$ (M= Ti, Fe): Absolute cationic ordering in NASICON-type phases, J. Am. Chem. Soc. 133 (2011) 11900-11903. https://doi.org/10.1021/ja204321y

[17] C. Masquelier, L. Croguennec, Polyanionic (phosphates, silicates, sulfates) frameworks as electrode materials for rechargeable Li (or Na) batteries, Chem. Rev. 113 (2013) 6552-6591. https://doi.org/10.1021/cr3001862

[18] Y. Ono, Y. Yui, M. Hayashi, K. Asakura, H. Kitabayashi, K.I. Takahashi, Electrochemical properties of $NaCuO_2$ for sodium-ion secondary batteries, ECS Trans. 58 (2014) 33-39. https://doi.org/10.1149/05812.0033ecst

[19] I. Hasa, J. Hassoun, Y.K. Sun, B. Scrosati, Sodium-ion battery based on an electrochemically converted $NaFePO_4$ cathode and nanostructured tin–carbon anode, Chem. Phys. Chem. 15 (2014) 2152-2155. https://doi.org/10.1002/cphc.201400088

[20] R. Sharma, A. Dhillon, D. Kumar, Biosorbents from agricultural by-products: Updates after 2000s, In bio-and nanosorbents from natural resources, Springer, Cham. 2018, pp. 1-20. https://doi.org/10.1007/978-3-319-68708-7_1

[21] R. Sharma, D. Kumar, Nanocomposites: An approach towards pollution control, Nanocomposites for pollution control, Pan Stanford, 2018, pp. 3-46. https://doi.org/10.1201/b22390-1

[22] R. Sharma, D. Kumar, Nanoadsorbents: An approach towards wastewater treatment, Nanotechnology for sustainable water resources, Wiley-Scrivener Book, 2018, pp. 371-405. https://doi.org/10.1002/9781119323655.ch12

[23] S. Nehra, R. Sharma, D. Kumar, Chitosan-based membranes for wastewater desalination and heavy metal detoxification, Nanoscale materials in water purification, Elsevier, 2019, pp. 799-814. https://doi.org/10.1016/B978-0-12-813926-4.00037-9

[24] R. Sharma, D. Kumar, Adsorption of Cr(III) and Cu(II) on hydrothermally synthesized graphene oxide–calcium–zinc nanocomposite, J. Chem. Eng. Data. 63 (2018) 4560-4572. https://doi.org/10.1021/acs.jced.8b00637

[25] R. Sharma, A. Dhillon, D. Kumar, Mentha-stabilized silver nanoparticles for high-performance colorimetric detection of Al (III) in aqueous systems, Sci. rep. 8 (2018) 5189-5202. https://doi.org/10.1038/s41598-018-23469-1

[26] R. Sharma, S. Raghav, M. Nair, D. Kumar, Kinetics and adsorption studies of mercury and lead by ceria nanoparticles entrapped in tamarind powder, ACS omega 3 (2018) 14606-14619. https://doi.org/10.1021/acsomega.8b01874

[27] R., Joshi, R. Sharma, A. Kuila, Lipase production from Fusarium incarnatum KU377454 and its immobilization using Fe_3O_4 NPs for application in waste cooking oil degradation, Bioresource Technol. Rep. 5 (2019) 134-140. https://doi.org/10.1016/j.biteb.2019.01.005

[28] S. Nehra, R. Sharma, D. Kumar, Nanomaterials as an emerging opportunity to procure safe drinking water, Biopolymers: Structure, performance and applications, NOVA Science Publishers, 2018, pp. 67-94.

[29] N. Yabuuchi, M. Kajiyama, J. Iwatate, H. Nishikawa, S. Hitomi, R. Okuyama, R. Usui, Y. Yamada, S. Komaba, P2-type Na_x [$Fe_{1/2}Mn_{1/2}$]O_2 made from earth-abundant elements for rechargeable Na batteries, Nat. Mater. 11 (2012) 512. https://doi.org/10.1038/nmat3309

[30] Y. Cao, L. Xiao, W. Wang, D. Choi, Z. Nie, J. Yu, L.V. Saraf, Z. Yang, J. Liu, Reversible sodium ion insertion in single crystalline manganese oxide nanowires with long cycle life, Adv. Mater. 23 (2011) 3155-3160. https://doi.org/10.1002/adma.201100904

[31] H. Liu, H. Zhou, L. Chen, Z. Tang, W. Yang, Electrochemical insertion/deinsertion of sodium on NaV_6O_{15} nanorods as cathode material of rechargeable sodium-based batteries, J. Power Sources 196 (2011) 814-819. https://doi.org/10.1016/j.jpowsour.2010.07.062

[32] Sun, H.W. Lee, M. Pasta, Y. Sun, W. Liu, Y. Li, H.R. Lee, N. Liu, Y. Cui, Carbothermic reduction synthesis of red phosphorus-filled 3D carbon material as a high-capacity anode for sodium ion batteries, Energy Storage Mater. 4 (2016) 130-136. https://doi.org/10.1016/j.ensm.2016.04.003

[33] W.J. Li, S.L. Chou, J.Z. Wang, H.K. Liu, S.X. Dou, Simply mixed commercial red phosphorus and carbon nanotube composite with exceptionally reversible sodium-ion storage, Nano Lett. 13 (2013) 5480-5484. https://doi.org/10.1021/nl403053v

[34] X. Xie, S. Chen, B. Sun, C. Wang, G. Wang, 3D Networked tin oxide/graphene aerogel with a hierarchically porous architecture for high-rate performance sodium-ion batteries, ChemSusChem 8 (2015) 2948-2955. https://doi.org/10.1002/cssc.201500149

[35] X. Xie, D. Su, J. Zhang, S. Chen, A.K. Mondal, G. Wang, A comparative investigation on the effects of nitrogen-doping into graphene on enhancing the electrochemical performance of SnO_2/graphene for sodium-ion batteries, Nanoscale 7 (2015) 3164-3172. https://doi.org/10.1039/C4NR07054B

[36] J. Wang, C. Luo, T. Gao, A. Langrock, A.C. Mignerey, C. Wang, An advanced MoS_2/carbon anode for high-performance sodium-ion batteries, Small 11 (2015) 473-481. https://doi.org/10.1002/smll.201401521

[37] Q. Sun, Q.Q. Ren, H. Li, Z.W. Fu, High capacity Sb_2O_4 thin film electrodes for rechargeable sodium battery, Electrochem. Commun. 13 (2011) 1462-1464. https://doi.org/10.1016/j.elecom.2011.09.020

[38] Y. Wang, D. Su, C. Wang, G. Wang, SnO_2@ MWCNT nanocomposite as a high capacity anode material for sodium-ion batteries, Electrochem. Commun. 29 (2013) 8-11. https://doi.org/10.1016/j.elecom.2013.01.001

[39] H. Xiong, M.D. Slater, M. Balasubramanian, C.S. Johnson, T. Rajh, Amorphous TiO_2 nanotube anode for rechargeable sodium ion batteries, J. Phys. Chem. Lett. 2 (2011) 2560-2565. https://doi.org/10.1021/jz2012066

[40] X. Xie, Z. Ao, D. Su, J. Zhang, G. Wang, MoS_2/Graphene composite anodes with enhanced performance for sodium-ion batteries: The role of the two-dimensional heterointerface, Adv. Funct. Mater. 25 (2015) 1393-1403. https://doi.org/10.1002/adfm.201404078

[41] C. Chen, Y. Wen, X. Hu, X. Ji, M. Yan, L. Mai, P. Hu, B. Shan, Y. Huang, Na^+ intercalation pseudo capacitance in graphene-coupled titanium oxide enabling ultra-fast sodium storage and long-term cycling, Nat. Commun. 6 (2015) 6929. https://doi.org/10.1038/ncomms7929

[42] X. Xie, D. Su, S. Chen, J. Zhang, S. Dou, G. Wang, SnS_2 Nanoplatelet@graphene nanocomposites as high-capacity anode materials for sodium-ion batteries, Chem. Asian J. 9 (2014) 1611-1617. https://doi.org/10.1002/asia.201400018

[43] J. Fullenwarth, A. Darwiche, A. Soares, B. Donnadieu, L. Monconduit, NiP_3: A promising negative electrode for Li-and Na-ion batteries, J. Mater. Chem. A 2 (2014) 2050-2059. https://doi.org/10.1039/C3TA13976J

[44] C. Wu, P. Kopold, Y.L. Ding, P.A. Van Aken, J. Maier, Y. Yu, Synthesizing porous $NaTi_2(PO_4)_3$ nanoparticles embedded in 3D graphene networks for high-rate and long cycle-life sodium electrodes, ACS Nano 9 (2015) 6610-6618. https://doi.org/10.1021/acsnano.5b02787

[45] G. Pang, C. Yuan, P. Nie, B. Ding, J. Zhu, X. Zhang, Synthesis of NASICON-type structured $NaTi_2(PO_4)_3$–graphene nanocomposite as an anode for aqueous rechargeable Na-ion batteries, Nanoscale 6 (2014) 6328-6334. https://doi.org/10.1039/C3NR06730K

[46] J.M. Fan, J.J. Chen, Q. Zhang, B.B. Chen, J. Zang, M.S. Zheng, Q.F. Dong, An amorphous carbon nitride composite derived from ZIF-8 as anode material for sodium-ion batteries, ChemSusChem 8 (2015) 1856-1861. https://doi.org/10.1002/cssc.201500192

[47] W. Luo, Z. Jian, Z. Xing, W. Wang, C. Bommier, M.M. Lerner, X. Ji, Electrochemically expandable soft carbon as anodes for Na-ion batteries, ACS Cent. Sci. 1 (2015) 516-522. https://doi.org/10.1021/acscentsci.5b00329

[48] S. Komaba, Y. Matsuura, T. Ishikawa, N. Yabuuchi, W. Murata, S. Kuze, Redox reaction of Sn-polyacrylate electrodes in aprotic Na cell, Electrochem. Commun. 21 (2012) 65-68. https://doi.org/10.1016/j.elecom.2012.05.017

[49] M. He, K. Kravchyk, M. Walter, M.V. Kovalenko, Monodisperse antimony nanocrystals for high-rate Li-ion and Na-ion battery anodes: Nano versus bulk, Nano Lett. 14 (2014) 1255-1262. https://doi.org/10.1021/nl404165c

[50] B. Farbod, K. Cui, W.P. Kalisvaart, M. Kupsta, B. Zahiri, A. Kohandehghan, E.M. Lotfabad, Z. Li, E.J. Luber, D. Mitlin, Anodes for sodium ion batteries based on tin–germanium–antimony alloys, ACS Nano 8 (2014) 4415-4429. https://doi.org/10.1021/nn4063598

Materials Research Forum LLC
https://doi.org/10.21741/9781644900833-1

[51] L. Baggetto, J.K. Keum, J.F. Browning, G.M. Veith, Germanium as negative electrode material for sodium-ion batteries, Electrochem. Commun. 34 (2013) 41-44. https://doi.org/10.1016/j.elecom.2013.05.025

[52] P.R. Abel, M.G. Fields, A. Heller, C.B. Mullins, Tin–Germanium alloys as anode materials for sodium-ion batteries, ACS Appl. Mater. Interfaces 6 (2014) 15860-15867. https://doi.org/10.1021/am503365k

[53] Y.M. Lin, P.R. Abel, A. Gupta, J.B. Goodenough, A. Heller, C.B. Mullins, Sn–Cu nanocomposite anodes for rechargeable sodium-ion batteries, ACS Appl. Mater. Interfaces 5 (2013) 8273-8277. https://doi.org/10.1021/am4023994

[54] S. Li, J. Qiu, C. Lai, M. Ling, H. Zhao, S. Zhang, Surface capacitive contributions: Towards high rate anode materials for sodium ion batteries, Nano Energy 12 (2015) 224-230. https://doi.org/10.1016/j.nanoen.2014.12.032

[55] S.H. Lee, P. Liu, C.E. Tracy, D.K. Benson, All-solid-state rocking chair lithium battery on a flexible al substrate, Electrochem. Solid State Lett. 2 (1999) 425-427. https://doi.org/10.1149/1.1390859

[56] M. Sawicki, L.L. Shaw, Advances and challenges of sodium ion batteries as post lithium ion batteries, RSC Adv. 5 (2015) 53129-53154. https://doi.org/10.1039/C5RA08321D

[57] N. Anantharamulu, K.K. Rao, G. Rambabu, B. Vijaya Kumar, V. Radha, M. Vithal, A wide-ranging review on NASICON type materials, J. Mater. Sci. 46 (2011) 2821-2837. https://doi.org/10.1007/s10853-011-5302-5

[58] S. Patoux, G. Rousse, J.B. Leriche, C. Masquelier, Structural and electrochemical studies of rhombohedral $Na_2TiM(PO_4)_3$ and $Li_{1.6}Na_{0.4}TiM (PO_4)_3$ (M= Fe, Cr) phosphates, Chem. Mater. 15 (2003) 2084-2093. https://doi.org/10.1021/cm020479p

[59] Q. Sun, Q.Q. Ren, Z.W. Fu, NASICON-type $Fe_2(MoO_4)_3$ thin film as cathode for rechargeable sodium ion battery, Electrochem. Commun. 23 (2012) 145-148. https://doi.org/10.1016/j.elecom.2012.07.023

[60] C. Zhu, P. Kopold, P.A. Van Aken, J. Maier, Y. Yu, High power–high energy sodium battery based on threefold interpenetrating network, Adv. Mater. 28 (2016) 2409-2416. https://doi.org/10.1002/adma.201505943

[61] W. Song, X. Ji, Z. Wu, Y. Zhu, Y. Yang, J. Chen, M. Jing, F. Li, C.E. Banks, First exploration of Na-ion migration pathways in the NASICON structure $Na_3V_2(PO_4)_3$, J. Mater. Chem. A, 2 (2014) 5358-5362. https://doi.org/10.1039/c4ta00230j

[62] H.P. Hong, Crystal structures and crystal chemistry in the system $Na_{1+x}Zr_2Si_xP_{3-x}O_{12}$, Mater. Res. Bull. 11 (1976) 173-182. https://doi.org/10.1016/0025-5408(76)90073-8

[63] J.B. Goodenough, Y. Kim, Challenges for rechargeable Li batteries, Chem. Mater. 22 (2009) 587-603. https://doi.org/10.1021/cm901452z

[64] K.D. Kreuer, H. Kohler, J. Maier, In high conductivity solid ionic conductors: Recent trends and applications (Ed. T. Takahashi), Singapore (1989) 242. https://doi.org/10.1142/9789814434294_0011

[65] J. Maier, U. Warhus, Thermodynamic investigations of Na_2ZrO_3 by electrochemical means, J. Chem. Thermodyn. 18 (1986) 309-316. https://doi.org/10.1016/0021-9614(86)90075-3

[66] J. Maier, Point-defect thermodynamics and size effects, Solid State Ionics 131 (2000) 13-22. https://doi.org/10.1016/S0167-2738(00)00618-4

[67] J. Maier, Thermodynamics of electrochemical lithium storage, Angew. Chem. Int. Ed. 52 (2013) 4998-5026. https://doi.org/10.1002/anie.201205569

[68] J. Maier, Mass transport in the presence of internal defect reactions—concept of conservative ensembles: I, chemical diffusion in pure compounds, J. Am. Ceram. Soc. 76 (1993) 1212-1217. https://doi.org/10.1111/j.1151-2916.1993.tb03743.x

[69] P.P. Prosini, M. Lisi, D. Zane, M. Pasquali, Determination of the chemical diffusion coefficient of lithium ion $LiFePO_4$, Solid State Ionics 148(2002) 45-51. https://doi.org/10.1016/S0167-2738(02)00134-0

[70] X.H. Rui, N. Yesibolati, S.R. Li, C.C. Yuan, C.H. Chen, Determination of the chemical diffusion coefficient of Li^+ in intercalation-type $Li_3V_2(PO_4)_3$ anode material, Solid State Ionics 187 (2011) 58-63. https://doi.org/10.1016/j.ssi.2011.02.013

[71] D.H. Lee, J. Xu, Y.S. Meng, An advanced cathode for Na-ion batteries with high rate and excellent structural stability, Phys. Chem. Chem. Phys. 15 (2013) 3304-3312. https://doi.org/10.1039/c2cp44467d

[72] P. Hu, J. Ma, T. Wang, B. Qin, C. Zhang, C. Shang, J. Zhao, G. Cui, NASICON-structured $NaSn_2(PO_4)_3$ with excellent high-rate properties as anode material for lithium ion batteries, Chem. Mater. 27 (2015) 6668-6674. https://doi.org/10.1021/acs.chemmater.5b02471

[73] W. Shen, H. Li, C. Wang, Z. Li, Q. Xu, H. Liu, Y. Wang, Improved electrochemical performance of the $Na_3V_2(PO_4)_3$ cathode by B-doping of the carbon coating layer for sodium-ion batteries, J. Mater. Chem. A, 3 (2015) 15190-15201. https://doi.org/10.1039/C5TA03519H

[74] F. Sagane, Synthesis of NaTi$_2$(PO$_4$)$_3$ Thin-film electrodes by sol-gel method and study on the kinetic behavior of Na$^+$-ion insertion/extraction reaction in aqueous solution, J. Electrochem. Soc. 163 (2016) A2835-A2839. https://doi.org/10.1149/2.0161614jes

[75] E. de la Llave, V. Borgel, K.J. Park, J.Y. Hwang, Y.K. Sun, P. Hartmann, F.F. Chesneau, D. Aurbach, Comparison between Na-Ion and Li-Ion cells: Understanding the critical role of the cathode's stability and the anodes pretreatment on the cells behavior, ACS Appl. Mater. Interfaces 8 (2016) 1867-1875. https://doi.org/10.1021/acsami.5b09835

[76] S. Song, H.M. Duong, A.M. Korsunsky, N. Hu, L. Lu, A Na$^+$ superionic conductor for room-temperature sodium batteries, Sci. Rep. 6 (2016) 32330. https://doi.org/10.1038/srep32330

[77] M.W. Verbrugge, B.J. Koch, Modeling lithium intercalation of single-fiber carbon microelectrodes, J. Electrochem. Soc. 143 (1996) 600-608. https://doi.org/10.1149/1.1836486

[78] Y. Niu, M. Xu, Y. Zhang, J. Han, Y. Wang, C.M. Li, Detailed investigation of a NaTi$_2$(PO$_4$)$_3$ anode prepared by pyro-synthesis for Na-ion batteries, RSC Adv. 6 (2016) 45605-45611. https://doi.org/10.1039/C6RA06533C

[79] D.J. Kim, R. Ponraj, A.G. Kannan, H.W. Lee, R. Fathi, R. Ruffo, C.M. Mari, D.K. Kim, Diffusion behavior of sodium ions in Na$_{0.44}$MnO$_2$ in aqueous and non-aqueous electrolytes, J. Power Sources 244 (2013) 758-763. https://doi.org/10.1016/j.jpowsour.2013.02.090

[80] K. Zhong, R. Hu, G. Xu, Y. Yang, J.M. Zhang, Z. Huang, Adsorption and ultrafast diffusion of lithium in bilayer graphene ab initio and kinetic Monte Carlo simulation study, (2019). https://doi.org/10.1103/PhysRevB.99.155403

[81] S. Guo, J. Yi, Y. Sun, H. Zhou, Recent advances in titanium-based electrode materials for stationary sodium-ion batteries, Energy Environ. Sci. 9 (2016) 2978-3006. https://doi.org/10.1039/C6EE01807F

[82] P. Yu, B.N. Popov, J.A. Ritter, R.E. White, Determination of the lithium ion diffusion coefficient in graphite, J. Electrochem. Soc. 146 (1999) 8-14. https://doi.org/10.1149/1.1391556

[83] D. Wang, Q. Liu, C. Chen, M. Li, X. Meng, X. Bie, Y. Wei, Y. Huang, F. Du, C. Wang, G. Chen, NASICON-structured NaTi$_2$(PO$_4$)$_3$@C nanocomposite as the low operation-voltage anode material for high-performance sodium-ion batteries, ACS Appl. Mater. Interfaces 8 (2016) 2238-2246. https://doi.org/10.1021/acsami.5b11003

[84] N. Böckenfeld, A. Balducci, Determination of sodium ion diffusion coefficients in sodium vanadium phosphate, J. Solid State Electrochem. 18 (2014) 959-964. https://doi.org/10.1007/s10008-013-2342-6

[85] Y. Zhu, Y. Xu, Y. Liu, C. Luo, C. Wang, Comparison of electrochemical performances of olivine $NaFePO_4$ in sodium-ion batteries and olivine $LiFePO_4$ in lithium-ion batteries, Nanoscale 5 (2013) 780-787. https://doi.org/10.1039/C2NR32758A

[86] R. Amin, P. Balaya, J. Maier, Anisotropy of electronic and ionic transport in $LiFePO_4$ single crystals, Electrochem. Solid State Lett. 10 (2007) A13-A16. https://doi.org/10.1149/1.2388240

[87] J.Y. Shin, D. Samuelis, J. Maier, Sustained lithium-storage performance of hierarchical, nanoporous anatase TiO_2 at high rates: Emphasis on interfacial storage phenomena, Adv. Funct. Mater. 21 (2011) 3464-3472. https://doi.org/10.1002/adfm.201002527

[88] J. Sheng, H. Zang, C. Tang, Q. An, Q. Wei, G. Zhang, L. Chen, C. Peng, L. Mai, Graphene wrapped NASICON-type $Fe_2(MoO_4)_3$ nanoparticles as an ultra-high rate cathode for sodium ion batteries, Nano Energy 24 (2016) 130-138. https://doi.org/10.1016/j.nanoen.2016.04.021

[89] S.A. Novikova, R.V. Larkovich, A.A. Chekannikov, T.L. Kulova, A.M. Skundin, A.B. Yaroslavtsev, Electrical conductivity and electrochemical characteristics of $Na_3V_2(PO_4)_3$-based NASICON-type materials, Inorg. Mater. 54 (2018) 794-804. https://doi.org/10.1134/S0020168518080149

[90] Q. Zheng, H. Yi, X. Li, H. Zhang, Progress and prospect for NASICON-type $Na_3V_2(PO_4)_3$ for electrochemical energy storage, J. Energy Chem. (2018). https://doi.org/10.1016/j.jechem.2018.05.001

[91] W. Song, Z. Wu, J. Chen, Q. Lan, Y. Zhu, Y. Yang, C. Pan, H. Hou, M. Jing, X. Ji, High-voltage NASICON sodium ion batteries: Merits of fluorine insertion, Electrochim. Acta 146 (2014) 142-150. https://doi.org/10.1016/j.electacta.2014.09.068

[92] H. Zhang, B. Qin, D. Buchholz, S. Passerini, High-efficiency sodium-ion battery based on NASICON electrodes with high power and long lifespan, ACS Appl. Energy Mater. 1 (2018) 6425-6432. https://doi.org/10.1021/acsaem.8b01390

[93] S.A. Needham, G.X. Wang, H.K. Liu, Synthesis of NiO nanotubes for use as negative electrodes in lithium ion batteries, J. Power Sources 159 (2006) 254-257. https://doi.org/10.1016/j.jpowsour.2006.04.025

[94] J.Z. Wang, L. Lu, M. Lotya, J.N. Coleman, S.L. Chou, H.K. Liu, A.I. Minett, J. Chen, Development of MoS_2–CNT composite thin film from layered MoS_2 for lithium batteries, Adv. Energy Mater. 3 (2013) 798-805. https://doi.org/10.1002/aenm.201201000

[95] J. Zheng, L. Liu, G. Ji, Q. Yang, L. Zheng, J. Zhang, Hydrogenated anatase TiO_2 as lithium-ion battery anode: Size–reactivity correlation, ACS Appl. Mater. Interfaces, 8 (2016) 20074-20081. https://doi.org/10.1021/acsami.6b05993

[96] M. Yoshio, H. Wang, K. Fukuda, T. Umeno, T. Abe, Z. Ogumi, Improvement of natural graphite as a lithium-ion battery anode material, from raw flake to carbon-coated sphere, J. Mater. Chem. 14 (2004) 1754-1758. https://doi.org/10.1039/b316702j

[97] Y. Zhao, Z. Wei, Q. Pang, Y. Wei, Y. Cai, Q. Fu, F. Du, A. Sarapulova, H. Ehrenberg, B. Liu, G. Chen, NASICON-Type $Mg_{0.5}Ti_2(PO_4)_3$ Negative electrode material exhibits different electrochemical energy storage mechanisms in Na-ion and Li-ion batteries, ACS Appl. Mater. Interfaces 9 (2017) 4709-4718. https://doi.org/10.1021/acsami.6b14196

[98] V.T. Nguyen, Y.L. Liu, S.A. Hakim, S.Y. Amr Rady Radwan, W. Chen, Synthesis and electrochemical performance of $Fe_2(MoO_4)_3$/RGO nanocomposite cathode material for sodium-ion batteries, Int. J. Electrochem. Sci. 10 (2015) 10565-10575. https://doi.org/10.1149/2.0011505jss

[99] F. Sauvage, E. Quarez, J.M. Tarascon, E. Baudrin, Crystal structure and electrochemical properties vs. Na^+ of the sodium fluorophosphate $Na_{1.5}VOPO_4F_{0.5}$, Solid State Sci. 8 (2006) 1215-1221. https://doi.org/10.1016/j.solidstatesciences.2006.05.009

[100] R. Essehli, I. Belharouak, H. Ben Yahia, K. Maher, A. Abouimrane, B. Orayech, S. Calder, X.L. Zhou, Z. Zhou, Y.K. Sun, Alluaudite $Na_2Co_2Fe(PO_4)_3$ as an electroactive material for sodium ion batteries, Dalton Trans. 44 (2015) 7881-7886. https://doi.org/10.1039/C5DT00971E

[101] J. Isasi, A. Daidouh, Synthesis, structure and conductivity study of new monovalent phosphates with the langbeinite structure, Solid State Ionics 133 (2000) 303-313. https://doi.org/10.1016/S0167-2738(00)00677-9

[102] M. Xu, C.J. Cheng, Q.Q. Sun, S.J. Bao, Y.B. Niu, H. He, Y. Li, J. Song, A 3D porous interconnected $NaVPO_4F$/C network: Preparation and performance for Na-ion batteries, RSC Adv. 5 (2015) 40065-40069. https://doi.org/10.1039/C5RA05161D

[103] S. Zhou, G. Barim, B.J. Morgan, B.C. Melot, R.L. Brutchey, Influence of rotational distortions on Li^+-and Na^+-intercalation in anti-NASICON $Fe_2(MoO_4)_3$, Chem. Mater. 28 (2016) 4492-4500. https://doi.org/10.1021/acs.chemmater.6b01806

[104] Y. Qi, L. Mu, J. Zhao, Y.S. Hu, H. Liu, S. Dai, Superior Na-storage performance of low-temperature-synthesized $Na_3(VO_{1-x}PO_4)_2F_{1+2x}$ ($0 \leq x \leq 1$) nanoparticles for Na-ion batteries, Angew. Chem. Int. Ed. 54 (2015) 9911-9916. https://doi.org/10.1002/anie.201503188

[105] J. Zhao, L. Mu, Y. Qi, Y.S. Hu, H. Liu, S. Dai, A phase-transfer assisted solvo-thermal strategy for low-temperature synthesis of $Na_3(VO_{1-x}PO_4)_2F_{1+2x}$ cathodes for sodium-ion batteries, Chem. Commun. 51 (2015) 7160-7163. https://doi.org/10.1039/C5CC01504A

[106] C. Masquelier, C. Wurm, J. Rodríguez-Carvajal, J. Gaubicher, L. Nazar, A powder neutron diffraction investigation of the two rhombohedral NASICON analogues: γ-$Na_3Fe_2(PO_4)_3$ and $Li_3Fe_2(PO_4)_3$, Chem. Mater. 12 (2000) 525-532. https://doi.org/10.1021/cm991138n

[107] D. Li, J. Xue, M. Liu, Synthesis of $Fe_2(MoO_4)_3$ microspheres by self-assembly and photocatalytic performances, New J. Chem. 39 (2015) 1910-1915. https://doi.org/10.1039/C4NJ01731E

[108] Y. Niu, M. Xu, Reduced graphene oxide and $Fe_2(MoO_4)_3$ composite for sodium-ion batteries cathode with improved performance, J. Alloy. Compd. 674 (2016) 392-398. https://doi.org/10.1016/j.jallcom.2016.02.223

[109] Y. Song, H. Wang, Z. Li, N. Ye, L. Wang, Y. Liu, $Fe_2(MoO_4)_3$ nanoparticle-anchored MoO_3 nanowires: Strong coupling via the reverse diffusion of heteroatoms and largely enhanced lithium storage properties, RSC Adv. 5 (2015) 16386-16393. https://doi.org/10.1039/C4RA15655B

[110] V. Nguyen, Y. Liu, Y. Li, S.A. Hakim, X. Yang, W. Chen, Synthesis and electrochemical performance of $Fe_2(MoO_4)_3$/carbon nanotubes nanocomposite cathode material for sodium-ion battery, ECS J. Solid State Sci. Technol. 4 (2015) M25-M29. https://doi.org/10.1149/2.0011505jss

[111] V. Nguyen, Y. Liu, X. Yang, W. Chen, $Fe_2(MoO_4)_3$/nanosilver composite as a cathode for sodium-ion batteries, ECS Electrochem. Lett. 4 (2015) A29-A32. https://doi.org/10.1149/2.0021503eel

[112] S. Kajiyama, J. Kikkawa, J. Hoshino, M. Okubo, E. Hosono, Assembly of $Na_3V_2(PO_4)_3$ nanoparticles confined in a one-dimensional carbon sheath for enhanced sodium-ion cathode properties, Chem. Eur. J. 20 (2014) 12636-12640. https://doi.org/10.1002/chem.201403126

[113] X. Rui, W. Sun, C. Wu, Y. Yu, Q. Yan, An advanced sodium-ion battery composed of carbon coated $Na_3V_2(PO_4)_3$ in a porous graphene network, Adv. Mater. 27 (2015) 6670-6676. https://doi.org/10.1002/adma.201502864

[114] W. Duan, Z. Zhu, H. Li, Z. Hu, K. Zhang, F. Cheng, J. Chen, $Na_3V_2(PO_4)_3$@C core–shell nanocomposites for rechargeable sodium-ion batteries, J. Mater. Chem. A, 2 (2014) 8668-8675. https://doi.org/10.1039/C4TA00106K

[115] W. Shen, H. Li, C. Wang, Z. Li, Q. Xu, H. Liu, Y. Wang, Improved electrochemical performance of the $Na_3V_2(PO_4)_3$ cathode by B-doping of the carbon coating layer for sodium-ion batteries, J. Mater. Chem. A 3 (2015) 15190-15201. https://doi.org/10.1039/C5TA03519H

[116] J. Liu, K. Tang, K. Song, P.A. Van Aken, Y. Yu, J. Maier, Electrospun $Na_3V_2(PO_4)_3$/C nanofibers as stable cathode materials for sodium-ion batteries, Nanoscale 6 (2014) 5081-5086. https://doi.org/10.1039/c3nr05329f

[117] Q. Liu, D. Wang, X. Yang, N. Chen, C. Wang, X. Bie, Y. Wei, G. Chen, F. Du, Carbon-coated $Na_3V_2(PO_4)_2F_3$ nanoparticles embedded in a mesoporous carbon matrix as a potential cathode material for sodium-ion batteries with superior rate capability and long-term cycle life, J. Mater. Chem. A 3 (2015) 21478-21485. https://doi.org/10.1039/C5TA05939A

[118] S. Li, Y. Dong, L. Xu, X. Xu, L. He, L. Mai, Effect of carbon matrix dimensions on the electrochemical properties of $Na_3V_2(PO_4)_3$ nanograins for high-performance symmetric sodium-ion batteries, Adv. Mater. 26 (2014) 3545-3553. https://doi.org/10.1002/adma.201305522

[119] J. Geng, F. Li, S. Ma, J. Xiao, M. Sui, First principle study of $Na_3V_2(PO_4)_2F_3$ for Na batteries application and experimental investigation, Int. J. Electrochem. Sci. 11 (2016) 3815-3823. https://doi.org/10.20964/110483

[120] J. Fang, S. Wang, Z. Li, H. Chen, L. Xia, L. Ding, H. Wang, Porous $Na_3V_2(PO_4)_3$@C nanoparticles enwrapped in three-dimensional graphene for high performance sodium-ion batteries, J. Mater. Chem. A 4 (2016) 1180-1185. https://doi.org/10.1039/C5TA08869K

[121] W. Shen, H. Li, Z. Guo, C. Wang, Z. Li, Q. Xu, H. Liu, Y. Wang, Y. Xia, Double-nanocarbon synergistically modified $Na_3V_2(PO_4)_3$: An advanced cathode for high-rate and long-life sodium-ion batteries, ACS Appl. Mater. Interfaces 8 (2016) 15341-15351. https://doi.org/10.1021/acsami.6b03410

[122] C. Zhu, K. Song, P.A. Van Aken, J. Maier, Y. Yu, Carbon-coated $Na_3V_2(PO_4)_3$ embedded in porous carbon matrix: an ultrafast Na-storage cathode with the potential

of outperforming Li cathodes, Nano Lett. 14 (2014) 2175-2180.
https://doi.org/10.1021/nl500548a

[123] W. Song, X. Ji, J. Chen, Z. Wu, Y. Zhu, K. Ye, H. Hou, M. Jing, C.E. Banks,
Mechanistic investigation of ion migration in $Na_3V_2(PO_4)_2F_3$ hybrid-ion batteries,
Phys. Chem. Chem. Phys. 17 (2015) 159-165. https://doi.org/10.1039/C4CP04649H

[124] W. Ren, Z. Zheng, C. Xu, C. Niu, Q. Wei, Q. An, K. Zhao, M. Yan, M. Qin, L.
Mai, Self-sacrificed synthesis of three-dimensional $Na_2V_2(PO_4)_3$ nanofiber network
for high-rate sodium–ion full batteries, Nano Energy 25 (2016) 145-153.
https://doi.org/10.1016/j.nanoen.2016.03.018

[125] W. Shen, C. Wang, Q. Xu, H. Liu, Y. Wang, Nitrogen-doping-induced defects of a
carbon coating layer facilitate Na-storage in electrode materials, Adv. Energy Mater.
5 (2015) 1400982. https://doi.org/10.1002/aenm.201400982

[126] W. Song, X. Ji, Z. Wu, Y. Zhu, F. Li, Y. Yao, C.E. Banks, Multifunctional dual
$Na_3V_2(PO_4)_2F_3$ cathode for both lithium-ion and sodium-ion batteries, RSC Adv. 4
(2014) 11375-11383. https://doi.org/10.1039/C3RA47878E

[127] Y. Jiang, Z. Yang, W. Li, L. Zeng, F. Pan, M. Wang, X. Wei, G. Hu, L. Gu, Y.
Yu, Nanoconfined carbon-coated $Na_3V_2(PO_4)_3$ particles in mesoporous carbon
enabling ultralong cycle life for sodium-ion batteries, Adv. Energy Mater. 5 (2015)
1402104. https://doi.org/10.1002/aenm.201402104

[128] M. Chen, K. Kou, M. Tu, J. Hu, B. Yang, Fabrication of multi-walled carbon
nanotubes modified $Na_3V_2(PO_4)_3$/C and its application to high-rate lithium-ion
batteries cathode, Solid State Ionics 274 (2015) 24-28.
https://doi.org/10.1016/j.ssi.2015.02.021

[129] W. Song, X. Ji, Z. Wu, Y. Yang, Z. Zhou, F. Li, Q. Chen, C.E. Banks, Exploration
of ion migration mechanism and diffusion capability for $Na_3V_2(PO_4)_2F_3$ cathode
utilized in rechargeable sodium-ion batteries, J. Power Sources 256 (2014) 258-263.
https://doi.org/10.1016/j.jpowsour.2014.01.025

[130] S. Chen, Z. Ao, B. Sun, X. Xie, G. Wang, Porous carbon nanocages encapsulated
with tin nanoparticles for high performance sodium-ion batteries, Energy Storage
Mater. 5 (2016) 180-190. https://doi.org/10.1016/j.ensm.2016.07.001

[131] J. Park, J.S. Kim, J.W. Park, T.H. Nam, K.W. Kim, J.H. Ahn, G. Wang, H.J. Ahn,
Discharge mechanism of MoS_2 for sodium ion battery: Electrochemical
measurements and characterization, Electrochim. Acta 92 (2013) 427-432.
https://doi.org/10.1016/j.electacta.2013.01.057

[132] X. Xie, K. Kretschmer, J. Zhang, B. Sun, D. Su, G. Wang, Sn@ CNT nanopillars grown perpendicularly on carbon paper: A novel free-standing anode for sodium ion batteries, Nano Energy 13 (2015) 208-217. https://doi.org/10.1016/j.nanoen.2015.02.022

[133] Y. Kim, K.H. Ha, S.M. Oh, K.T. Lee, High-capacity anode materials for sodium-ion batteries, Chem. Eur. J. 20 (2014) 11980-11992. https://doi.org/10.1002/chem.201402511

[134] S. Hariharan, K. Saravanan, P. Balaya, α-MoO$_3$: A high performance anode material for sodium-ion batteries, Electrochem. Commun. 31 (2013) 5-9. https://doi.org/10.1016/j.elecom.2013.02.020

[135] M.R. Palacin, Recent advances in rechargeable battery materials: A chemist's perspective, Chem. Soc. Rev. 38 (2009) 2565-2575. https://doi.org/10.1039/b820555h

[136] F. Lalère, J.B. Leriche, M. Courty, S. Boulineau, V. Viallet, C. Masquelier, V. Seznec, An all-solid state NASICON sodium battery operating at 200 C, J. Power Sources 247 (2014) 975-980. https://doi.org/10.1016/j.jpowsour.2013.09.051

[137] S.I. Park, I. Gocheva, S. Okada, J.I. Yamaki, Electrochemical properties of NaTi$_2$(PO$_4$)$_3$ anode for rechargeable aqueous sodium-ion batteries, J. Electrochem. Soc. 158 (2011) A1067-A1070. https://doi.org/10.1149/1.3611434

[138] Y. Fang, L. Xiao, J. Qian, Y. Cao, X. Ai, Y. Huang, H. Yang, 3D Graphene decorated NaTi$_2$(PO$_4$)$_3$ microspheres as a superior high-rate and ultracycle-stable anode material for sodium ion batteries, Adv. Energy Mater. 6 (2016) 1502197. https://doi.org/10.1002/aenm.201502197

[139] P. Senguttuvan, G. Rousse, M.E. Arroyoy de Dompablo, H. Vezin, J.M. Tarascon, M.R. Palacín, Low-potential sodium insertion in a NASICON-type structure through the Ti(III)/Ti(II) redox couple, J. Am. Chem. Soc. 135 (2013) 3897-3903. https://doi.org/10.1021/ja311044t

[140] G. Pang, P. Nie, C. Yuan, L. Shen, X. Zhang, H. Li, C. Zhang, Mesoporous NaTi$_2$(PO$_4$)$_3$/CMK-3 nanohybrid as anode for long-life Na-ion batteries, J. Mater. Chem. A 2 (2014) 20659-20666. https://doi.org/10.1039/C4TA04732J

[141] W. Wang, B. Jiang, L. Hu, S. Jiao, NASICON material NaZr$_2$(PO$_4$)$_3$: A novel storage material for sodium-ion batteries, J. Mater. Chem. A, 2 (2014) 1341-1345. https://doi.org/10.1039/C3TA14310D

[142] I. Hasa, S. Passerini, J. Hassoun, A rechargeable sodium-ion battery using a nanostructured Sb–C anode and P2-type layered Na$_{0.6}$Ni$_{0.22}$Fe$_{0.11}$Mn$_{0.66}$O$_2$ cathode, RSC Adv. 5 (2015) 48928-48934. https://doi.org/10.1039/C5RA06336A

[143] Y. Zhang, H. Zhao, Y. Du, Symmetric full cells assembled by using self-supporting $Na_3V_2(PO_4)_3$ bipolar electrodes for superior sodium energy storage, J. Mater. Chem. A, 4 (2016) 7155-7159. https://doi.org/10.1039/C6TA02218A

[144] G.B. Xu, L.W. Yang, X.L. Wei, J.W. Ding, J.X. Zhong, P.K. Chu, Hierarchical porous nanocomposite architectures from multi-wall carbon nanotube threaded mesoporous $NaTi_2(PO_4)_3$ nanocrystals for high-performance sodium electrodes, J. Power Sources 327 (2016) 580-590. https://doi.org/10.1016/j.jpowsour.2016.07.089

[145] Z. Huang, L. Liu, L. Yi, W. Xiao, M. Li, Q. Zhou, G. Guo, X. Chen, H. Shu, X. Yang, X. Wang, Facile solvothermal synthesis of $NaTi_2(PO_4)_3$/C porous plates as electrode materials for high-performance sodium ion batteries, J. Power Sources 325 (2016) 474-481. https://doi.org/10.1016/j.jpowsour.2016.06.066

[146] Y. Niu, M. Xu, C. Guo, C.M. Li, Pyro-synthesis of a nanostructured $NaTi_2(PO_4)_3$/C with a novel lower voltage plateau for rechargeable sodium-ion batteries, J. Colloid Interf. Sci. 474 (2016) 88-92. https://doi.org/10.1016/j.jcis.2016.04.021

[147] Z. Jian, Y. Sun, X. Ji, A new low-voltage plateau of $Na_3V_2(PO_4)_3$ as an anode for Na-ion batteries, Chem. Commun. 51 (2015) 6381-6383. https://doi.org/10.1039/C5CC00944H

[148] Q. Zhang, C. Liao, T. Zhai, H. Li, A high rate 1.2 V aqueous sodium-ion battery based on all NASICON structured $NaTi_2(PO_4)_3$ and $Na_3V_2(PO_4)_3$, Electrochim. Acta 196 (2016) 470-478. https://doi.org/10.1016/j.electacta.2016.03.007

[149] C. Masquelier, Solid electrolytes: Lithium ions on the fast track, Nat. Mater. 10 (2011) 649. https://doi.org/10.1038/nmat3105

[150] H.K. Roh, H.K. Kim, M.S. Kim, D.H. Kim, K.Y. Chung, K.C. Roh, K.B. Kim, In situ synthesis of chemically bonded $NaTi_2(PO_4)_3$/rGO 2D nanocomposite for high-rate sodium-ion batteries, Nano Res. 9 (2016) 1844-1855. https://doi.org/10.1007/s12274-016-1077-y

Sodium-Ion Batteries: Materials and Applications
Materials Research Foundations **76** (2020) 31-72

Materials Research Forum LLC
https://doi.org/10.21741/9781644900833-2

Chapter 2

Carbon Anodes for Sodium-Ion Batteries

Syed Mustansar Abbas[1]*, Muhammad Iftikhar[1,2], Ata-ur-Rehman[2]

[1] Nanoscience and Technology Department, National Centre for Physics, Islamabad, Pakistan

[2] Department of Chemistry, Quaid-e-Azam University, Islamabad, Pakistan

*qau_abbas@yahoo.com

Abstract

Current attempts have shown great prospect of substituting lithium-ion batteries (LIBs) with their rival named as Sodium-ion batteries (SIBs), as both share similar chemistry while lithium being scarce and expensive in comparison to the earth crust rich sodium. The poor performance of SIBs anode has restricted its development in the past. Recently, a significant amount of research has been focused on anode materials for SIBs. Carbonaceous anodes have become viable SIB anodes providing high safety, abundant resources, and nontoxicity. In this chapter, the prominent sodium storage capabilities of some potent carbonaceous anodes namely hard carbon, graphite, carbon nanofibers, graphene, biomass derivatized carbon and heteroatom-doped carbon, have been discussed.

Keywords

Sodium-Ion Battery, Anode, Carbon Nanofiber, Hard Carbon, Graphene

Contents

1.   Introduction

The rapid development of modern society in various fields has exponentially increased energy consumption. Because of the depletion in fossil fuel, there is larger interest in sustainable green renewable energy and non-exhaustive sources like solar, waves and wind etc. However, for consistent use of this sustainable energy in an electrical grid, a suitable energy storage device is essential. In this context, a surge has risen for advanced energy storage and conversion technologies that can fulfil globally increasing energy needs and prevent environmental pollution. Rechargeable batteries or secondary batteries are one of the most suitable among all available technologies because of their tuneability, high energy storage and conversion capabilities, and ease of fabrication and maintenance [1, 2]. Today, lithium-ion batteries (LIBs), as well as sodium-ion batteries (SIBs), are the most widely studied rechargeable battery systems.

LIBs are the most popular and globally recognized power sources of the present time pertaining to the several advantages that they offer like high energy density, lasting cycle life, eco-friendly and lightweight [3, 4]. In last decade or so, there is a vibrant swing towards SIBs research due to some drawbacks associated with LIBs like high cost, limited lithium resources, safety problems and complex cycling protocols [2, 5-9]. Being a part of alkali metals in the periodic table both lithium, and sodium share similar electrochemistry. Owing to the low cost of sodium, (only ~3% compared with Li source) SIBs are expected to be more suitable for large-scale grid storage applications [10], geographically even distribution (23600 ppm compared with 20 ppm of Li) [11] and can be drained to zero charge, making it easy for shipment and storage while LIBs must retain some charge for transportation and storage [12]. SIBs offer another economical potential in the form of using aluminium current collector for anode as it does not form an alloy with aluminium and thus eliminating the risk of corrosion while LIBs utilize more costly copper anode [13, 14]. However, some inherent shortcomings associated with SIBs like the apparently larger ionic radius of Na^+ (1.02 Å as compared to 0.76 Å for Li^+), heavier mass of Na (23 g mol^{-1} as equated to 6.9 g mol^{-1} for Li) make SIBs to demonstrate low energy and power density in comparison to LIBs [12, 15]. Therefore, it is highly demanded that suitable anode, cathode, separators and electrolytes may be searched to increase the efficiency of most cost-effective storage devices in the form of SIBs. A brief comparison is tabulated between sodium and lithium (Table 1).

Table 1 *Evaluation of physical characteristics for LIBs and SIBs. Reproduced with permission [13]. Copyright 2019 The American Chemical Society*

Property	LIBs	SIBs
Relative atomic mass	6.94	23.00
Mass-to-electron ratio	6.94	23.00
Shannon's ionic radii/Å	0.76	1.02
E° (vs SHE)/V	-3.04	-2.71
Melting point/°C	180.5	97.7
Theoretical capacity of metal electrodes/mAh g^{-1}	3861	1166
Theoretical capacity of metal electrodes/mAh cm^{-3}	2062	1131
Theoretical capacity of $ACoO_2$/mAh g^{-1}	274	235
Theoretical capacity of $ACoO_2$/mAh cm^{-3}	1378	1193
Molar conductivity in $AClO4$/PC/S cm^2 mol^{-1}	6.54	7.16
Desolvation energy in PC/kJ mol^{-1}	218.0	157.3
Coordination preference	octahedral and tetrahedral	octahedral and prismatic

2.   Overview of SIBs electrode materials

Fig. 1 shows a simplified schematic diagram of a working SIB consisting of an anode and cathode mounted on their respective current collectors, electrolyte, and separator film [16]. The working principle is almost the same as that for LIBs following reversible intercalation/deintercalation between anode and cathode. During charging, the desodiation takes place from the cathode as it is oxidized and Na^+ ions are inserted into anode as reduction takes place. During the discharging cycle, the anode gets oxidized releasing the Na^+ ions that travel through electrolyte towards the cathode, which is reduced during sodiation process and electrons flow through the external circuit [17].

Over the past few decades, SIBs have been recorded with many distinct anode and cathode materials. Some of the most widely reported cathodes include phosphates, sulfides, polyanions, sulphates, layered oxides, fluorides, Prussian blue and various types of organic polymers. Layered oxides provide high potency as cathode material for SIBs, however, their multiphase and irreversible transitions during cycling result in lower rate performance and capacity decay [18]. Several types of layered oxides have been reported that are generally classified into O3 or P2 type depending on the accommodation of Na^+ ions at octahedral or prismatic sites, respectively. Some of the representative layered oxides include O3-$NaNi_{1/3}Mn_{1/3}Co_{1/3}O_2$ ($\sim$120 mAh g^{-1} discharge capacity at C/10 up to 50th cycle) [19] P2-$Na_{0.67}Ni_{0.25}Mg_{0.1}Mn_{0.65}O_2$, [20] (140 mAh g^{-1} discharge capacity and energy density of 335 Wh kg^{-1}) and P2-Na$_{2/3}$[Fe$_{1/2}$Mn$_{1/2}$]O$_2$ [21] (~149 mAh g^{-1} discharge capacity at 12 mA g^{-1} at the end of 30 cycles). Due to their rigid, perovskite-like framework structure and ambient temperature synthetic protocols, the Prussian blue and its structural analogues have proved suitable insertion hosts for SIBs [22-25]. Some of the representative materials in this class include KFe$_2$(CN)$_6$ [22], KMnFe(CN)$_6$ (80−120 mAh g^{-1}) [13, 22], Na$_4$Fe(CN)$_6$ (~90 mAh g^{-1}) [26]. The defects in the framework structure, together with low electronic conductivity, hamper their use as standard cathode for SIBs [27, 28]. Polyanion compounds are considered as research hotspot for SIBs cathode because of their high operating voltage, diversity in structure and excellent cycling performance, however, they also suffer from capacity fading upon prolonged cycling [29]. Some of the representative cathodes of this class include, NASICAON-type Na$_3$V$_2$(PO$_4$)$_3$-C [30] (~90 mAh g^{-1} at 2340 mA g^{-1} after 10000 cycles), NaFe(SO$_4$)$_2$ [31] (~80 mAh g^{-1} at 0.1 C for 80 cycles), NaFePO$_4$ [32] (144.3 mAh g^{-1} after 300 cycles at 0.1 C). Recently organic compounds have received special emphasis owing to their resource abundance, eco-friendly, recycling potential, good designability and low cost [33]. Various combination compounds of organic tetrasodium salts (Na$_4$C$_8$H$_2$O$_6$, Na$_2$C$_8$H$_2$O$_6$/Na$_4$C$_8$H$_2$O$_6$, and Na$_4$C$_8$H$_2$O$_6$/Na$_6$C$_8$H$_2$O$_6$) [34] can deliver specific capacities around 180 mAh g^{-1} with excellent Coulombic efficiencies (C.E), however, organic compounds have poor inherent conductivity and they are prone to be dissolved in organic electrolytes, which limits their practical applicability. Fig. 2a shows a comparative summary of the representative cathode materials used for SIBs [35].

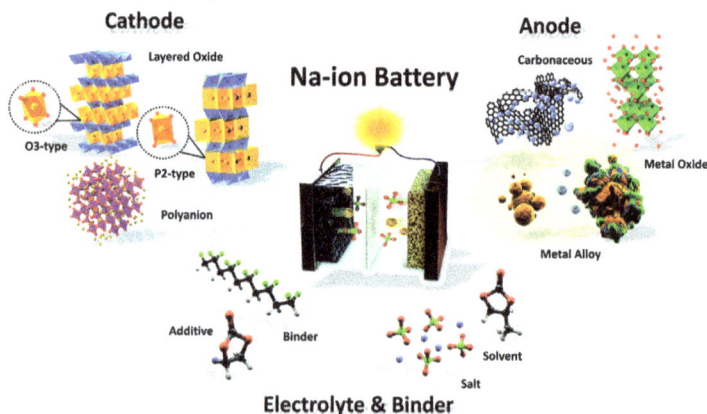

Figure 1. Scheme showing the operating principle of SIBs and most usual anode, cathode, electrolyte, binders and additive used. Reproduced with permission [16]. Copyright 2019 Royal Society of Chemistry.

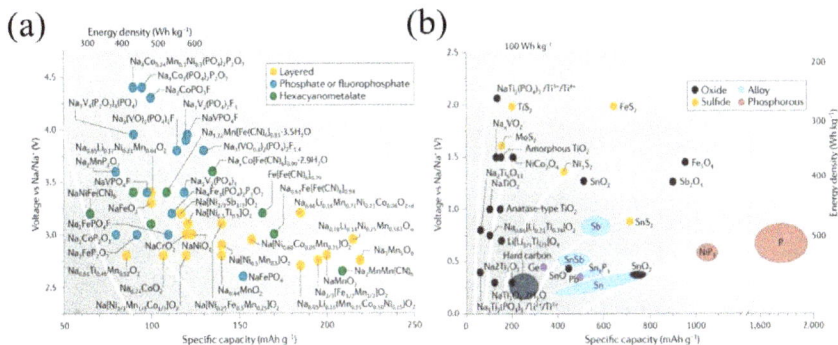

Figure 2. Relationships between specific capacity and voltage for present (a) cathode and (b) anode materials in SIBs. Reproduced with permission [35]. Copyright 2019 Nature Publishing Group.

In comparison to cathodes, less attention has been given to anode materials for SIBs. Presently most of the SIB anode research is focused on carbonaceous materials, metal oxides, carbon-based composites and alloy based materials. Conjugated aromatic polymers are also a strong contender for SIB anode but their complicated synthesis protocols limit their use in the commercial cell. A layered 2D conjugated polymer with substantial porosity was prepared by polymerization of tetrabromopolyaromatic monomers and has been used as an anode for SIBs with successful capacity outcomes ($\sim$79 mAh g^{-1} (70%)) at 5 A g^{-1} after 7700 cycles) [36]. Various metal oxides [37-40] and metal sulfides [41-43] have been widely searched in this direction and they present high capacities with good cycle life, however, the usual volume variation problem is much more pronounced for metal oxides and metal sulfides for SIBs anode. Titanium-based oxides, [15, 44, 45] especially anatase TiO_2 is reported as a potential anode for SIBs whereby it presents high capacities at higher current rates with good stability (134 mAh g^{-1} (95%) at 3.35 A g^{-1} after 4500 cycles). Similarly, Sn [38, 44, 46] is well known for its high potential as anode materials and reported $Sn_{10}Bi_{10}Sb_{80}$ alloy [47] shows 620 mAh g^{-1} of discharge capacity at the end of the 100th cycle. Some iron-based sulfide material also display value as anode like Fe_3S_4 [48] is reported to undergo conversion reaction yielding FeS_2, and FeS quantum size particles that help to gain the extra electrochemical stability and hence the discharge capacities achieved are quite encouraging (275 mAh g^{-1} at 20 A g^{-1} after 3500 cycles). Fig. 2b shows a comparative summary of the representative anode materials used for SIBs [35]. Among these anodes, the highest attention is paid to different carbon-based materials, therefore, this chapter focusses on different kinds of carbon-based materials that have been explored as SIB anodes.

3.   Carbon anode materials for advanced SIBs

3.1   Graphite as anode for SIBs

Graphite is one of the major carbon allotropes found naturally and can be prepared from petroleum or coal at high temperatures. Graphite has the layered framework in an ABAB-hexagonal stacking with graphene layers which are packed in combination with the weak Van der Waals forces. The interlayers are separated at a distance of which is suitable for hosting guest metal ions especially Li^+ by forming graphite-intercalation compounds (GICs). It is regarded as a semi-metal because of its effective electronic composition that offers both thermal and electrical conductivity [49].

Commercially graphite has been the most commonly applied LIB anode caused by the formation of LiC_6 (372 mAh g^{-1}) during intercalation however; its structure seems less favourable for forming Na-C binary compounds when used as anode for SIBs [50] and in this case, it may give only 31 mAh g^{-1} theoretical capacity for NaC_{70} [51]. Several computational studies have been conducted to explore the reason for this low capacity and density functional theory (DFT) studies have revealed the formation of NaC_6 and NaC_8 intercalated structures that suffer high instability due to the incompatibility of the larger and heavier Na^+ ions to accommodate in to the graphitic layers, stretching in C-C bonds lengths and high Na/Na$^+$ redox potential [52, 53].

Scientists have attempted to tune up the interlayer lattice distance of graphite to accommodate more Na^+ ions and have reported graphitic network with an interlayer spacing of 4.3 Å (typical graphite d-spacing = 3.354 Å) and they have reported capacities as good as 284 mAh g^{-1} (20 mA g^{-1}) for 2000 cycles [54]. The Fig. 3 illustrates an interlayer expansion of graphite electrode during sodium storage process [54]. Some other studies have suggested that choosing an appropriate solvent with graphite can help enhance Na^+ ion intercalation by co-intercalation effect following equation 1 [55].

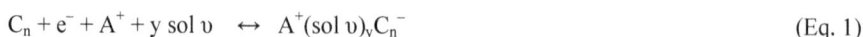

$$C_n + e^- + A^+ + y \text{ sol } \upsilon \quad \leftrightarrow \quad A^+(\text{sol } \upsilon)_y C_n^- \tag{Eq. 1}$$

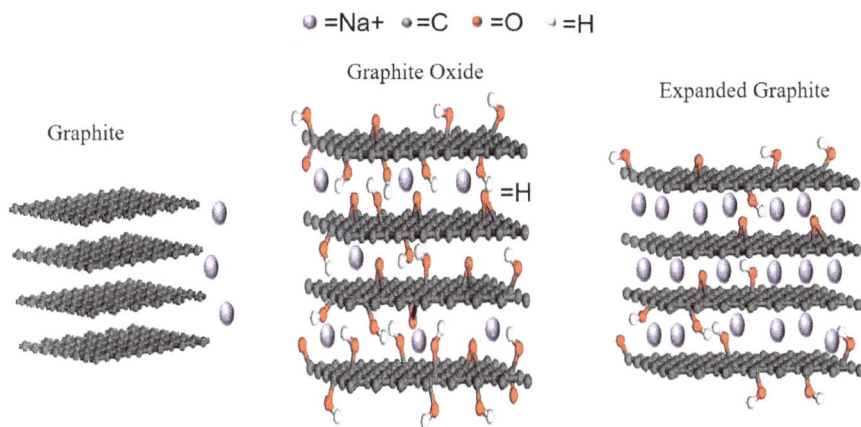

Figure 3. A scheme showing the possible sodium storage in graphite-based materials. Reproduced with permission [54]. Copyright 2019 Nature Publishing Group.

For example, using diglyme as a solvent for electrolyte the capacity observed was 100 mAh g^{-1} under the current rate of 37 mA g^{-1} for 1000 cycles [55] and by using ether-based electrolyte the capacity increased to 150 mAh g^{-1} under 100 mA g^{-1} current and retain to 127 mAh g^{-1} at the completion of 300 cycles [56]. Several other solvent combinations (DME, EC/DEC, TEGDME and DEGDME) with sodium electrolytes (NaClO$_4$, NaPF$_6$, NaCF$_3$SO$_3$) were also evaluated and their phase transitions were measured by operando X-ray diffraction technique [57].

The electrochemical performance of SIBs seems to be considerably affected by the electrolyte components with graphite and it has been found that the co-intercalation effect of electrolyte helps reduce the usual repulsive interactions of Na$^+$ ions and graphite together with the higher solvation energy and increased stability of the Na-solvent complex [50, 57]. Along with the favourable effects of co-intercalation the drawbacks include; the higher consumption of electrolyte, high intercalation voltage leading to the low energy density of full sodium ion cell and large volume expansions (~350 %) [58]. Till now the insertion/intercalation of Na$^+$ ions within graphite layers has proved impractical, so alternative schemes for bulk power storage systems are needed to augment the thermodynamic stability of sodium-graphite intercalation compounds.

3.2 Hard carbon as anode for SIBs

Hard carbon is the material most frequently recorded for SIBs anode. Amorphous carbon materials that are generally obtained from the pyrolysis (T < 1500 °C) of hydrocarbon and organic polymer-based precursors that are divided into two groups, namely graphitizable (soft carbon) and non-graphitizable (hard carbon) carbons [59]. These types of carbons exhibit a structure bearing planar hexagonal layers with partial crosslinking of sp^3 hybridized carbon atoms and area with graphitic-like structure bearing sp^2 carbon atom layers but lacking long-range order in the c direction (Fig. 4a,b) [59, 60]. Soft carbon is the one with reduced crosslinking as a result when these are subjected to heat treatment in the temperature range between 1500 to 3000 °C, the interlayer carbon atoms become mobilized and develop into graphite-like crystallites. While hard carbon has a high degree of crosslinking, therefore, their carbon layers remain immobilized even at high temperature and hence they will never transform towards a true graphitic structure [61]. Hard carbons receive their popular name as they are mechanically much harder in comparison to soft carbons. The degree of crosslinking is related to the accretion status of the transitional state achieved in the course of carbonization process as soft carbons are generally shaped from the fluid or gaseous state, unlike the hard carbon that originates from the solid phase. The general type of precursors used for hard carbon includes oxygen-rich organic materials, polymers such as phenol-formaldehyde resins together with plant-derived materials including cellulose, sugar, charcoal, and coconut shells etc. On the other hand, soft carbons are prepared from hydrocarbons and specifically aromatic based products [49]. The pyrolysis conditions and the composition of precursor play a primary role in the purity and yield of hard carbon. The two bands at 1350 (D-band) and 1580 cm^{-1} (G-band) in Raman spectra demonstrate the varying degree of defect-induced mode and ordered graphitic mode, respectively. The degree of graphitization reduces when the rate of carbonization rises owing to increased relative order and the higher angle shift of (002) planes in XRD along with I_D/I_G ratio in Raman can provide and experimental evidence [62]. During carbonization process of hard carbon precursors the release of gases lead to materials with excellent properties like its exceptional porosity (1000 m^2 g^{-1}) [63], low particle density (1.4–1.8 g cm^{-3}) [59, 64-66] low bulk powder density (0.35 g cm^{-1}) [66], and the significant preservation of the initial polymeric microstructure.

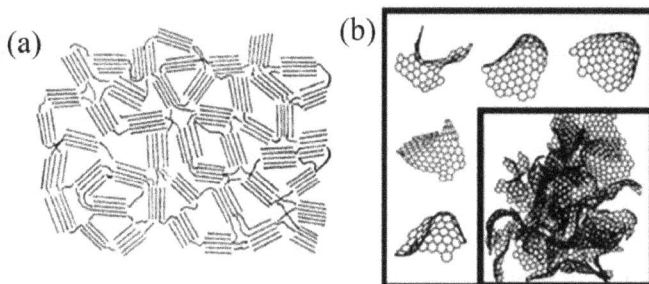

Figure 4. A variety of schemes put forward by scientists in order to explain the actual design of hard carbon. Reproduced with permission [59, 60]. Copyright 2019 Royal Society of Chemistry.

Hard carbon is the most successfully used carbon anode for SIBs however there are numerous challenges related to their use. The precursor used for preparing hard carbon together with synthesis procedures, particle sizes, porosity, and vacancy defects plays an important role in achieving higher reversible capacities for SIBs. The first ever possibility of non-graphitic carbon as anode for SIBs was reported in 1993 where disordered soft carbon derived from pyrolysis of petroleum coke was used [51]. Later Stevens and Dahn explained the falling cards or house of cards mechanism of sodium storage in hard carbon [67-69]. According to which the first stage involves the addition of Na^+ ions between the graphene layers causing an interlayer expansion resulting in the sloping voltage region as shown in Fig. 5a [69] followed by a second stage in which Na^+ ion adsorb on the nanopores of the hard carbon structure producing a plateau region. The mechanism was named on cards due to the inherent structure of graphene layers placed one over the other like playing cards. The experimental evidence was provided by using small angle X-ray scattering (SAXS) where the diffraction peak shifts to lower 2theta suggesting an interlayer expansion of graphene layers (d-spacing = 3.8 Å) upon insertion of Na^+ ions [70]. ^{23}Na NMR studies further revealed the sodiation mechanism in hard carbon where the two resonance peaks at 5.2 and 9.9 ppm are because of the presence of Na^+ ion between the layers of hard carbon while a broader resonance peaks at about 9.0 and 16.0 ppm suggest the presence of Na^+ ion in randomly distributed nanopores in hard carbon [71]. Templated carbon with hierarchical porosity with a pore volume of 0.798 cm^3 g^{-1} and an exposed surface of 346 m^2 g^{-1} was developed by Wenzel et al., and found to present a superior rate capability of 100 mAh g^{-1} at C/5 [72]. They also demonstrated the relationship between the electrochemical and physicochemical properties with commercially available high surface area (1041 m^2 g^{-1}) porous (1.008 cm^3 g^{-1} pore volume) and non-porous carbon samples. They achieved improved capacity and rate performance unlike the high surface area commercial porous carbon, certifying that surface area and pore volume do not directly impact on increasing performance of SIBs anode [72]. Later some *ab initio* studies have indicated that instead of increased interlayer distance the generation of vacancy defects in hard carbon helps increase the specific capacity by developing stronger binding with Na^+ ions and overtaking the Van der Waal forces [73]. Bommier et al., later deviated from the card-house model by putting forward three-stage Na^+ ions intercalation in hard carbon as shown in Fig. 5b. According to their study, the surface assembly of Na^+ ions at defect bearing sites represented by the slope of the voltage region in Fig. 5b is followed by intercalation and finally adsorbed at the pore sites in hard carbon lattice during the plateau region [73, 74].

3.3 Graphene as anode for SIBs

To overcome the world energy demands, scientists are trying to regenerate the renewable resources. Rechargeable energy storage devices can be a surprise if their specific energy capacity and stability related problems get resolved. The unique structure of graphene makes it the ideal candidate for anode material in energy storage devices [75-80]. Graphene, as well as its modified forms like GO & rGO, are extensively utilized for SIB applications. Chou et al., first used graphene as an anode in SIBs in 2013 and achieved promising results [81]. Since then various composites of graphene, GO and rGO have been applied as promising anode for SIBs e.g., metal oxides graphene composites [82, 83].

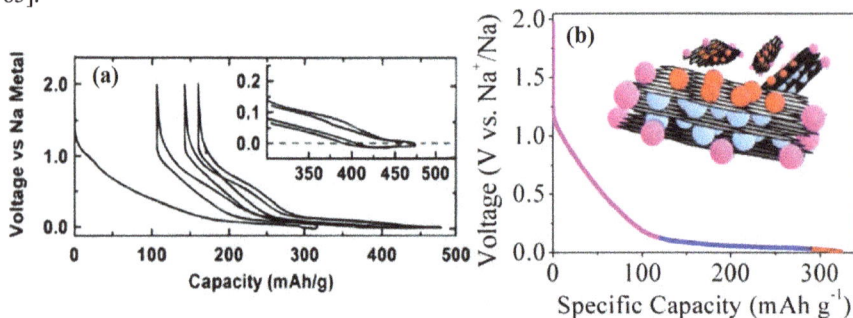

Figure 5(a) The voltage profile of glucose pyrolyzed to 1000 °C vs Na/Na⁺; the details of the low-potential region are shown as an inset. Reproduced with permission [69]. Copyright 2019, Electrochemical Society (b) Visual representation of the card-house model on sodium-ion storage in hard carbon. The two distinct phases: pore filling and intercalation inside TNs are seen. Reproduced with permission [74] Copyright (2019) The American Chemical Society

Among various composites, SnO_2 graphene composite showed very interesting properties due to their p-type nature, unidentified valence and large band gap for rechargeable energy storage devices [84-87]. SnO has a high theoretical capacity, large interlayer spacing and the layered morphology helps in minimizing the volumetric expansions during intercalation and de-intercalation. In this regard, Hang and coworkers fabricated 2D SnO anode that offers superior Na^+ ion shuttling [88]. Present SnO electrodes, however, suffer from major capacity fading issues, mostly because of their volumetric changes, so various groups have opted carbon supports to composite the tin oxide (SnO) so to have better anode characteristics. The sluggish charge conduction issue of SnO can be mitigated by incorporating carbon cloth [89], carbon nanotubes (CNTs) [90], and graphene [90, 91]. Chen et al., [92] fabricated a flexible anode for SIBs composed of SnO nanoflake supported over 3D graphene/CNTs. Ultimately the 3D graphene/CNTs@SnO_2 showed high rate performance and excellent steadiness during repeated cycling. When subjected to current rates starting from 100 to 1000 mA g^{-1}, the resulting specific capacity of the graphene/CNTs@SnO_2 declined from 584 to 390 mA g^{-1}, respectively.

In continuation of this work on metal oxide/graphene composites various transition metals have been used as bare or composited to make them compatible in energy storage application. In 2014, Jian et al., [82] fabricated Fe_2O_3/GNS for SIBs which showed cycling performance of 400 mAh g^{-1} (100 mA g^{-1}) above 200 cycles. Because of its suitable theoretical capacity, the ferric oxide is also being considered as potential electrode material for SIBs, because of its natural abundance, low processing cost and non-toxicity but for practical application in energy storage devices the volume expansion, particle agglomeration and capacity fading issue must be mitigated [93-98]. To overcome the former issues making composites with carbonaceous material like graphene, porous carbon and CNTs may be a good addition. GO/rGO@Fe_2O_3 with different composite ratios exhibited promising electrochemical results.

Among other transition metals, CuO has gathered much attention principally due to its chemical stability, nontoxic nature, abundant sources and its low cost. CuO has undergone through many for its affability as an anode in secondary batteries [99]. However, due to high volume expansion and low conductivity, CuO is not suitable for SIBs. To eliminate these problems Klein et al., proposed a conversion type reaction for CuO in SIBs including building composites, optimizing nanoparticles structure and tough architecture [99-103]. Wang et al., reported an electrospinning technique to synthesize CuO quantum dots which were wrapped in carbon nanofibers. They utilized it as binder-free SIB anode achieving higher gravimetric capacity. Similarly, rGO with excellent electrical and robust mechanical flexibility was incorporated into CuO anodes. The synergistic effect of CuO and rGO is ascertained from the high performance of CuO/rGO composite ranging from 470 (100 mA g^{-1}) to 350 mAh g^{-1} (2000 mA g^{-1}) [104].

Due to the suitable theoretical capacity of 718 mAh g^{-1}, abundance and eco-friendly nature, NiO is one of the most well-known anode materials for SIBs. The drawback of NiO is the sluggish Na^+ ion kinetics offered by the NiO and high particle agglomeration produced by volume changes that lead to irreversible capacity fading and lower cyclic stability issues [105]. Electrochemical properties of NiO for SIBs are even not seen in some cases [106]. He et al., [107] followed mechanistic approach using a series of co-related techniques like synchrotron studies, real-time electron spectroscopy and X-ray diffraction to observe stepwise reaction products and it was found that NiO continuously gets transformed into Ni and Na_2O when bond with Na^+. A NiO/NiO-graphene hybrid fabricated from a Ni derived MOF (metal organic framework) has been utilized that offered high capacity as SIB anode [108]. As displayed in Fig. 6a-d, the cycle stability of the NiO/Ni/graphene anode was highly favorable with low fading capacities (0.2% / cycle) and the rate capabilities reported at 0.2, 0.5, 1 and 2 A g^{-1} were respective of 295, 385, 207 and 248 mAh g^{-1}.

Various reports suggest the possibility of CoO as a potential substitute for the graphite anode of LIBs mainly because of its modest theoretical capacity (715 mAh g^{-1}) [109] but its applicability for the SIBs has not been much reported [110]. Chang et al., revealed a low-temperature fabrication of CoO microsphere films on to a Cu foil with an effective surface area of 103 m^2 g^{-1} using a hydrothermal method. As an anode for SIBs, these CoO microspheres exhibited a capacity of 172 mAh g^{-1} (100 mA g^{-1}) at 100th cycle [111]. In order to mitigate the problems of large volumetric changes and enhancing the electrical conductivity, composites with graphene can provide positive solutions to these drawbacks.

Titanium dioxide (TiO_2) has proved itself a versatile candidate with applications in photocatalysis, fuel cells and batteries etc. undoubtfully because of its nontoxic nature, abundant resources, cost-effectiveness and exceptional stability. TiO_2 has already been proven an effective electrode for SIBs and has various phases such as anatase, brookite, rutile and hollandite [112, 113]. Owing to different structures TiO_2 can offer excellent stable capacities when utilized as anode for SIBs in its various polymorphs mainly anatase, rutile and brookite. Anatase has been reported to be more favoured and can offer 2D diffusion channels for Na^+ ion insertion [114]. But the main problem for TiO_2 as anode for SIBs belong to its low conductivity, sluggish kinetics for Na^+, low capacity with large capacity loss, especially at higher current densities. These problems can be minimized by forming composites, heteroatom doping and combination with carbonaceous materials. Generally, CNTs, porous carbon and graphene are regarded as superior additive [115, 116].

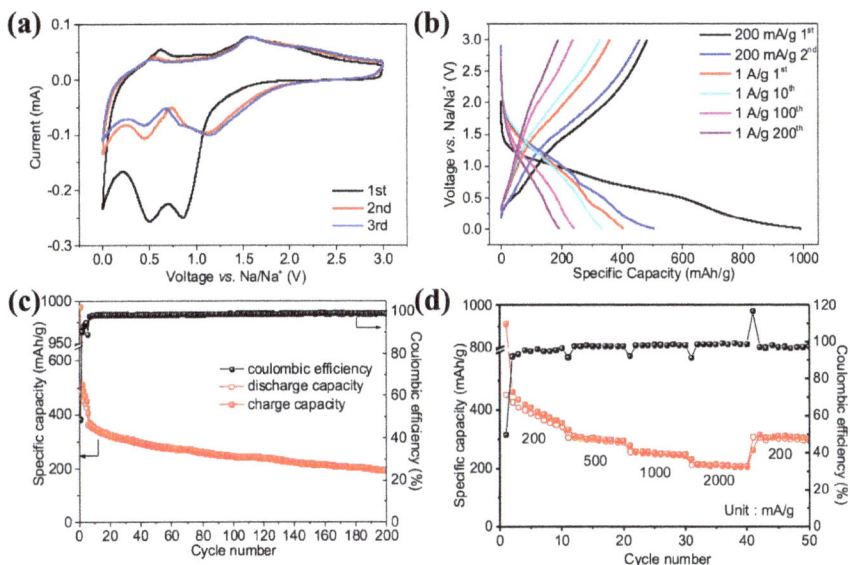

Figure 6. (a) Cyclic voltammogram curves for first three cycles of NiO/Ni/Graphene composite vs Na/Na$^+$ (b) Voltage profile of NiO/Ni/Graphene composite vs Na/Na$^+$ for selected cycles at different current densities. (c) Cyclic charge-discharge performance of NiO/Ni/Graphene composite with corresponding columbic efficiencies. (d) Rate performance of NiO/Ni/Graphene composite at various current densities. Reproduced with permission [108], Copyright 2019 The American Chemical Society.

Due to its well renowned electrical conductivity, price compatibility and high chemical stabilization, MoO_2 has gained more attention among various transition metals and has exhibited promising results as an electrode for SIBs [117, 118]. But MoO_2 is not an upright candidate for energy storage devices owing to volume expansion problem. Many trials are being carried out to overcome this issue especially by applying conductive carbon coatings to prevent volume expansion beyond a certain limit [119, 120]. Huang et al., investigated MoO_2/GO composite for SIB anode that is capable to deliver a gravimetric capacity of 483 mAh g^{-1} (100 mA g^{-1}) at 10th cycle [120].

Materials Research Forum LLC
https://doi.org/10.21741/9781644900833-2

VO_2 is yet another transition metal oxide much explored for being a potential anodic material in LIBs whereby it offers high capacity while being abundant in sources with low cost, however for SIB anode its results are not so impressive especially as anode [121]. Man thiram et al., [122] investigated electrochemical properties of VO_2/rGO in energy storage systems however, owing to the large particle size of VO_2 higher capacities were not achieved. More studies are required to understand the reaction mechanism at low potential regions. Due to its layered structure, V_2O_5 is commonly used as a catalyst and in power storage devices [123-127]. The intercalating distance of orthorhombic V_2O_5 (4.37 Å) is not considered suitable for SIBs as owing to layer type its capacity fades rapidly [128]. Moretti's group first time investigated the use of amorphous V_2O_5 as anode for SIBs. To increase the cycle stability, rate capacity and electronic conductivity GO/rGO can be used as an additive. Only a few results have been reported as cathode for SIBs using V_2O_5/graphene composite [128, 129].

Transition metal dichalcogenides, especially metal sulfides, have recently gained researchers interest with their distinctive physicochemical characteristics and potential applications in energy storage devices [130, 131]. WS_2/graphene nanocomposites fabricated via simple hydrothermal process [132] have been used as an anode in SIBs exhibiting the reversible specific capacity of 594 mAh g^{-1} and after 500 cycles 283 mAh g^{-1} (40 mA g^{-1}) capacity was still retained. The proposed electrochemical reaction may be as,

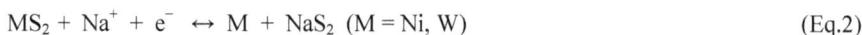

$$MS_2 + Na^+ + e^- \leftrightarrow M + NaS_2 \ (M = Ni, W) \tag{Eq.2}$$

In 2014, various investigations have been carried out into MoS_2/rGO nanocomposites. The exfoliated MoS_2/rGO composite revealed a specific capacity of 165 mAh g^{-1} at the end of the 50th cycle under 20 mA g^{-1} current rate [133].

In the preparation of SbOx/rGO composite, Zhou et al., [134] used wet ball milling technique and it delivered 352 mAh g^{-1} (5 A g^{-1}) capacity at 100th cycle, while 409 mAh g^{-1} capacity is offered while using current density of 1 A g^{-1}, with more than 95% retention in capacity for SIB. Similarly, Sb_2S_3/rGO composite based SIB anodes were investigated offering 700 mAh g^{-1} at 0.05 A g^{-1} with smaller capacity fading till the end of 50 cycles [135]. The following reactions were proposed to be taking place,

$$Sb_2S_3 + 6Na^+ + 6e^- \leftrightarrow 2Sb + 3Na_2S \tag{Eq.3}$$

$$2Sb \ + \ 6Na^+ \ + \ 6e^- \ \leftrightarrow \ 2Na_3Sb \qquad\qquad (Eq.4)$$

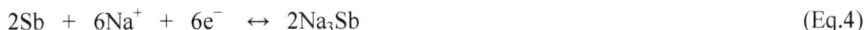

Sulphides have been opted as SIB anode because of their suitably higher capacity, natural abundance and economy. But unfortunately, their sodium intercalating mechanism controlled fabrication and cycling performance still matters of concern.

3.4   Carbon nanofibers as anode for SIBs

Carbon nanofibers belong to 1D (one-dimensional) carbon-based materials consisting of discontinuous cylindrical filaments having aspect ratios greater than 100. They contain sp^2-hybridized carbon atoms but differ from CNTs as they do not have graphene layers [136-138]. Carbon nanofibers possess excellent electronic conductivity making them well suited for applications as an electrode for both LIBs and SIBs to enhance their electrochemical performance. Carbon nanofibers have been structured by several synthetic protocols including electrospinning, catalytic chemical vapour deposition growth, biomass methods and template-based synthesis [139-145].

ID carbon nanofibers that are synthesized from cracking of polymers display encouraging electrochemical performance as SIBs anode due to the abundance of defect sites for Na^+ ion storage [146-159]. To see the effect of porous morphology Li et al., used Pluronic F-127, a triblock copolymer as a precursor to synthesize porous carbon nanofibers that were found to be highly effective as SIB anode [148]. As shown in Fig. 7a,b the porous carbon fibre shows a good rate capability of 60 mAh g^{-1} (10,000 mA g^{-1}) with an excellent reversible capacity of 266 mAh g^{-1} when cycled at 50 mA g^{-1} current density. These distinctive results have been credited to the 3D interconnected microspores that not only increase the surface available for a reaction but also increase the structural stability during cycling to prevent any cracking. Motivated by this work, several other efforts have been carried out to synthesize porous carbon nanofibers together with nitrogen and/or oxygen doping to induce defects to increase ion-storage sites. For this purpose, usually, some rich nitrogen-containing polymers (e.g. polyamide, polypyrrole) is used and subjected to carbonization to induce pores during decomposition and possible heteroatom doping [146, 152, 154, 156-158]. A practical example is a preparation of dual doped porous carbon nanofiber that is prepared from the breakdown of bacterial cellulose and polyaniline [160]. The porous cellulose@PANI nanofibers based material presents 545 mAh g^{-1} (100 mA g^{-1}) capacity after 100 cycles with improved rate performance. The high reversible capacity is ascribed to 3D interconnected pores and heteroatom doping that can together enhance electronic conductivity, penetration of electrolyte, fast electron transfer pathway and increased storage sites.

Materials Research Forum LLC
https://doi.org/10.21741/9781644900833-2

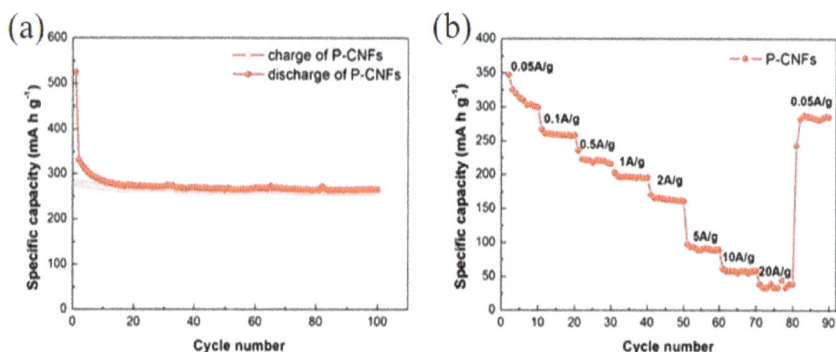

Figure 7. (a) Cyclic charge-discharge performance of P-CNFs for 100 cycles at 50 mA g^{-1} and (b) Rate capability of P-CNFs at various current densities. Reproduced with permission, [147] Copyright 2019 Royal Society of Chemistry.

Other than fabricating various morphologies of carbon nanofibers, scientists have attempted to prepare nanocomposite structures to get more benefited from the capabilities of carbon nanofibers as anode for SIBs. Using both outer surface anchoring and encapsulation methodologies various carbon nanofiber based composites are reported. Hou et al., prepared polypyrrole derived carbon nanofiber that was decorated with Sb nanoparticles to tune up the surface of carbon nanofiber [143]. An improved cycling performance with a capacity retention of 96.7% was observed when the materials electrode was cycled at 100 mA g^{-1} for 100 cycles. When the materials electrode was cycled at a current density of 100 mA g^{-1} for 100 cycles an improved cycling performance with a capacity retention of 96.7% was observed. In order to circumvent the huge volume changes of Sb, an encapsulation strategy was proposed by Zhu et al., and they prepared 0.4 nm thick MoS_2 nanoplates that were encapsulated in carbon nanofiber [161]. They noticed an increased reversible capacity of up to 484 mAh g^{-1} at a moderately low current of 1000 mA g^{-1} and 253 mAh g^{-1} even at a very high rate of 10,000 mA g^{-1}. Various researchers have adopted similar strategy using other metal oxides, sulfides and selenides such as $Li_4Ti_5O_{12}$ [162], SnO_2 [163], $MnFe_2O_4$ [164], TiO_2 [115], MoS_2 [165, 166], $FeSe_2$ [167], and $NiSe_2$ [168] for improving electronic and ionic conductivities also adopt an almost similar strategy. Later, porous carbon nanofibers bearing electrodeposited SnO_2 with carbon coating ($PCNF@SnO_2@C$) was also applied as composite carbon fibre anode for SIBs and the results were quite impressive with the high storage capability of 374 mAh g^{-1} up to 100 cycles at 50 mA g^{-1} of current density and high rate capability [169]. The reason for this electrochemical activity was attributed to the porous and elastic nature of carbon nanofibers and protective effect of the carbon coating layer together with SnO_2 as active redox metal. There are numerous other research articles highlighting the benefits of porous carbon nanofibers, doping with some heteroatom, surface anchoring with metals and encapsulation structures for the enhancement of structural feature that can ultimately effect sodium storage performance as displayed in Table 2 [146, 148, 149, 151-154, 156-158, 170] and Table 3 [58, 115, 143, 161, 171-181].

Table 2   *List of reported carbon nanofibers based anodes for SIBs. Reproduced with permission [138]. Copyright 2019 Royal Society of Chemistry*

Materials	Preparation strategy	Diameter & Surface area	Electrochemical performance (mAh g^{-1})	Rate (mA g^{-1})
CNFs	Biomass method	50 ~ 100 nm / 377 m^2 g^{-1}	176 after 600 cycles	200
CNFs	Electrospinning	200 ~ 300 nm /NA	233	50
Porous CNFs	Electrospinning, F127 template	~280 nm / 74.59 m^2 g^{-1}	266 after 100 cycles	50
CNFs@N-doped porous carbon	Biomass (CNFs), Oxidative template assembly (porous carbon)	80 ~ 110 nm / 300.2 m^2 g^{-1}	240 after 100 cycles	100
N-doped porous CNFs	Oxidative template assembly	50 ~ 70 nm / 1508 m^2 g^{-1}	243 after 100 cycles	50
N-doped CNFs	Electrospinning	100 ~ 300 nm / 564.4 m^2 g^{-1}	377 after 100 cycles	100
N-doped hollow CNFs	Oxidative template assembly, KOH activation	~300 nm / 868 m^2 g^{-1}	160 after 100 cycles	50
N-doped CNFs	Electrospinning, N2 treatment	~150 nm / 513.89 m^2 g^{-1}	254 after 100 cycles	50
N-doped CNFs	Electrospinning, urea treatment	~250 nm / 8.16 m^2 g^{-1}	354 after 100 cycles	50
N, O-doped porous CNFs	Biomass (CNFs), Oxidative template assembly (porous carbon), KOH activation	100 ~ 150 nm / 1426.1 m^2 g^{-1}	545 after 100 cycles	100
N, B-doped porous CNFs	Biomass (CNFs), NH4HB4O7·H2O treatment	30 ~ 80 nm / 1585 m^2 g^{-1}	581 after 120 cycles	100

Table 3   *List of reported carbon nanofibers based composite anodes for SIBs. Reproduced with permission [138]. Copyright 2019 Royal Society of Chemistry*

Materials	Preparation strategy	Diameter & Surface area	Active material & Weight Content	Electrochemical performance (mAh g^{-1})	Rate (mA g^{-1})
Sb anchored on CNFs	Biomass method (CNFs), Solution method (Sb)	~50 nm/ 160.3 m^2 g^{-1}	Sb, 80.8 wt%	542. 5 after 100 cycles	100
PCNF@SnO2@C	Electrospinning (CNFs), Electrodeposition (SnO$_2$), CVD (carbon coating layer)	~200 nm / N/A	SnO$_2$, 38.5 wt%	374 after 100 cycles	50
Sb@CNFs	Electrospinning	~200 nm / N/A	Sb, 38 wt%	446 after 400 cycles	200
Sb@CNFs	Electrospinning	~250 nm / N/A	Sb, 54 wt%	350 after 300 cycles	100

Sn nanodots@Porous N-doped CNFs	Electrospinning	120 nm / 316 $m^2\,g^{-1}$	Sn, 63 wt%	483 after 1300 cycles	2 A g^{-1}
SnSb@Porous CNFs	Electrospinning	~250 nm / N/A	SnSb, 60 wt%	345 after 200 cycles	0.2 C
SnSb@Porous CNFs	Electrospinning	100 nm / N/A	SnSb, 44 wt%	600 after 100 cycles	200
SnO$_x$@CNFs	Electrospinning	100 ~ 300 nm / N/A	SnO$_x$, 17 wt%	210 after 300 cycles	500
Li$_4$Ti$_5$O$_{12}$@CNFs	Electrospinning	100 nm / N/A	Li$_4$Ti$_5$O$_{12}$, 90.5 wt%	162.5 after 100 cycles	0.2 C
TiO$_2$@CNFs	Electrospinning	~120 nm /15.7 $m^2\,g^{-1}$	TiO$_2$, 81 wt%	237.1 after 1000 cycles	200
MnFe$_2$O$_4$@CNFs	Electrospinning	180 nm/239.6 $m^2\,g^{-1}$	MnFe$_2$O$_4$, 66.8 wt%	360 after 4200 cycles	2000
MoS2@CNFs	Electrospinning	~150 nm / N/A	MoS$_2$, 83.2 wt%	283.9 after 600 cycles	100
MoS$_2$@CNFs	Electrospinning	~50 nm / N/A	MoS$_2$, 62 wt%	484 after 100 cycles	1 A g^{-1}
NiSe$_2$-rGO-CNFs	Electrospinning, Selenization	~1 µm / 119 $m^2\,g^{-1}$	NiSe$_2$, 75 wt%	468 after 100 cycles	200
Hollow nanospheres FeSe$_2$-rGO-graphitic CNFs	Electrospinning , Selenization	~2 µm / 34 $m^2\,g^{-1}$	FeSe$_2$, 73 wt%	412 after 150 cycles	1 A g^{-1}

3.5   Biomass-derived carbon as anode for SIBs

Biomass is an organic material mainly composed of C, H and O. After burning process of biomass the derived carbon can be a useful raw material for SIBs anode. Usually, carbon obtained from biomass has an amorphous structure which is more suitable for batteries due to its low production cost and easy synthesis process [182-185]. Biomass-derived carbon resources are found abundantly and eco-friendly, so these are attracting considerable attention towards them. Biomass carbon is commonly obtained through followings steps (a) washing, (b) drying to remove moisture, (c) calcination at elevated temperature. Previously, a diversity of carbonaceous materials have been obtained through biomass and used in storage devices [186]. Carbonaceous material obtained through carbonization of woods have current collector free and binder free properties [183]. Ding et al., [182] fabricated less graphitic carbon nanosheets micropores frameworks possessing a wider distance between the layers (0.388 nm) as compared to that for ordered graphite (0.335 nm). When applied as an anode in SIB the insertion and deinsertion capability of Na$^+$ ion is 306 and 532 mAh g^{-1} at varying current densities. While at a current of 100 mA g^{-1} the sodium storing capacity dropped to 255 mAh g^{-1} upon extended cycling of up to 200. Mitlin et al., [187] obtained carbon from banana peels that have a lesser surface area and this material showed 336 mAh g^{-1} as capacity with 11% drop in capacity after repeatedly undergoing 300 cycles of charge and discharge. At low current density 500 mA g^{-1} the achievable gravimetric capacity is 221 mAh g^{-1} and after cycling it merely reduces by 7%. Yang and co-workers [184] obtained

N-doped carbon from okra through carbonization and exfoliated the sheets to get the high specific area with facile sodiation and desodiation into carbon sheets during cycling. The remarkably high surface area yielded a capacity of 292.2 mAh g^{-1} with a steady C.E for more than 2000 cycles. Hu et al., [188] demonstrated various biomass-derived hard carbon-based potential materials for SIBs anode. Similarly, it is reported that the hard carbon-based micro tubes derived from carbonization of natural carbon at 1000, 1300, and 1600 °C are capable to offer a temperature dependent, sufficiently high SIB anode performance. The electrochemical activity has also increased with an increase in the temperatures of carbonization from 1000, 1300 to 1600 °C, demonstrating that the electrochemistry is seriously linked to the carbonizing temperature of the hard carbon microtubes [189]. Apple waste derived hard carbonaceous anode material [190] showed high SIB performance and has the advantage of incorporation of nitrogen and sulfur from apple proteins, that assists sodiation and desodiation reactions.

Zhao et al., [191] presented carbon nanoparticles derived from the flame deposition of coconut oil and demonstrated a second cycle sodiation and desodiation capacities of 277 and 217 mAh g^{-1} at 100 mA g^{-1} after 20 cycles. Doped carbon microsphere were also produced through carbonization from oatmeal [192]. The electrochemical results showed that after 50 cycles N-doped carbon microsphere delivered a capacity of 336 mAh g^{-1} when cycled at 50 mA g^{-1} current. Huang et al., [185] described a hard carbon SIB anode material derived from pomelo peels by pyrolyzing in H_3PO_4 to yield a high surface area hard carbon containing O/P containing functional groups that can show a pseudocapacitive effect from exterior redox reaction. After 220 cycles the reversible capacity of SIBs remained 181 mAh g^{-1} (200 mA g^{-1}) but, regrettably, the initial C.E of the material was only 27%. Hu et al. [193] suggested a single stage heat treatment to get carbon sheath from natural leaves and showed promising electrochemical performances with a large first C.E of 74.8% together with a high reversible capacity of 360 mAh g^{-1} **(Fig. 8a-i)**. Li and coworkers [183] fabricated an ultra-thin mesoporous carbon by simple carbonization of wood and the as-synthesized mesoporous carbon had well-aligned pores in the horizontal and vertical directions. In SIBs, mesoporous carbon is proposed to be applied with high outcomes in terms of capacity (13.6 mAg cm^{-3}) which is greater than that for orthodox LIBs together with benefits of being binder and current collector free anode. Zheng et al., [194] fabricated a dual-natured hard carbon by hydrothermal and subsequent pyrolytic treatment of holly leaf and the resulting carbon possess both mesopores and macropores with tiny stranded graphitic domains. It has been proposed that the hydrothermal treatment is vital for the lamella-like structuring and enlarged pores along with nanoporous graphitic portions obtained directly by pyrolyzing the holly leaves. Lamellar fabricated carbon showed the high sodium storage of 318 mAh g^{-1} (20 mA g^{-1})

with superb rate performance. It is worth to conclude that to choose the biomass precursor consideration should be made that it must not be obtained from highly consumed food to avoid its scarcity and price issues. It should be significant that biomass contains unique chemical composition and microstructure which may directly define the composition and structure of obtained carbonaceous material, most importantly electrochemical features of sodium storage performances [195].

3.6   Heteroatom-doped carbon materials as anode for SIBs

Heteroatom doped (e.g. N, S, P and B) carbon anode materials have been explored for SIBs and have proved to boost the capacity, electronic wettability and electronic conductivity which can ease the electrode/electrolyte interactions and charge transfer [196, 197]. Thus, various heteroatom based carbonaceous materials have come up with improved performance as conceivable anodes for SIBs while noteworthy results have been shown by some candidates. So far, N-doped carbon anodic materials have come up with great potential that enhances the electronic conductivity and reaction activity by producing extrinsic imperfections. Huang et al., [198] in 2013, reported carbon nanofiber doped with nitrogen showing 134 mAh g^{-1} reversible capacity after 200 cycles at 0.2 A g^{-1} while the anodic material has been shown to deliver a 72 mAh g^{-1} capacity at a current of 20 A g^{-1}. Afterwards, a number of carbon-based materials with different morphologies have been doped with nitrogen for enhanced performance as SIBs anode. For instance, porous carbon-based nanosheets, mesoporous carbon, carbon films, CNTs and carbon spheres with N-doping have been reported in addition to the various graphene-based anodes with nitrogen doping, etc. [146, 158, 196-200]. From these 3d graphene foams [197] doped with nitrogen have shown a higher N-content of 6.8 % displaying a capacity of 1057.1 mAh g^{-1} (0.1 A g^{-1}) and 594 mAh g^{-1} (150 A g^{-1}). They have also explored rGO and GO-based nitrogen doped anode for SIBs with improved sodiation capacity and they proposed that the synergistic effect of 3d foam like morphology and N-doping impart role in improving the electrochemical cycling behaviour of the electrodes. Lou et al. [154] found that after extending their cycling up to 7000 cycles, the reversible storage capacities of free-standing carbon nanofiber films doped with nitrogen (7.15 wt%) were around 210 mAh g^{-1} at 15 mA g^{-1} while at the higher current rate of 5000 mA g^{-1} it was still maintained at a level of 154 mAh g^{-1}. The cyclic capability has been designated to high mechanical flexibility with improved structural stability.

Recent attempts to improve the reversible sodium ion capacity by using sulfur-doped carbonaceous anodes have also proved somewhat fruitful. Furthermore, the introduction of S can increase the intercalating distance between the layers which not only increases capacity but the speed of Na$^+$ intercalation/de-intercalation. Therefore, S-doping may significantly enhance sodium storage characteristics of carbonaceous material. Chen et al. [201] have outlined the electrochemical features of sulfur covalently linked to graphene with the specific sodium storage capability of 291 mAh g^{-1} (0.05 A g^{-1}) and outstanding durability after 200 continuous cycles. Jiang et al., [202] discovered Na storage performance of S-doped disarranged carbon with 26.91% sulfur concentration achieved by pyrolizing sulfur and NTCDA (1,4,5,8-Naphthalenetetracarboxylic dianhydride). The maximum sodium storage of 516 mAh g^{-1} (20 mA g^{-1}) and 271 mAh g^{-1} (1000 mA g^{-1}) was achieved upon 1000 cycles corresponding to capacity maintenance of 85.9% **(Fig. 9a-c).** Similarly, the effect of sulfur doping on carbon as a function of sodium storage performance has also been found by Huang et al., [203]. The resulting doped material with a 15.17 wt% sulfur revealed a wider 0.386 nm interlayer spacing showing an elevated capacity of 303.2 mAh g^{-1} (500 mA g^{-1}) and a high C.E of 73.6% after 700 cycles.

Similarly, The phosphorous doped carbon nanosheets as SIB anode holding a capacity of 328 mAh g^{-1} (100 mA g^{-1}) have been reported with wider interplanar distance of 0.42 nm [204] and after 5000 cycles it is fascinating to observe that the as-doped material presented a storage competence of up to 149 mAh g^{-1} (5 A g^{-1}). This prominent efficiency behaviour is credited to the following arguments: a) smaller Na$^+$ path length and more active sites produced from owing to the ultrathin nanosheets accompanied with large surface area, b) the proficient intercalating distance which can ease the insertion/de-insertion of Na$^+$, c) transformation of the electronic states upon doping that also facilitate the adsorption of ions in the electrolyte.

Boron came up as a significant dopant for the carbonaceous anode in SIB applications, particularly of its similarity in size with carbon, along with lower electronegativity. Wang's group [205] reported the role of boron as a linker with the oxygen that helps in the expansion of the intercalating distance in the rGO along with the generation of defects that promote the sodium ion storage ability of rGO. As anticipated the rGO incorporated with boron produced 280 mAh g^{-1} (0.02 mA g^{-1}) capacity that is greater as compared to bare rGO matching nicely with the outcomes of the results reported by Mizuno and Ling [206]. Furthermore, as a next step, various groups have exploited the effect of doping of more than one heteroatoms to revive the sodiation capacity of the carbon-based anodes. To additionally increase the properties of carbon both N, S- and N, P-codoping were fabricated as N-doping alone improves the electronic conductivity and S or P-doping alone can affect the gaps between the layers of carbon. Zhang et al., [207] fabricated graded nitrogen and sulfur co-doped carbon nanospheres by pyrolysis of polyaniline and cellulose which exhibited a steady 3400 electrochemical cycles displaying a sodium storage capacity of 150 mAh g^{-1} which is substantially superior to S-doped carbon microsphere, bare and N-doped carbon sphere. Zhou and coworkers [208] used a gas-solid phase synthetic approach to replace nitrogen atoms from the doped nitrogen-rich carbon by sulfur atoms and noted the efficiency of this codoped material for SIB anode. The incorporation of the sulfur has a prominent effect on the carbon matrix producing large specific area, distorted structure and greater interlayer distance. Moreover, they used DFT calculations to show the S-doping influence, revealing that computed d-spacing of N-C and S-N/C are 3.47 and 3.73 Å, respectively. The N-doped carbon and S/N-codoped cyclic voltammetry curves show a pair of other redox spikes (1.85/1.05 V) that are unlike the Na-S battery, showing that during the Na$^+$ intercalation/deintercalation process the C-S-C-bonds are not reconstructing and broken and the capacity may be partially enhanced by the Faradaic reaction between S and Na$^+$ ions that are associated with carbon. S/N co-doped carbon improved electrochemical behaviour and they can deliver a capacity of 350 mAh g^{-1} (500 mA g^{-1}) and 110 mAh g^{-1} (10,000 mA g^{-1}), respectively in SIB.

Li et al., [209] fabricated a P/N-codoped carbon microsphere using the solvothermal process and checked its activity as SIB anode. Their results showed the exceptional sodium storing capacity of the N/P-codoped carbon electrode with the sodium storage capability of 305 mAh g^{-1} (100 mA g^{-1}) and 136 mAh g^{-1} (5000 mA g^{-1}). N/O-codoped carbon network was fabricated by Yu and co-workers [149] by thermally blending a mixture of bacteria-contaminated cellulose and polyaniline through KOH. The interlinked carbon nanofibrous network is able to facilitate the electron passage through channels and intensive nanopores in the carbon nanofibers can make available a number of lively spots for storing more sodium ions. The co-doping of N and O would improve the surface wettability and electrical conductivity of anode material based on carbon. Furthermore, it facilitates the contact between the electrode and the electrolyte and reduces the propagation range of Na$^+$ with its wonderful 3D carbon framework and large available surface. The NOC (N/O codoped carbon) demonstrated outstanding sodium storage characteristics with elevated capacities and cycle stabilization taking advantage of these properties. A stable capacity of 545 mAh g^{-1} (100 mA g^{-1}) was maintained at 100th cycle along with a nominal capacity of 240 mAh g^{-1} (2000 mA g^{-1}) after 2000 cycles.

References

[1] B. Dunn, H. Kamath, J.M. Tarascon, Electrical energy storage for the grid: A battery of choices, Science 334 (2011) 928-935. https://doi.org/10.1126/science.1212741

[2] H. Pan, Y.-S. Hu, L. Chen, Room-temperature stationary sodium-ion batteries for large-scale electric energy storage, Energy Environ. Sci. 6 (2013) 2338-2360. https://doi.org/10.1039/c3ee40847g

[3] S. Goriparti, E. Miele, F. De Angelis, E. Di Fabrizio, R.P. Zaccaria, C. Capiglia, Review on recent progress of nanostructured anode materials for Li-ion batteries. J. Power sources, 257 (2014) 421-443. https://doi.org/10.1016/j.jpowsour.2013.11.103

[4] C. De las Casas, W. Li, A review of application of carbon nanotubes for lithium ion battery anode material, J. Power sources 208 (2012) 74-85. https://doi.org/10.1016/j.jpowsour.2012.02.013

[5] V. Palomares, P. Serras, I. Villaluenga, K.B. Hueso, J. Carretero-González, T. Rojo, Na-ion batteries, recent advances and present challenges to become low cost energy storage systems, Energy Environ. Sci. 5 (2012) 5884-5901. https://doi.org/10.1039/c2ee02781j

[6] L.P. Wang, L. Yu, X. Wang, M. Srinivasan, Z.J. Xu, Recent developments in electrode materials for sodium-ion batteries, J. Mater. Chem. A 3 (2015) 9353-9378. https://doi.org/10.1039/C4TA06467D

[7] D. Kundu, E. Talaie, V. Duffort, L.F. Nazar, The emerging chemistry of sodium ion batteries for electrochemical energy storage, Angew. Chem. Int. Ed. 54 (2015) 3431-3448. https://doi.org/10.1002/anie.201410376

[8] C. Nithya, S. Gopukumar, Sodium ion batteries: A newer electrochemical storage, Wiley Interdisciplinary Reviews: Energy Environ. 4 (2015) 253-278. https://doi.org/10.1002/wene.136

[9] B.L. Ellis, L.F. Nazar, Sodium and sodium-ion energy storage batteries, Curr.Opinion in Solid State Mater. Sci. 16 (2012) 168-177. https://doi.org/10.1016/j.cossms.2012.04.002

[10] T.B. Reddy, Linden's handbook of batteries, Vol. 4. 2011: Mcgraw-hill New York.

[11] R.S. Carmichael, Practical Handbook of Physical Properties of Rocks and Minerals (1988). 2017: CRC press. https://doi.org/10.1201/9780203710968

[12] Y. Luo, Y. Tang, S. Zheng, Y. Yan, H. Xue, H. Pang, Dual anode materials for lithium-and sodium-ion batteries, J. Mater. Chem. A. 6 (2018) 4236-4259. https://doi.org/10.1039/C8TA00107C

[13] N. Yabuuchi, K. Kubota, M. Dahbi, S. Komaba, Research development on sodium-ion batteries, Chem. Rev. 114 (2014) 11636-11682. https://doi.org/10.1021/cr500192f

[14] M.D. Slater, D. Kim, E. Lee, C.S. Johnson, Sodium-ion batteries, Adv. Funct. Mater. 23 (2013) 947-958. https://doi.org/10.1002/adfm.201200691

[15] G. Ali, A. Badshah, K.Y. Chung, K.-W. Nam, M. Jawad, M. Arshad, S.M. Abbas, Superior shuttling of lithium and sodium ions in manganese-doped titania@ functionalized multiwall carbon nanotube anodes, Nanoscale 9 (2017) 9859-9871. https://doi.org/10.1039/C7NR01417A

[16] J.Y. Hwang, S.T. Myung, Y.K. Sun, Sodium-ion batteries: Present and future, Chem. Soc. Rev. 46 (2017) 3529-3614. https://doi.org/10.1039/C6CS00776G

[17] K. Kubota, S. Komaba, Practical issues and future perspective for Na-ion batteries, J. Electrochem. Soc. 162 (2015) A2538-A2550. https://doi.org/10.1149/2.0151514jes

[18] M.H. Han, E. Gonzalo, G. Singh, T. Rojo, A comprehensive review of sodium layered oxides: powerful cathodes for Na-ion batteries. Energy Environ. Sci. 8 (2015) 81-102. https://doi.org/10.1039/C4EE03192J

[19] M. Sathiya, K. Hemalatha, K. Ramesha, J.-M. Tarascon, A. Prakash, Synthesis, structure, and electrochemical properties of the layered sodium insertion cathode material: $NaNi_{1/3}Mn_{1/3}Co_{1/3}O_2$, Chem. Mater. 24 (2012) 1846-1853. https://doi.org/10.1021/cm300466b

[20] K. Hemalatha, M. Jayakumar, P. Bera, A. Prakash, Improved electrochemical performance of $Na_{0.67}MnO_2$ through Ni and Mg substitution. J. Mater. Chem. A. 3 (2015) 20908-20912. https://doi.org/10.1039/C5TA06361B

[21] N. Yabuuchi, M. Kajiyama, J. Iwatate, H. Nishikawa, S. Hitomi, R. Okuyama, R. Usui, Y. Yamada, S. Komaba, P2-type $Na_x[Fe_{1/2}Mn_{1/2}]O_2$ made from earth-abundant elements for rechargeable Na batteries, Nat. Mater. 11 (2012) 512. https://doi.org/10.1038/nmat3309

[22] Y. Lu, L. Wang, J. Cheng, J.B. Goodenough, Prussian blue: A new framework of electrode materials for sodium batteries, Chem. Commun. 48 (2012) 6544-6546. https://doi.org/10.1039/c2cc31777j

[23] Y. You, X.-L. Wu, Y.-X. Yin, Y.-G. Guo, High-quality Prussian blue crystals as superior cathode materials for room-temperature sodium-ion batteries, Energy Environ. Sci. 7 (2014) 1643-1647. https://doi.org/10.1039/C3EE44004D

[24] Y. Yue, A.J. Binder, B. Guo, Z. Zhang, Z.A. Qiao, C. Tian, S. Dai, Mesoporous prussian blue analogues: Template-free synthesis and sodium-ion battery applications, Angew. Chem. Int. Ed. 53 (2014) 3134-3137. https://doi.org/10.1002/anie.201310679

[25] D. Yang, J. Xu, X.Z. Liao, H. Wang, Y.S. He, Z.-F. Ma, Prussian blue without coordinated water as a superior cathode for sodium-ion batteries. Chem. Commun. 51 (2015) 8181-8184. https://doi.org/10.1039/C5CC01180A

[26] J. Qian, M. Zhou, Y. Cao, X. Ai, H. Yang, Nanosized $Na_4Fe(CN)_6$/C composite as a low-cost and high-rate cathode material for sodium-ion batteries. Adv. Energy Mater. 2 (2012) 410-414. https://doi.org/10.1002/aenm.201100655

[27] K. Hurlbutt, S. Wheeler, I. Capone, M. Pasta, Prussian Blue analogs as battery materials, Joule 2 (2018) 1950-1960. https://doi.org/10.1016/j.joule.2018.07.017

[28] Y. Xu, S. Zheng, H. Tang, X. Guo, H. Xue, H. Pang, Prussian blue and its derivatives as electrode materials for electrochemical energy storage, Energy Storage Mater. 9 (2017) 11-30. https://doi.org/10.1016/j.ensm.2017.06.002

[29] S.P. Guo, J.C. Li, Q.T. Xu, Z. Ma, H. G. Xue, Recent achievements on polyanion-type compounds for sodium-ion batteries: syntheses, crystal chemistry and electrochemical performance, J. Power sources 361 (2017) 285-299. https://doi.org/10.1016/j.jpowsour.2017.07.002

[30] X. Jiang, L. Yang, B. Ding, B. Qu, G. Ji, J.Y. Lee, Extending the cycle life of $Na_3V_2(PO_4)_3$ cathodes in sodium-ion batteries through interdigitated carbon scaffolding, J. Mater. Chem. A. 4 (2016) 14669-14674. https://doi.org/10.1039/C6TA05030A

[31] P. Singh, K. Shiva, H. Celio, J.B. Goodenough, Eldfellite, $NaFe(SO_4)_2$: An intercalation cathode host for low-cost Na-ion batteries. Energy Environ. Sci. 8 (2015) 3000-3005. https://doi.org/10.1039/C5EE02274F

[32] K. Zaghib, J. Trottier, P. Hovington, F. Brochu, A. Guerfi, A. Mauger, C. Julien, Characterization of Na-based phosphate as electrode materials for electrochemical cells, J. Power sources 196 (2011) 9612-9617. https://doi.org/10.1016/j.jpowsour.2011.06.061

[33] C. Yuan, Q. Wu, Q. Shao, Q. Li, B. Gao, Q. Duan, H.-g. Wang, Free-standing and flexible organic cathode based on aromatic carbonyl compound/carbon nanotube composite for lithium and sodium organic batteries, J. Colloid Interface Sci. 517 (2018) 72-79. https://doi.org/10.1016/j.jcis.2018.01.095

[34] S. Wang, L. Wang, Z. Zhu, Z. Hu, Q. Zhao, J. Chen, All Organic Sodium-Ion Batteries with $Na_4C_8H_2O_6$, Angew. Chem. Int. Ed. 53 (2014) 5892-5896. https://doi.org/10.1002/anie.201400032

[35] J.W. Choi, D. Aurbach, Promise and reality of post-lithium-ion batteries with high energy densities, Nature Rev. Mater. 1 (2016) 16013. https://doi.org/10.1038/natrevmats.2016.13

[36] W. Liu, X. Luo, Y. Bao, Y.P. Liu, G.-H. Ning, I. Abdelwahab, L. Li, C.T. Nai, Z.G. Hu, D. Zhao, A two-dimensional conjugated aromatic polymer via C–C coupling reaction, Nature Chem. 9 (2017) 563. https://doi.org/10.1038/nchem.2696

[37] X. Deng, Z. Chen, Y. Cao, Transition metal oxides based on conversion reaction for sodium-ion battery anodes, Mater. Today Chem. 9 (2018) 114-132. https://doi.org/10.1016/j.mtchem.2018.06.002

[38] B. Huang, Z. Pan, X. Su, L. An, Tin-based materials as versatile anodes for alkali (earth)-ion batteries, J. Power sources, 395 (2018) 41-59. https://doi.org/10.1016/j.jpowsour.2018.05.063

[39] J. Tang, A.D. Dysart, V.G. Pol, Advancement in sodium-ion rechargeable batteries, Current opinion in chemical engineering, 9 (2015) 34-41. https://doi.org/10.1016/j.coche.2015.08.007

[40] L. Wang, Z. Wei, M. Mao, H. Wang, Y. Li, J. Ma, Metal oxide/graphene composite anode materials for sodium-ion batteries, Energy Storage Mater. 16 (2019) 434-454. https://doi.org/10.1016/j.ensm.2018.06.027

[41] Y. Liu, C. Yang, Q. Zhang, M. Liu, Recent Progress in the Design of Metal Sulfides as Anode Materials for Sodium Ion Batteries, Energy Storage Mater. (2019) https://doi.org/10.1016/j.ensm.2019.01.001

[42] J. Chen, S. Li, K. Qian, P.S. Lee, NiMn layered double hydroxides derived multiphase Mn-doped Ni sulfides with reduced graphene oxide composites as anode materials with superior cycling stability for sodium ion batteries, Mater. Today Energy 9 (2018) 74-82. https://doi.org/10.1016/j.mtener.2018.02.008

[43] L. Wang, J. Wang, F. Guo, L. Ma, Y. Ren, T. Wu, P. Zuo, G. Yin, J. Wang, Understanding the initial irreversibility of metal sulfides for sodium-ion batteries via operando techniques, Nano Energy 43 (2018) 184-191. https://doi.org/10.1016/j.nanoen.2017.11.029

[44] S. Nie, L. Liu, J. Liu, J. Xia, Y. Zhang, J. Xie, M. Li, X. Wang, TiO_2-Sn/C composite nanofibers with high-capacity and long-cycle life as anode materials for sodium ion batteries, J. Alloys Compds. 772 (2019) 314-323. https://doi.org/10.1016/j.jallcom.2018.09.044

[45] Z. Diao, D. Zhao, C. Lv, H. Liu, D. Yang, S. Shen, Ultrafine polycrystalline titania nanofibers for superior sodium storage, J. Energy Chem. 38 (2019) 153-161. https://doi.org/10.1016/j.jechem.2018.12.009

[46] X. Li, X. Li, L. Fan, Z. Yu, B. Yan, D. Xiong, X. Song, S. Li, K.R. Adair, D. Li, Rational design of Sn/SnO2/porous carbon nanocomposites as anode materials for

sodium-ion batteries, Applied Surf. Sci. 412 (2017) 170-176.
https://doi.org/10.1016/j.apsusc.2017.03.203

[47] H. Xie, W.P. Kalisvaart, B.C. Olsen, E.J. Luber, D. Mitlin, J.M. Buriak, Sn–Bi–Sb
alloys as anode materials for sodium ion batteries, J. Mater. Chem. A. 5 (2017) 9661-
9670. https://doi.org/10.1039/C7TA01443K

[48] Q. Li, Q. Wei, W. Zuo, L. Huang, W. Luo, Q. An, V.O. Pelenovich, L. Mai, Q.
Zhang, Greigite Fe_3S_4 as a new anode material for high-performance sodium-ion
batteries, Chem. Sci. 8 (2017) 160-164. https://doi.org/10.1039/C6SC02716D

[49] C. Bommier, D. Mitlin, X. Ji, Internal structure–Na storage mechanisms–
electrochemical performance relations in carbons, Prog. Mater. Sci. 97 (2018) 170-
203. https://doi.org/10.1016/j.pmatsci.2018.04.006

[50] G. Yoon, H. Kim, I. Park, K. Kang, Conditions for reversible Na intercalation in
graphite: theoretical studies on the interplay among guest ions, solvent, and graphite
host. Adv. Energy Mater.7 (2017) 1601519. https://doi.org/10.1002/aenm.201601519

[51] M.M. Doeff, Y. Ma, S.J. Visco, L.C. De Jonghe, Electrochemical insertion of
sodium into carbon, J. Electrochem. Soc. 140 (1993) L169-L170.
https://doi.org/10.1149/1.2221153

[52] K. Nobuhara, H. Nakayama, M. Nose, S. Nakanishi, H. Iba, First-principles study of
alkali metal-graphite intercalation compounds, J. Power Sources 243 (2013) 585-587.
https://doi.org/10.1016/j.jpowsour.2013.06.057

[53] Y. Okamoto, Density functional theory calculations of alkali metal (Li, Na, and K)
graphite intercalation compounds, J. Phys. Chem. C. 118 (2014) 16-19.
https://doi.org/10.1021/jp4063753

[54] Y. Wen, K. He, Y. Zhu, F. Han, Y. Xu, I. Matsuda, Y. Ishii, J. Cumings, C. Wang,
Expanded graphite as superior anode for sodium-ion batteries, Nature Commun. 5
(2014) 4033. https://doi.org/10.1038/ncomms5033

[55] B. Jache, P. Adelhelm, Use of graphite as a highly reversible electrode with superior
cycle life for sodium-ion batteries by making use of co-intercalation phenomena,
Angew. Chem. Int. Ed. 53 (2014) 10169-10173.
https://doi.org/10.1002/anie.201403734

[56] H. Kim, J. Hong, Y.U. Park, J. Kim, I. Hwang, K. Kang, Sodium storage behavior in
natural graphite using ether-based electrolyte Systems. Adv. Funct. Mater. 25 (2015)
534-541. https://doi.org/10.1002/adfm.201402984

[57] H. Kim, J. Hong, G. Yoon, H. Kim, K.-Y. Park, M.-S. Park, W.-S. Yoon, K. Kang, Sodium intercalation chemistry in graphite, Energy Environ. Sci. 8 (2015) 2963-2969. https://doi.org/10.1039/C5EE02051D

[58] Y. Liu, N. Zhang, C. Yu, L. Jiao, J. Chen, $MnFe_2O_4$@C nanofibers as high-performance anode for sodium-ion batteries, Nano Lett. 16 (2016) 3321-3328. https://doi.org/10.1021/acs.nanolett.6b00942

[59] R.E. Franklin, Crystallite growth in graphitizing and non-graphitizing carbons, Proceedings of the Royal Society of London. Series A. Mathematical and Physical Sciences 209 (1951) 196-218. https://doi.org/10.1098/rspa.1951.0197

[60] A.P. Terzyk, S. Furmaniak, P.J. Harris, P.A. Gauden, J. Włoch, P. Kowalczyk, G. Rychlicki, How realistic is the pore size distribution calculated from adsorption isotherms if activated carbon is composed of fullerene-like fragments? Phys. Chem. Chem. Phys. 9 (2007) 5919-5927. https://doi.org/10.1039/b710552e

[61] H. Marsh, W. Wynne-Jones, The surface properties of carbon-I The effect of activated diffusion in the determination of surface area, Carbon, 1 (1964) 269-279. https://doi.org/10.1016/0008-6223(64)90281-7

[62] H. Hou, X. Qiu, W. Wei, Y. Zhang, X. Ji, Carbon anode materials for advanced sodium-ion batteries, Adv. Energy Mater.7 (2017) 1602898. https://doi.org/10.1002/aenm.201602898

[63] R. Gray, Coal to coke conversion, in Introduction to carbon science 1989, Elsevier. p. 285-321. https://doi.org/10.1016/B978-0-408-03837-9.50014-2

[64] Z. Li, C. Bommier, Z.S. Chong, Z. Jian, T.W. Surta, X. Wang, Z. Xing, J.C. Neuefeind, W.F. Stickle, M. Dolgos, Mechanism of Na-ion storage in hard carbon anodes revealed by heteroatom doping, Adv. Energy Mater.7 (2017) 1602894. https://doi.org/10.1002/aenm.201602894

[65] J. Kipling, J. Sherwood, P. Shooter, N. Thompson, The pore structure and surface area of high-temperature polymer carbons, Carbon 1 (1964) 321-328. https://doi.org/10.1016/0008-6223(64)90286-6

[66] A. Martynenko, True, particle, and bulk density of shrinkable biomaterials: evaluation from drying experiments. Drying Technol. 32 (2014) 1319-1325. https://doi.org/10.1080/07373937.2014.894522

[67] D. Stevens, J. Dahn, The mechanisms of lithium and sodium insertion in carbon materials, J. Electrochem. Soc. 148 (2001) A803-A811. https://doi.org/10.1149/1.1379565

[68] D. Stevens, J. Dahn, An In Situ Small-Angle X-Ray Scattering Study of Sodium Insertion into a Nanoporous Carbon Anode Material within an Operating Electrochemical Cell. J. Electrochem. Soc.147 (2000) 4428-4431. https://doi.org/10.1149/1.1394081

[69] D. Stevens, J. Dahn, High capacity anode materials for rechargeable sodium-ion batteries, J. Electrochem. Soc.147 (2000) 1271-1273. https://doi.org/10.1149/1.1393348

[70] S. Komaba, W. Murata, T. Ishikawa, N. Yabuuchi, T. Ozeki, T. Nakayama, A. Ogata, K. Gotoh, K. Fujiwara, Electrochemical Na insertion and solid electrolyte interphase for hard-carbon electrodes and application to Na-Ion batteries, Adv. Funct. Mater. 21 (2011) 3859-3867. https://doi.org/10.1002/adfm.201100854

[71] K. Gotoh, T. Ishikawa, S. Shimadzu, N. Yabuuchi, S. Komaba, K. Takeda, A. Goto, K. Deguchi, S. Ohki, K. Hashi, NMR study for electrochemically inserted Na in hard carbon electrode of sodium ion battery, J. Power Sources, 225 (2013) 137-140. https://doi.org/10.1016/j.jpowsour.2012.10.025

[72] S. Wenzel, T. Hara, J. Janek, P. Adelhelm, Room-temperature sodium-ion batteries: Improving the rate capability of carbon anode materials by templating strategies, Energy Environ. Sci. 4 (2011) 3342-3345. https://doi.org/10.1039/c1ee01744f

[73] P.C. Tsai, S.-C. Chung, S.-k. Lin, A. Yamada, Ab initio study of sodium intercalation into disordered carbon. J. Mater. Chem. A. 3 (2015) 9763-9768. https://doi.org/10.1039/C5TA01443C

[74] C. Bommier, T.W. Surta, M. Dolgos, X. Ji, New mechanistic insights on Na-ion storage in nongraphitizable carbon, Nano Lett. 15 (2015) 5888-5892. https://doi.org/10.1021/acs.nanolett.5b01969

[75] Y. Zhang, Y. Zhao, Z. Bakenov, A simple approach to synthesize nanosized sulfur/graphene oxide materials for high-performance lithium/sulfur batteries, Ionics 20 (2014) 1047-1050. https://doi.org/10.1007/s11581-014-1165-5

[76] M. Pumera, Graphene-based nanomaterials for energy storage. Energy Environ. Sci. 4 (2011) 668-674. https://doi.org/10.1039/C0EE00295J

[77] C. Xu, B. Xu, Y. Gu, Z. Xiong, J. Sun, X. Zhao, Graphene-based electrodes for electrochemical energy storage, Energy Environ. Sci. 6 (2013) 1388-1414. https://doi.org/10.1039/c3ee23870a

[78] J. Zhu, D. Yang, Z. Yin, Q. Yan, H. Zhang, Graphene and graphene-based materials for energy storage applications. Small, 10 (2014) 3480-3498. https://doi.org/10.1002/smll.201303202

[79] S. Wu, R. Xu, M. Lu, R. Ge, J. Iocozzia, C. Han, B. Jiang, Z. Lin, Graphene-containing nanomaterials for lithium-ion batteries, Adv. Energy Mater.5 (2015) 1500400. https://doi.org/10.1002/aenm.201500400

[80] S. Wu, R. Ge, M. Lu, R. Xu, Z. Zhang, Graphene-based nano-materials for lithium–sulfur battery and sodium-ion battery, Nano Energy 15 (2015) 379-405. https://doi.org/10.1016/j.nanoen.2015.04.032

[81] Y.X. Wang, S.L. Chou, H.-K. Liu, S.-X. Dou, Reduced graphene oxide with superior cycling stability and rate capability for sodium storage, Carbon 57 (2013) 202-208. https://doi.org/10.1016/j.carbon.2013.01.064

[82] Z. Jian, B. Zhao, P. Liu, F. Li, M. Zheng, M. Chen, Y. Shi, H. Zhou, Fe_2O_3 nanocrystals anchored onto graphene nanosheets as the anode material for low-cost sodium-ion batteries, Chem. Commun. 50 (2014) 1215-1217. https://doi.org/10.1039/C3CC47977C

[83] W. Wang, L. Hu, J. Ge, Z. Hu, H. Sun, H. Sun, H. Zhang, H. Zhu, S. Jiao, In situ self-assembled $FeWO_4$/graphene mesoporous composites for Li-ion and Na-ion batteries, Chem. Mater. 26 (2014) 3721-3730. https://doi.org/10.1021/cm501122u

[84] K.J. Saji, K. Tian, M. Snure, A. Tiwari, 2D tin monoxide—an unexplored p-type van der waals semiconductor: material characteristics and field effect transistors, Adv. Electron. Mater. 2 (2016) 1500453. https://doi.org/10.1002/aelm.201500453

[85] H. Yamaguchi, S. Nakanishi, H. Iba, T. Itoh, Amorphous polymeric anode materials from poly (acrylic acid) and tin (II) oxide for lithium ion batteries, J. Power Sources 275 (2015) 1-5. https://doi.org/10.1016/j.jpowsour.2014.10.071

[86] L.Y. Liang, Z.M. Liu, H.T. Cao, X.Q. Pan, Microstructural, optical, and electrical properties of SnO thin films prepared on quartz via a two-step method, ACS Appl. Mater. Interfaces 2 (2010) 1060-1065. https://doi.org/10.1021/am900838z

[87] Y. Zhang, Z. Ma, D. Liu, S. Dou, J. Ma, M. Zhang, Z. Guo, R. Chen, S. Wang, p-Type SnO thin layers on n-type SnS 2 nanosheets with enriched surface defects and embedded charge transfer for lithium ion batteries, J. Mater. Chem. A. 5 (2017) 512-518. https://doi.org/10.1039/C6TA09748K

[88] F. Zhang, J. Zhu, D. Zhang, U. Schwingenschlögl, H.N. Alshareef, Two-dimensional SnO anodes with a tunable number of atomic layers for sodium ion batteries, Nano Lett.17 (2017) 1302-1311. https://doi.org/10.1021/acs.nanolett.6b05280

[89] R. Raccichini, A. Varzi, S. Passerini, B. Scrosati, The role of graphene for electrochemical energy storage, Nat. Mater. 14 (2015) 271. https://doi.org/10.1038/nmat4170

[90] N. Mahmood, C. Zhang, H. Yin, Y. Hou, Graphene-based nanocomposites for energy storage and conversion in lithium batteries, supercapacitors and fuel cells, J. Mater. Chem. A. 2 (2014) 15-32. https://doi.org/10.1039/C3TA13033A

[91] M. Reddy, G. Subba Rao, B. Chowdari, Metal oxides and oxysalts as anode materials for Li ion batteries, Chem. Rev. 113 (2013) 5364-5457. https://doi.org/10.1021/cr3001884

[92] M. Chen, D. Chao, J. Liu, J. Yan, B. Zhang, Y. Huang, J. Lin, Z.X. Shen, Rapid pseudocapacitive sodium-ion response induced by 2D ultrathin tin monoxide nanoarrays, Adv. Funct. Mater. 27 (2017) 1606232. https://doi.org/10.1002/adfm.201606232

[93] J. Zhu, D. Deng, Single-crystalline α-Fe_2O_3 void@ frame microframes for rechargeable batteries, J. Mater. Chem. A. 4 (2016) 4425-4432. https://doi.org/10.1039/C6TA00870D

[94] D. Kong, C. Cheng, Y. Wang, B. Liu, Z. Huang, H.Y. Yang, Seed-assisted growth of α-Fe_2O_3 nanorod arrays on reduced graphene oxide: A superior anode for high-performance Li-ion and Na-ion batteries, J. Mater. Chem. A. 4 (2016) 11800-11811. https://doi.org/10.1039/C6TA04370D

[95] G.D. Park, J.S. Cho, J.K. Lee, Y.C. Kang, Na-ion storage performances of $FeSe_x$ and Fe_2O_3 hollow nanoparticles-decorated reduced graphene oxide balls prepared by nanoscale Kirkendall diffusion process, Sci. Rep. 6 (2016) 22432. https://doi.org/10.1038/srep22432

[96] H. Li, L. Xu, H. Sitinamaluwa, K. Wasalathilake, C. Yan, Coating Fe_2O_3 with graphene oxide for high-performance sodium-ion battery anode, Composites Commun. 1 (2016) 48-53. https://doi.org/10.1016/j.coco.2016.09.004

[97] Z.J. Zhang, Y.-X. Wang, S.-L. Chou, H.-J. Li, H.-K. Liu, J.-Z. Wang, Rapid synthesis of α-Fe_2O_3/rGO nanocomposites by microwave autoclave as superior anodes for sodium-ion batteries, J. Power Sources 280 (2015) 107-113. https://doi.org/10.1016/j.jpowsour.2015.01.092

[98] T. Li, A. Qin, L. Yang, J. Chen, Q. Wang, D. Zhang, H. Yang, In situ grown Fe2O3 single crystallites on reduced graphene oxide nanosheets as high performance conversion anode for sodium-ion batteries, ACS Appl. Mater. Interfaces 9 (2017) 19900-19907. https://doi.org/10.1021/acsami.7b04407

[99] W. Guo, W. Sun, Y. Wang, Multilayer CuO@NiO hollow spheres: microwave-assisted metal–organic-framework derivation and highly reversible structure-matched stepwise lithium storage. ACS Nano 9 (2015) 11462-11471. https://doi.org/10.1021/acsnano.5b05610

[100] S. Ko, J.I. Lee, H.S. Yang, S. Park, U. Jeong, Mesoporous CuO particles threaded with CNTs for high-performance lithium-ion battery anodes, Adv. Mater. 24 (2012) 4451-4456. https://doi.org/10.1002/adma.201201821

[101] B. Wang, X.L. Wu, C.Y. Shu, Y.G. Guo, C.R. Wang, Synthesis of CuO/graphene nanocomposite as a high-performance anode material for lithium-ion batteries, J. Mater. Chem. 20 (2010) 10661-10664. https://doi.org/10.1039/c0jm01941k

[102] R. Sahay, P. Suresh Kumar, V. Aravindan, J. Sundaramurthy, W. Chui Ling, S.G. Mhaisalkar, S. Ramakrishna, S. Madhavi, High aspect ratio electrospun CuO nanofibers as anode material for lithium-ion batteries with superior cycleability, J. Phys. Chem. C, 116 (2012) 18087-18092. https://doi.org/10.1021/jp3053949

[103] D. Yin, G. Huang, Z. Na, X. Wang, Q. Li, L. Wang, CuO nanorod arrays formed directly on Cu foil from MOFs as superior binder-free anode material for lithium-ion batteries, ACS Energy Lett., 2 (2017) 1564-1570. https://doi.org/10.1021/acsenergylett.7b00215

[104] D. Li, D. Yan, X. Zhang, J. Li, T. Lu, L. Pan, Porous CuO/reduced graphene oxide composites synthesized from metal-organic frameworks as anodes for high-performance sodium-ion batteries, J. Colloid Interface Sci. 497 (2017) 350-358. https://doi.org/10.1016/j.jcis.2017.03.037

[105] J. Sun, C. Lv, F. Lv, S. Chen, D. Li, Z. Guo, W. Han, D. Yang, S. Guo, Tuning the shell number of multishelled metal oxide hollow fibers for optimized lithium-ion storage, ACS Nano 11 (2017) 6186-6193. https://doi.org/10.1021/acsnano.7b02275

[106] Y. Jiang, M. Hu, D. Zhang, T. Yuan, W. Sun, B. Xu, M. Yan, Transition metal oxides for high performance sodium ion battery anodes, Nano Energy, 5 (2014) 60-66. https://doi.org/10.1016/j.nanoen.2014.02.002

[107] K. He, F. Lin, Y. Zhu, X. Yu, J. Li, R. Lin, D. Nordlund, T.C. Weng, R.M. Richards, X.Q. Yang, Sodiation kinetics of metal oxide conversion electrodes: A comparative study with lithiation, Nano Lett. 15 (2015) 5755-5763. https://doi.org/10.1021/acs.nanolett.5b01709

[108] F. Zou, Y.-M. Chen, K. Liu, Z. Yu, W. Liang, S.M. Bhaway, M. Gao, Y. Zhu, Metal organic frameworks derived hierarchical hollow NiO/Ni/graphene composites for lithium and sodium storage, ACS Nano 10 (2015) 377-386. https://doi.org/10.1021/acsnano.5b05041

[109] C. Peng, B. Chen, Y. Qin, S. Yang, C. Li, Y. Zuo, S. Liu, J. Yang, Facile ultrasonic synthesis of CoO quantum dot/graphene nanosheet composites with high lithium storage capacity, ACS Nano 6 (2012) 1074-1081. https://doi.org/10.1021/nn202888d

[110] L. Chang, K. Wang, L.-a. Huang, Z. He, S. Zhu, M. Chen, H. Shao, J. Wang, Hierarchical CoO microflower film with excellent electrochemical lithium/sodium storage performance, J. Mater. Chem. A. 5 (2017) 20892-20902. https://doi.org/10.1039/C7TA05027E

[111] L. Chang, K. Wang, L. Huang, Z. He, H. Shao, J. Wang, Hierarchically porous CoO microsphere films with enhanced lithium/sodium storage properties, J. Alloys Compds, 725 (2017) 824-834. https://doi.org/10.1016/j.jallcom.2017.07.122

[112] D.V. Bavykin, J.M. Friedrich, F.C. Walsh, Protonated titanates and TiO_2 nanostructured materials: synthesis, properties, and applications, Adv. Mater. 18 (2006) 2807-2824. https://doi.org/10.1002/adma.200502696

[113] L. Wu, D. Bresser, D. Buchholz, G.A. Giffin, C.R. Castro, A. Ochel, S. Passerini, Unfolding the mechanism of sodium insertion in anatase TiO_2 nanoparticles, Adv. Energy Mater.5 (2015) 1401142. https://doi.org/10.1002/aenm.201401142

[114] D. Su, S. Dou, G. Wang, Anatase TiO_2: better anode material than amorphous and rutile phases of TiO2 for Na-ion batteries, Chem. Mater.27 (2015) 6022-6029. https://doi.org/10.1021/acs.chemmater.5b02348

[115] Y. Xiong, J. Qian, Y. Cao, X. Ai, H. Yang, Electrospun TiO_2/C Nanofibers As a High-Capacity and Cycle-Stable Anode for Sodium-Ion Batteries, ACS Appl. Mater. Interfaces 8 (2016) 16684-16689. https://doi.org/10.1021/acsami.6b03757

[116] J.Y. Hwang, S.T. Myung, J.H. Lee, A. Abouimrane, I. Belharouak, Y.K. Sun, Ultrafast sodium storage in anatase TiO_2 nanoparticles embedded on carbon nanotubes. Nano Energy 16 (2015) 218-226. https://doi.org/10.1016/j.nanoen.2015.06.017

[117] Y. Sun, X. Hu, W. Luo, Y. Huang, Self-assembled hierarchical MoO_2/graphene nanoarchitectures and their application as a high-performance anode material for lithium-ion batteries, ACS Nano 5 (2011) 7100-7107. https://doi.org/10.1021/nn201802c

[118] Y. Shi, B. Guo, S.A. Corr, Q. Shi, Y.-S. Hu, K.R. Heier, L. Chen, R. Seshadri, G.D. Stucky, Ordered mesoporous metallic MoO_2 materials with highly reversible lithium storage capacity, Nano Lett. 9 (2009) 4215-4220. https://doi.org/10.1021/nl902423a

[119] X. Zhao, H.-E. Wang, X. Chen, J. Cao, Y. Zhao, Z.G. Neale, W. Cai, J. Sui, G. Cao, Tubular MoO_2 organized by 2D assemblies for fast and durable alkali-ion storage, Energy Storage Mater. 11 (2018) 161-169. https://doi.org/10.1016/j.ensm.2017.10.010

[120] J. Huang, Z. Xu, L. Cao, Q. Zhang, H. Ouyang, J. Li, Tailoring MoO2/Graphene Oxide Nanostructures for Stable, High-Density Sodium-Ion Battery Anodes. Energy Technology, 3 (2015) 1108-1114. https://doi.org/10.1002/ente.201500160

[121]S. Yang, Y. Gong, Z. Liu, L. Zhan, D.P. Hashim, L. Ma, R. Vajtai, P.M. Ajayan, Bottom-up approach toward single-crystalline VO2-graphene ribbons as cathodes for ultrafast lithium storage, Nano Lett. 13 (2013) 1596-1601. https://doi.org/10.1021/nl400001u

[122] G. He, L. Li, A. Manthiram, VO_2/rGO nanorods as a potential anode for sodium- and lithium-ion batteries, J. Mater. Chem. A. 3 (2015) 14750-14758. https://doi.org/10.1039/C5TA03188E

[123] A.M. Cao, J.S. Hu, H.P. Liang, L.J. Wan, Self-assembled vanadium pentoxide (V_2O_5) hollow microspheres from nanorods and their application in lithium-Ion batteries, Angew. Chem. Int. Ed. 44 (2005) 4391-4395. https://doi.org/10.1002/anie.200500946

[124] J. Liu, H. Xia, D. Xue, L. Lu, Double-shelled nanocapsules of V_2O_5-based composites as high-performance anode and cathode materials for Li ion batteries, J. Am. Chem. Soc. 131 (2009) 12086-12087. https://doi.org/10.1021/ja9053256

[125] Z. Chen, V. Augustyn, J. Wen, Y. Zhang, M. Shen, B. Dunn, Y. Lu, High-performance supercapacitors based on intertwined CNT/V_2O_5 nanowire nanocomposites, Adv. Mater. 23 (2011) 791-795. https://doi.org/10.1002/adma.201003658

[126] G. Gu, M. Schmid, P.W. Chiu, A. Minett, J. Fraysse, G.T. Kim, S. Roth, M. Kozlov, E. Muñoz, R.H. Baughman, V_2O_5 nanofibre sheet actuators. Nat. Mater. 2 (2003) 316. https://doi.org/10.1038/nmat880

[127] T. Zhai, H. Liu, H. Li, X. Fang, M. Liao, L. Li, H. Zhou, Y. Koide, Y. Bando, D. Golberg, Centimeter-long V2O5 nanowires: From synthesis to field-emission, electrochemical, electrical transport, and photoconductive properties. Adv. Mater. 22 (2010) 2547-2552. https://doi.org/10.1002/adma.200903586

[128] M. Lee, S.K. Balasingam, H.Y. Jeong, W.G. Hong, B.H. Kim, Y. Jun, One-step hydrothermal synthesis of graphene decorated V_2O_5 nanobelts for enhanced electrochemical energy storage, Sci. Rep. 5 (2015) 8151. https://doi.org/10.1038/srep08151

[129] R. Kiruthiga, C. Nithya, R. Karvembu, Reduced graphene oxide embedded V_2O_5 nanorods and porous honey carbon as high performance electrodes for hybrid sodium-ion supercapacitors, Electrochim. Acta 256 (2017) 221-231. https://doi.org/10.1016/j.electacta.2017.10.049

[130] W. Wu, L. Wang, Y. Li, F. Zhang, L. Lin, S. Niu, D. Chenet, X. Zhang, Y. Hao, T.F. Heinz, Piezoelectricity of single-atomic-layer MoS_2 for energy conversion and piezotronics. Nature 514 (2014) 470. https://doi.org/10.1038/nature13792

[131] J. Kibsgaard, Z. Chen, B.N. Reinecke, T.F. Jaramillo, Engineering the surface structure of MoS_2 to preferentially expose active edge sites for electrocatalysis, Nat. Mater. 11 (2012) 963. https://doi.org/10.1038/nmat3439

[132] D. Su, S. Dou, G. Wang, $WS_2@$ graphene nanocomposites as anode materials for Na-ion batteries with enhanced electrochemical performances, Chem. Commun. 50 (2014) 4192-4195. https://doi.org/10.1039/c4cc00840e

[133] G.S. Bang, K.W. Nam, J.Y. Kim, J. Shin, J.W. Choi, S.-Y. Choi, Effective liquid-phase exfoliation and sodium ion battery application of MoS_2 nanosheets, ACS Appl. Mater. Interfaces 6 (2014) 7084-7089. https://doi.org/10.1021/am4060222

[134] X. Zhou, X. Liu, Y. Xu, Y. Liu, Z. Dai, J. Bao, An SbO_x/reduced graphene oxide composite as a high-rate anode material for sodium-ion batteries, J. Phys. Chem. C. 118 (2014) 23527-23534. https://doi.org/10.1021/jp507116t

[135] Y. Denis, P.V. Prikhodchenko, C.W. Mason, S.K. Batabyal, J. Gun, S. Sladkevich, A.G. Medvedev, O. Lev, High-capacity antimony sulphide nanoparticle-decorated graphene composite as anode for sodium-ion batteries, Nat. Commun. 4 (2013) 2922. https://doi.org/10.1038/ncomms3922

[136] R. Baughman, AA Zhakidov, WA de Heer, Science 297 (2002) 787. https://doi.org/10.1126/science.1060928

[137] N. Rodriguez, A review of catalytically grown carbon nanofibers, J. Mater. Res. 8 (1993) 3233-3250. https://doi.org/10.1557/JMR.1993.3233

[138] W. Li, M. Li, K.R. Adair, X. Sun, Y. Yu, Carbon nanofiber-based nanostructures for lithium-ion and sodium-ion batteries, J. Mater. Chem. A. 5 (2017) 13882-13906. https://doi.org/10.1039/C7TA02153D

[139] W. Li, L. Zeng, Y. Wu, Y. Yu, Nanostructured electrode materials for lithium-ion and sodium-ion batteries via electrospinning, Science China Mater. 59 (2016) 287-321. https://doi.org/10.1007/s40843-016-5039-6

[140] M. Inagaki, Y. Yang, F. Kang, Carbon nanofibers prepared via electrospinning, Adv. Mater. 24 (2012) 2547-2566. https://doi.org/10.1002/adma.201104940

[141] Z.J. Fan, J. Yan, T. Wei, G.Q. Ning, L.J. Zhi, J.C. Liu, D.X. Cao, G.L. Wang, F. Wei, Nanographene-constructed carbon nanofibers grown on graphene sheets by chemical vapor deposition: high-performance anode materials for lithium ion batteries, ACS Nano 5 (2011) 2787-2794. https://doi.org/10.1021/nn200195k

[142] X.Q. Zhang, Q. Sun, W. Dong, D. Li, A.-H. Lu, J.Q. Mu, W.C. Li, Synthesis of superior carbon nanofibers with large aspect ratio and tunable porosity for electrochemical energy storage, J. Mater. Chem. A. 1 (2013) 9449-9455. https://doi.org/10.1039/c3ta10660h

[143] H. Hou, M. Jing, Y. Yang, Y. Zhang, W. Song, X. Yang, J. Chen, Q. Chen, X. Ji, Antimony nanoparticles anchored on interconnected carbon nanofibers networks as advanced anode material for sodium-ion batteries, J. Power Sources 284 (2015) 227-235. https://doi.org/10.1016/j.jpowsour.2015.03.043

[144] G. Zheng, Q. Zhang, J.J. Cha, Y. Yang, W. Li, Z.W. Seh, Y. Cui, Amphiphilic surface modification of hollow carbon nanofibers for improved cycle life of lithium sulfur batteries, Nano Lett. 13 (2013) 1265-1270. https://doi.org/10.1021/nl304795g

[145] Y. Yao, F. Wu, Naturally derived nanostructured materials from biomass for rechargeable lithium/sodium batteries, Nano Energy 17 (2015) 91-103. https://doi.org/10.1016/j.nanoen.2015.08.004

[146] L. Fu, K. Tang, K. Song, P.A. van Aken, Y. Yu, J. Maier, Nitrogen doped porous carbon fibres as anode materials for sodium ion batteries with excellent rate performance, Nanoscale 6 (2014) 1384-1389. https://doi.org/10.1039/C3NR05374A

[147] Y. Cao, L. Xiao, M.L. Sushko, W. Wang, B. Schwenzer, J. Xiao, Z. Nie, L.V. Saraf, Z. Yang, J. Liu, Sodium ion insertion in hollow carbon nanowires for battery applications, Nano Lett. 12 (2012) 3783-3787. https://doi.org/10.1021/nl3016957

[148] W. Li, L. Zeng, Z. Yang, L. Gu, J. Wang, X. Liu, J. Cheng, Y. Yu, Free-standing and binder-free sodium-ion electrodes with ultralong cycle life and high rate performance based on porous carbon nanofibers, Nanoscale 6 (2014) 693-698. https://doi.org/10.1039/C3NR05022J

[149] M. Wang, Z. Yang, W. Li, L. Gu, Y. Yu, Superior sodium storage in 3D interconnected nitrogen and oxygen dual-doped carbon network, Small 12 (2016) 2559-2566. https://doi.org/10.1002/smll.201600101

[150] D. Nan, Z.H. Huang, R. Lv, L. Yang, J.G. Wang, W. Shen, Y. Lin, X. Yu, L. Ye, H. Sun, Nitrogen-enriched electrospun porous carbon nanofiber networks as high-performance free-standing electrode materials, J. Mater. Chem. A. 2 (2014) 19678-19684. https://doi.org/10.1039/C4TA03868A

[151] W. Luo, J. Schardt, C. Bommier, B. Wang, J. Razink, J. Simonsen, X. Ji, Carbon nanofibers derived from cellulose nanofibers as a long-life anode material for rechargeable sodium-ion batteries, J. Mater. Chem. A. 1 (2013) 10662-10666. https://doi.org/10.1039/c3ta12389h

[152] Z. Zhang, J. Zhang, X. Zhao, F. Yang, Core-sheath structured porous carbon nanofiber composite anode material derived from bacterial cellulose/polypyrrole as an anode for sodium-ion batteries, Carbon 95 (2015) 552-559. https://doi.org/10.1016/j.carbon.2015.08.069

[153] T. Chen, Y. Liu, L. Pan, T. Lu, Y. Yao, Z. Sun, D.H. Chua, Q. Chen, Electrospun carbon nanofibers as anode materials for sodium ion batteries with excellent cycle performance, J. Mater. Chem. A. 2 (2014) 4117-4121. https://doi.org/10.1039/c3ta14806h

[154] S. Wang, L. Xia, L. Yu, L. Zhang, H. Wang, X.W. Lou, Free-standing nitrogen-doped carbon nanofiber films: integrated electrodes for sodium-ion batteries with ultralong cycle life and superior rate capability, Adv. Energy Mater. 6 (2016) 1502217. https://doi.org/10.1002/aenm.201502217

[155] J. Jin, B.-j. Yu, Z.-q. Shi, C.-y. Wang, C.-b. Chong, Lignin-based electrospun carbon nanofibrous webs as free-standing and binder-free electrodes for sodium ion batteries, J. Power Sources 272 (2014) 800-807. https://doi.org/10.1016/j.jpowsour.2014.08.119

[156] L. Zeng, W. Li, J. Cheng, J. Wang, X. Liu, Y. Yu, N-doped porous hollow carbon nanofibers fabricated using electrospun polymer templates and their sodium storage properties, RSC Adv. 4 (2014) 16920-16927. https://doi.org/10.1039/C4RA01200C

[157] J. Zhu, C. Chen, Y. Lu, Y. Ge, H. Jiang, K. Fu, X. Zhang, Nitrogen-doped carbon nanofibers derived from polyacrylonitrile for use as anode material in sodium-ion batteries, Carbon 94 (2015) 189-195. https://doi.org/10.1016/j.carbon.2015.06.076

[158] C. Chen, Y. Lu, Y. Ge, J. Zhu, H. Jiang, Y. Li, Y. Hu, X. Zhang, Synthesis of nitrogen-doped electrospun carbon nanofibers as anode material for high-performance sodium-ion batteries, Energy Technol. 4 (2016) 1440-1449. https://doi.org/10.1002/ente.201600205

[159] M. Wang, Y. Yang, Z. Yang, L. Gu, Q. Chen, Y. Yu, Sodium-ion batteries: Improving the rate capability of 3D interconnected carbon nanofibers thin film by boron, nitrogen dual-doping, Adv. Sci. 4 (2017) 1600468. https://doi.org/10.1002/advs.201600468

[160] M. Wang, Z. Yang, W. Li, L. Gu, Y. Yu, Superior sodium storage in 3d interconnected nitrogen and oxygen dual-doped carbon network, Small, 12 (2016) 2559-2566. https://doi.org/10.1002/smll.201600101

[161] Y. Zhu, X. Han, Y. Xu, Y. Liu, S. Zheng, K. Xu, L. Hu, C. Wang, Electrospun Sb/C fibers for a stable and Fast sodium-ion battery anode, ACS Nano 7 (2013) 6378-6386. https://doi.org/10.1021/nn4025674

Materials Research Forum LLC
https://doi.org/10.21741/9781644900833-2

[162] J. Liu, K. Tang, K. Song, P.A. van Aken, Y. Yu, J. Maier, Tiny $Li_4Ti_5O_{12}$ nanoparticles embedded in carbon nanofibers as high-capacity and long-life anode materials for both Li-ion and Na-ion batteries, Phys. Chem. Chem. Phys. 15 (2013) 20813-20818. https://doi.org/10.1039/c3cp53882f

[163] B. Zhang, J. Huang, J.-K. Kim, Ultrafine amorphous SnO_x Embedded in carbon nanofiber/carbon nanotube composites for Li-Ion and Na-ion batteries. Adv. Funct. Mater. 25 (2015) 5222-5228. https://doi.org/10.1002/adfm.201501498

[164] Y. Liu, N. Zhang, C. Yu, L. Jiao, J. Chen, $MnFe_2O_4$@C nanofibers as high-performance anode for sodium-ion batteries, Nano Lett. 16 (2016) 3321-3328. https://doi.org/10.1021/acs.nanolett.6b00942

[165] C. Zhu, X. Mu, P.A. van Aken, Y. Yu, J. Maier, Single-layered ultrasmall nanoplates of MoS_2 embedded in carbon nanofibers with excellent electrochemical performance for lithium and sodium storage, Angew. Chem. Int. Ed. 53 (2014) 2152-2156. https://doi.org/10.1002/anie.201308354

[166] X. Xiong, W. Luo, X. Hu, C. Chen, L. Qie, D. Hou, Y. Huang, Flexible Membranes of MoS_2/C Nanofibers by Electrospinning as Binder-Free Anodes for High-Performance Sodium-Ion Batteries. Scientific Reports, 5 (2015) 9254. https://doi.org/10.1038/srep09254

[167] J.S. Cho, J.-K. Lee, Y.C. Kang, Graphitic carbon-coated $FeSe_2$ hollow nanosphere-decorated reduced graphene oxide hybrid nanofibers as an efficient anode material for sodium ion batteries, Sci. Rep. 6 (2016) 23699. https://doi.org/10.1038/srep23699

[168]. J.S. Cho, S.Y. Lee, Y.C. Kang, First Introduction of NiSe2 to Anode Material for Sodium-Ion Batteries: A Hybrid of Graphene-Wrapped NiSe2/C Porous Nanofiber. Scientific Reports, 6 (2016) 23338. https://doi.org/10.1038/srep23338

[169] M. Dirican, Y. Lu, Y. Ge, O. Yildiz, X. Zhang, Carbon-confined SnO2-electrodeposited porous carbon nanofiber composite as high-capacity sodium-ion battery anode material, ACS Appl. Mater. Interfaces 7 (2015) 18387-18396. https://doi.org/10.1021/acsami.5b04338

[170] M. Wang, Y. Yang, Z. Yang, L. Gu, Q. Chen, Y. Yu, Sodium-ion batteries: Improving the rate capability of 3D Interconnected carbon nanofibers thin film by boron, nitrogen dual-doping, Adv. Sci. 4 (2017) 1600468. https://doi.org/10.1002/advs.201600468

[171] B. Zhang, F. Kang, J.-M. Tarascon, J.-K. Kim, Recent advances in electrospun carbon nanofibers and their application in electrochemical energy storage, Prog. Mater. Sci. 76 (2016) 319-380. https://doi.org/10.1016/j.pmatsci.2015.08.002

[172] L. Wu, X. Hu, J. Qian, F. Pei, F. Wu, R. Mao, X. Ai, H. Yang, Y. Cao, Sb–C nanofibers with long cycle life as an anode material for high-performance sodium-ion batteries, Energy Environ. Sci. 7 (2014) 323-328. https://doi.org/10.1039/C3EE42944J

[173] Y. Liu, N. Zhang, L. Jiao, J. Chen, Tin Nanodots encapsulated in porous nitrogen-doped carbon nanofibers as a free-standing anode for advanced sodium-ion batteries, Adv. Mater. 27 (2015) 6702-6707. https://doi.org/10.1002/adma.201503015

[174] L. Ji, M. Gu, Y. Shao, X. Li, M.H. Engelhard, B.W. Arey, W. Wang, Z. Nie, J. Xiao, C. Wang, Controlling SEI formation on SnSb-porous carbon nanofibers for improved Na ion storage, Adv. Mater. 26 (2014) 2901-2908. https://doi.org/10.1002/adma.201304962

[175] K. Shiva, H.B. Rajendra, A.J. Bhattacharyya, Electrospun SnSb crystalline nanoparticles inside porous carbon fibers as a high stability and rate capability anode for rechargeable batteries, ChemPlusChem 80 (2015) 516-521. https://doi.org/10.1002/cplu.201402291

[176] B. Zhang, J. Huang, J.K. Kim, Ultrafine Amorphous SnO_x embedded in carbon nanofiber/carbon nanotube composites for Li-Ion and Na-ion batteries. Adv. Funct. Mater. 25 (2015) 5222-5228. https://doi.org/10.1002/adfm.201501498

[177] J. Liu, K. Tang, K. Song, P.A. van Aken, Y. Yu, J. Maier, Tiny $Li_4Ti_5O_{12}$ nanoparticles embedded in carbon nanofibers as high-capacity and long-life anode materials for both Li-ion and Na-ion batteries. Phys. Chem. Chem. Phys. 15 (2013) 20813-20818. https://doi.org/10.1039/c3cp53882f

[178] X. Xiong, W. Luo, X. Hu, C. Chen, L. Qie, D. Hou, Y. Huang, Flexible membranes of MoS_2/C nanofibers by electrospinning as binder-free anodes for high-performance sodium-ion batteries, Sci. Rep. 5 (2015) 9254. https://doi.org/10.1038/srep09254

[179] C. Zhu, X. Mu, P.A. van Aken, Y. Yu, J. Maier, Single-layered ultrasmall nanoplates of MoS_2 embedded in carbon nanofibers with excellent electrochemical performance for lithium and sodium storage, Angew. Chem. Int. Ed. 53 (2014) 2152-2156. https://doi.org/10.1002/anie.201308354

[180] J.S. Cho, S.Y. Lee, Y.C. Kang, First introduction of $NiSe_2$ to anode material for sodium-ion batteries: A hybrid of graphene-wrapped $NiSe_2$/C porous nanofiber, Sci. Rep. 6 (2016) 23338. https://doi.org/10.1038/srep23338

[181] J.S. Cho, J.-K. Lee, Y.C. Kang, Graphitic carbon-coated FeSe 2 hollow nanosphere-decorated reduced graphene oxide hybrid nanofibers as an efficient anode material for sodium ion batteries, Sci. Rep. 6 (2016) 23699. https://doi.org/10.1038/srep23699

[182] J. Ding, H. Wang, Z. Li, A. Kohandehghan, K. Cui, Z. Xu, B. Zahiri, X. Tan, E.M. Lotfabad, B.C. Olsen, Carbon nanosheet frameworks derived from peat moss as high performance sodium ion battery anodes, ACS Nano 7 (2013) 11004-11015. https://doi.org/10.1021/nn404640c

[183] F. Shen, W. Luo, J. Dai, Y. Yao, M. Zhu, E. Hitz, Y. Tang, Y. Chen, V.L. Sprenkle, X. Li, Ultra-Thick, Low-tortuosity, and mesoporous wood carbon anode for high-performance sodium-ion batteries, Adv. Energy Mater.6 (2016) 1600377. https://doi.org/10.1002/aenm.201600377

[184] T. Yang, T. Qian, M. Wang, X. Shen, N. Xu, Z. Sun, C. Yan, A sustainable route from biomass byproduct okara to high content nitrogen-doped carbon sheets for efficient sodium ion batteries, Adv. Mater. 28 (2016) 539-545. https://doi.org/10.1002/adma.201503221

[185] K.I. Hong, L. Qie, R. Zeng, Z.-q. Yi, W. Zhang, D. Wang, W. Yin, C. Wu, Q.J. Fan, W.X. Zhang, Biomass derived hard carbon used as a high performance anode material for sodium ion batteries, J. Mater. Chem. A. 2 (2014) 12733-12738. https://doi.org/10.1039/C4TA02068E

[186] J. Górka, C. Vix-Guterl, C. Matei Ghimbeu, Recent progress in design of biomass-derived hard carbons for sodium ion batteries, C J. Carbon Res. 2 (2016) 24. https://doi.org/10.3390/c2040024

[187] E.M. Lotfabad, J. Ding, K. Cui, A. Kohandehghan, W.P. Kalisvaart, M. Hazelton, D. Mitlin, High-density sodium and lithium ion battery anodes from banana peels, ACS Nano 8 (2014) 7115-7129. https://doi.org/10.1021/nn502045y

[188] Y. Li, S. Xu, X. Wu, J. Yu, Y. Wang, Y.-S. Hu, H. Li, L. Chen, X. Huang, Amorphous monodispersed hard carbon micro-spherules derived from biomass as a high performance negative electrode material for sodium-ion batteries. J. Mater. Chem. A. 3 (2015) 71-77. https://doi.org/10.1039/C4TA05451B

[189] Y. Li, Y.S. Hu, M.M. Titirici, L. Chen, X. Huang, Hard carbon microtubes made from renewable cotton as high-performance anode material for sodium-ion batteries, Adv. Energy Mater.6 (2016) 1600659. https://doi.org/10.1002/aenm.201600659

[190] L. Wu, D. Buchholz, C. Vaalma, G.A. Giffin, S. Passerini, Apple-biowaste-derived hard carbon as a powerful anode material for Na-ion batteries. ChemElectroChem, 3 (2016) 292-298. https://doi.org/10.1002/celc.201500437

[191] R.R. Gaddam, D. Yang, R. Narayan, K. Raju, N.A. Kumar, X. Zhao, Biomass derived carbon nanoparticle as anodes for high performance sodium and lithium ion batteries, Nano Energy 26 (2016) 346-352. https://doi.org/10.1016/j.nanoen.2016.05.047

[192] D. Yan, C. Yu, X. Zhang, W. Qin, T. Lu, B. Hu, H. Li, L. Pan, Nitrogen-doped carbon microspheres derived from oatmeal as high capacity and superior long life anode material for sodium ion battery, Electrochim. Acta 191 (2016) 385-391. https://doi.org/10.1016/j.electacta.2016.01.105

[193] H. Li, F. Shen, W. Luo, J. Dai, X. Han, Y. Chen, Y. Yao, H. Zhu, K. Fu, E. Hitz, Carbonized-leaf membrane with anisotropic surfaces for sodium-ion battery, ACS Appl. Mater. Interfaces 8 (2016) 2204-2210. https://doi.org/10.1021/acsami.5b10875

[194] P. Zheng, T. Liu, X. Yuan, L. Zhang, Y. Liu, J. Huang, S. Guo, Enhanced performance by enlarged nano-pores of holly leaf-derived lamellar carbon for sodium-ion battery anode, Sci. Rep. 6 (2016) 26246. https://doi.org/10.1038/srep26246

[195] T.H. Emaga, C. Robert, S.N. Ronkart, B. Wathelet, M. Paquot, Dietary fibre components and pectin chemical features of peels during ripening in banana and plantain varieties, Bioresource Technol. 99 (2008) 4346-4354. https://doi.org/10.1016/j.biortech.2007.08.030

[196] H.g. Wang, Z. Wu, F.l. Meng, D.l. Ma, X.l. Huang, L.m. Wang, X.b. Zhang, Nitrogen-doped porous carbon nanosheets as low-cost, high-performance anode material for sodium-ion batteries, ChemSusChem 6 (2013) 56-60. https://doi.org/10.1002/cssc.201200680

[197] J. Xu, M. Wang, N.P. Wickramaratne, M. Jaroniec, S. Dou, L. Dai, High-performance sodium ion batteries based on a 3D anode from nitrogen-doped graphene foams, Adv.Mater. 27 (2015) 2042-2048. https://doi.org/10.1002/adma.201405370

[198] Z. Wang, L. Qie, L. Yuan, W. Zhang, X. Hu, Y. Huang, Functionalized N-doped interconnected carbon nanofibers as an anode material for sodium-ion storage with excellent performance, Carbon 55 (2013) 328-334. https://doi.org/10.1016/j.carbon.2012.12.072

[199] D. Li, L. Zhang, H. Chen, L.-x. Ding, S. Wang, H. Wang, Nitrogen-doped bamboo-like carbon nanotubes: promising anode materials for sodium-ion batteries, Chem. Commun. 51 (2015) 16045-16048. https://doi.org/10.1039/C5CC06266G

[200] Z. Wang, Y. Li, X.-J. Lv, N-doped ordered mesoporous carbon as a high performance anode material in sodium ion batteries at room temperature, RSC Adv. 4 (2014) 62673-62677. https://doi.org/10.1039/C4RA09084E

[201] X. Wang, G. Li, F.M. Hassan, J. Li, X. Fan, R. Batmaz, X. Xiao, Z. Chen, Sulfur covalently bonded graphene with large capacity and high rate for high-performance sodium-ion batteries anodes, Nano Energy 15 (2015) 746-754. https://doi.org/10.1016/j.nanoen.2015.05.038

Materials Research Forum LLC
https://doi.org/10.21741/9781644900833-2

[202] W. Li, M. Zhou, H. Li, K. Wang, S. Cheng, K. Jiang, A high performance sulfur-doped disordered carbon anode for sodium ion batteries, Energy Environ. Sci. 8 (2015) 2916-2921. https://doi.org/10.1039/C5EE01985K

[203] L. Qie, W. Chen, X. Xiong, C. Hu, F. Zou, P. Hu, Y. Huang, Sulfur-doped carbon with enlarged interlayer distance as a high-performance anode material for sodium-ion batteries, Adv. Sci. 2 (2015) 1500195. https://doi.org/10.1002/advs.201500195

[204] H. Hou, L. Shao, Y. Zhang, G. Zou, J. Chen, X. Ji, Large-area carbon nanosheets doped with phosphorus: a high-performance anode material for sodium-ion batteries, Adv. Sci. 4 (2017) 1600243. https://doi.org/10.1002/advs.201600243

[205] Y. Wang, C. Wang, Y. Wang, H. Liu, Z. Huang, Boric acid assisted reduction of graphene oxide: a promising material for sodium-ion batteries, ACS Appl. Mater. Interfaces 8 (2016) 18860-18866. https://doi.org/10.1021/acsami.6b04774

[206] C. Ling, F. Mizuno, Boron-doped graphene as a promising anode for Na-ion batteries, Phys.Chem. Chem. Phys. 16 (2014) 10419-10424. https://doi.org/10.1039/C4CP01045K

[207] D. Xu, C. Chen, J. Xie, B. Zhang, L. Miao, J. Cai, Y. Huang, L. Zhang, A hierarchical N/S-codoped carbon anode fabricated facilely from cellulose/polyaniline microspheres for high-performance sodium-ion batteries, Adv. Energy Mater.6 (2016) 1501929. https://doi.org/10.1002/aenm.201501929

[208] J. Yang, X. Zhou, D. Wu, X. Zhao, Z. Zhou, S-doped N-rich carbon nanosheets with expanded interlayer distance as anode materials for sodium-ion batteries, Adv. Mater. 29 (2017) 1604108. https://doi.org/10.1002/adma.201604108

[209] Y. Li, Z. Wang, L. Li, S. Peng, L. Zhang, M. Srinivasan, S. Ramakrishna, Preparation of nitrogen-and phosphorous co-doped carbon microspheres and their superior performance as anode in sodium-ion batteries, Carbon 99 (2016) 556-563. https://doi.org/10.1016/j.carbon.2015.12.066

Sodium-Ion Batteries: Materials and Applications   Materials Research Forum LLC
Materials Research Foundations **76** (2020) 73-92   https://doi.org/10.21741/9781644900833-3

Chapter 3

Organic Electrode Material for Sodium-Ion Batteries

Aneela Sabir[1,*], Tahmina Zia[1], Muhammad Usman[1], Muhammad Shafiq[1], Rafi Ullah Khan[1,2,]
Karl I Jacob[3], Rajender Boddula[4]

[1]Department of Polymer Engineering and Technology, University of the Punjab, Lahore, 54590
Pakistan

[2]Institute of Chemical Engineering and Technology (ICET), University of the Punjab, Lahore,
54590 Pakistan

[3]School of Materials Science and Engineering, Georgia Institute of Technology, Atlanta,
GA , USA

[4]CAS Key Laboratory of Nano system and Hierarchical Fabrication, National Center for
Nanoscience and Technology, Beijing 100190, P.R. China

*aneela.pet.ceet@pu.edu.pk

Abstract

A lot of work is done on functionalizing organic electrodes (OE) and incorporation of
nanostructures to tune their electrochemical properties. In collation, OE exhibit merits
like high capacity and structural design ability. Here in, organic electrodes based on their
reactions are divided into three classes; C=O, C-N=O and doping reactions. The
conductivity issue can be resolved through increasing conjugation. Theoretical capacity
can be elevated by expanding active groups. Working voltage can be regulated by tuning
grafting overseeing lowest unoccupied molecular orbital (LUMO). Future of organic
electrode relies mainly on aprotic electrolyte based full NaIBs with long cycle life.

Keywords

Organic Electrodes, Sodium-Ion Batteries, Electrochemical Energy Storage, C=N Based
Reaction, C=O Based Reaction, Doping Reaction

Contents

1. Introduction

Exploiting from tremendously abundant and inexpensive sodium reservoirs, sodium-ion batteries (NaIBs) are estimated as reassuring candidate for electrochemical energy conservation and storage on large scale. Owing to larger radius and atomic mass of Na+ than conventionally used materials, NaIBs having inorganic electrode encounter with little capacity and inadequate cycle life. Development of environment friendly, renewable, abundant raw material based batteries are gaining much attention. Organic electrode-based sodium ion batteries are one of them. Sodium-ion batteries (NaIBs) come with an edge for extensive energy storage, low cost and highly abundant mineral and similarities with conventionally used batteries on electrochemistry. Nowadays, NaIBs are in full swing. Typically used anode materials are inorganic based giving limited recycling

Materials Research Forum LLC
https://doi.org/10.21741/9781644900833-3

strategies [1, 2]. Comparable with inorganic electrodes, organic electrodes are conferred with various superiority. Including, organic materials comprised of naturally abundant (C, H and O) having flexible molecular structures giving room for Na ions and facilitating facile Na+ insertion/extraction mechanism [3]. Giving structural diversity with tunable functional groups resulting in tailored electrochemical properties [4]. To upgrade the performance of rechargeable Na-ion batteries anode material with greater energy storage capacity and improved cycle life are considerable [3, 5, 6]. Among numerous electrochemical energy storage (EES) technologies [7], lithium-ion batteries (LiIBs) are contender and drastically increased interests in research field and portable electronics [8, 9]. Although, geographically uneven distribution of lithium reservoir and its cost have created major concerns indeed, NaIBs go hand in hand with LiIBs since 1980s but left aside due to commercial success of LiIBs [10].

Presently anode materials used for NaIBs are carbon nano sheets, graphene materials, porous carbon nano sheets, biomass-derived carbon nano-spherules, intercalation-based compounds (e.g. $Na_2Ti_3O_7$), alloying materials (e.g. NaSn, Na_3P, Na_xSb). Including organic materials and polymers (e.g. pteridine, quinone, sodium terephthalates, polyimides) [11-15]. Organic materials act as electrodes are attracting much concern in EES system owing to their design ability, renewability, diversity and environmental benignity [16]. Although low capacity, inferior rate capability and limited cycling stability obstruct their application [17]. Na ion batteries consists of negatively charged anode and positively charged cathode where Na^+ transports through electrolyte and electrons travel through external circuit as shown in Fig. 1. During charging Na^+ moves from cathode to anode and vice versa.

Organic anodes confront three major challenges:

➢ Reduced reaction kinetics because of low conductivity of organic materials.

➢ Na ion insertion extraction motivates capacity decay on account of particle pulverization instigated by huge volume alterations.

➢ Organic Electrode material deprivation upon cycling in organic electrolytes.

Strategies to overcome above challenges are:

➢ Low conductivity is overlooked by adding 20-30% conductive carbon and reducing the particle size of organic compound in nanometer for enhancing active surface area, further enhancing reaction kinetics [18, 19].

➢ Nanosized organic material truncate particle pulverization, thus improving cycling stability.

➤Increasing the polarity through salt formation reduces solubility of organic material hindering loss of active material [17].

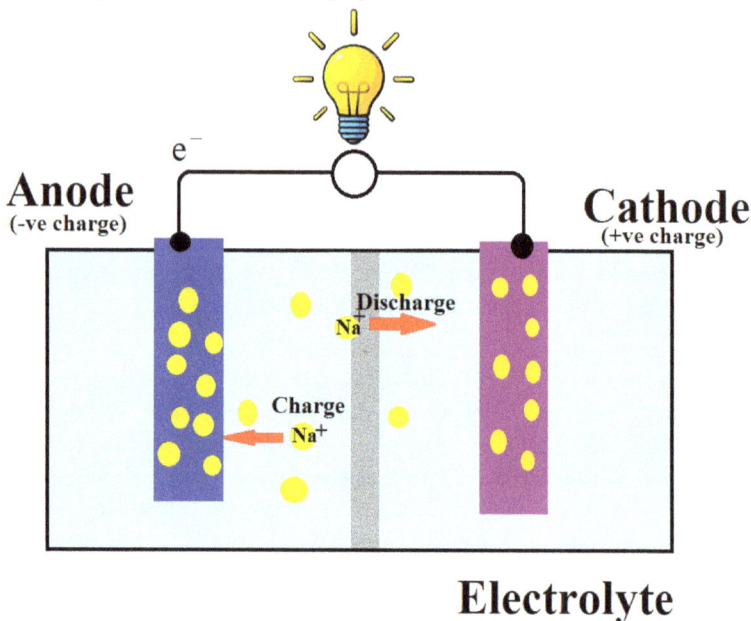

Anode
(-ve charge)

e^-

Cathode
(+ve charge)

Discharge
Na^+

Charge
Na^+

Electrolyte

Figure 1. Schematic representation of sodium-ion battery.

2.   Molecular design of electrodes for organic sodium ion batteries

2.1   Organic electrodes constituting of C=O based reaction

2.1.1 Carbonyl compounds

Carbonyl group is frequently used as organic functionality due to their efficient kinetics, structural diversity, high capacity and high reactivity. Virtually all carbonyl compounds are n-type organics with reversible single electron reduction per carbonyl giving Na based monovalent anion stabilized by other structural moieties. Depending upon stabilizing mechanism, carbonyl based organic molecules are divided in 3 types [20].

Type 1: Vicinal carbonyls employing stable enolates shown in Fig. 2.

Materials Research Forum LLC
https://doi.org/10.21741/9781644900833-3

Type 2: Carbonyl groups directly attached to aromatic ring giving delocalization stabilized well dispersed negative charge as shown in Fig. 3. Aromatic carboxylates, imides, anhydrides fall in this category [21].

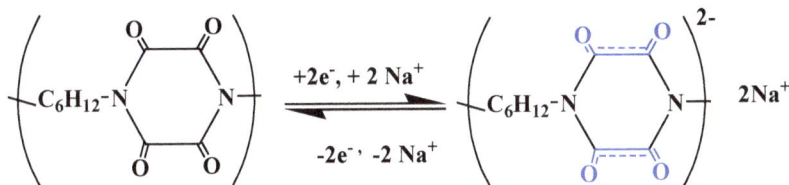

Figure 2. Stabilizing mechanism of vicinal carbonyls (type 1).

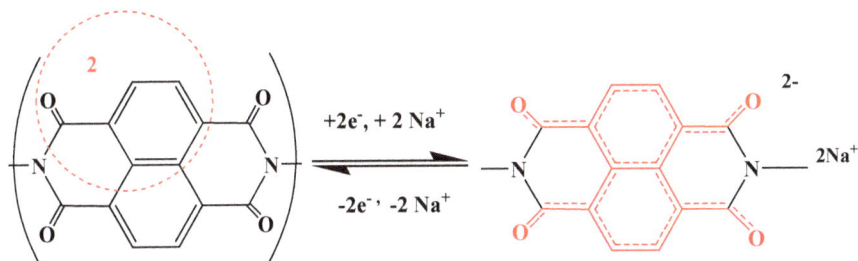

Figure 3. Stabilizing mechanism of carbonyls in conjugation with aromatic ring (type 2).

Type 3: Quinones and ketones belongs to this category .They show combined characteristics with Type 1 and 2 compounds as shown in Fig 4. The main stabilizing force comes from reduction induced structural moieties. Inductive and resonance effect can be used to tune redox behavior of carbonyls. Major pitfall is their solubility in organic electrolytes.

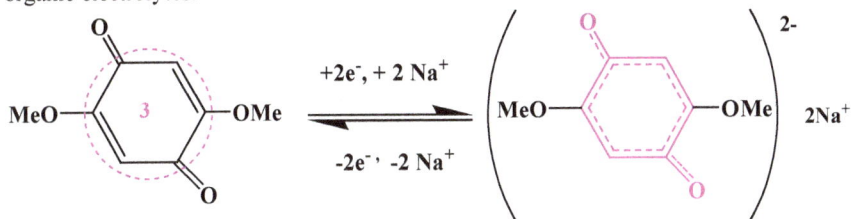

Figure 4. Stabilizing mechanism of hybrid of type 1 and 2 carbonyls (type 3).

Materials Research Forum LLC
https://doi.org/10.21741/9781644900833-3

2.1.2 Polyimides

In organic polyimides carbonyl group (C=O) is directly bonded with nitrogen and aromatic rings. Enolization reaction followed by association and disassociation of Na^+ to oxygen atom describes the redox mechanism of imides. Polyimides and their derivatives can be used as electrode material for NaIBs. Due to their inert stable structures and insolubility they exhibit long cycle life and highly constant electrochemical performance. Predominantly used are dianhydride derived polyimides with 1,4,5,8-naphthalenetetracarboxylic dianhydride (NTCDA), 1,2,4,5-benzenetetracarboxylic dianhydride (PMDA) and 3,4,9,10-perylenetetracarboxylic dianhydride (PTCDA) as monomer units as represented in Fig.5 [22].

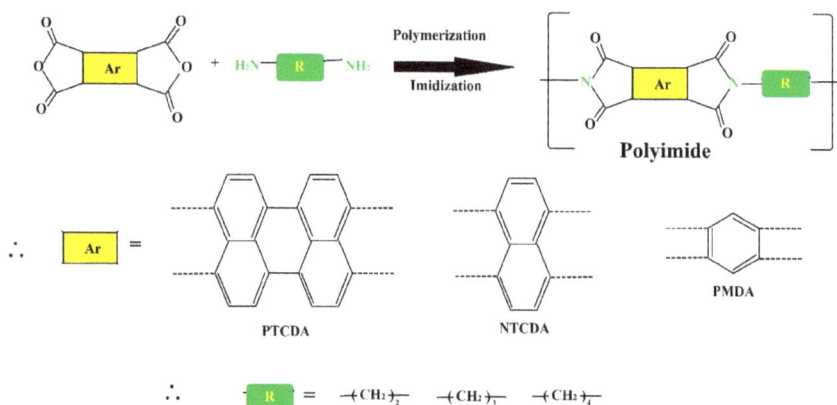

Figure 5. Imidization of polyimides.

NTCDA derived polyimide PNTCDA is used as electrode material for NaIBs. It gives 2 electrons transfer per molecule with discharge capacity of about 140mAh g^{-1}, discharge voltage of 1.7-2.4V with 60% capacity deliverance [23].

To overcome unwanted dissolution for small imides, polyimides with electro active center and inactive linker was used giving additional thermal stability and mechanical properties. By fluctuating the electro active core i.e., increasing aromaticity there was an increase in average discharge voltage [22].

Redox-inactive groups can be replaced by redox-active groups to avoid capacity loss. Redox active groups like anthraquinone were used with PMDA /NTCDA to form anthraquinonyl imide. It showed a capacity of ~200mAhg^{-1} at 5omAg^{-1}. Both polyimides undergo reversible, 4 electron transfer in 1.2-3.2 voltage range as shown in Fig. 6 [24].

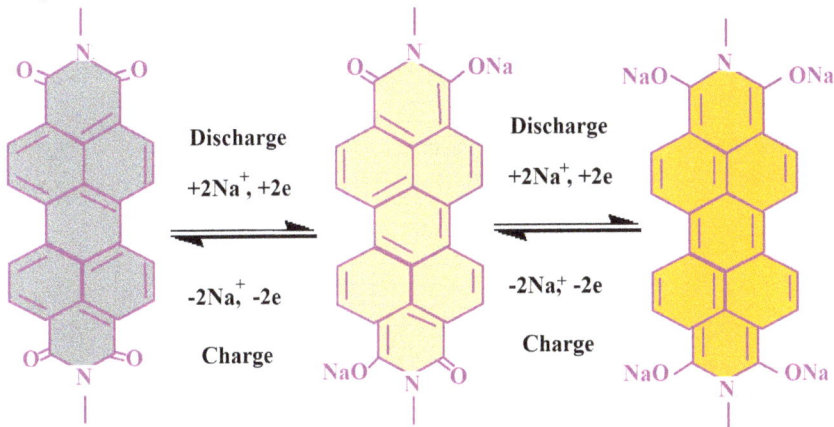

Figure 6. Schematic representation of working principles of PTCDA based polyimide.

2.1.3 Quinones

In quinone the stabilization mechanism follows type 3 mechanism i.e. stability of negative charge is done by formation of additional aromatic system. They undergo solubility issues in organic electrolytes [25]. The dissolved species travel through cathode and anode area resulting into shuttle effect and capacity loss giving low columbic efficiency. To overcome such issues, physiosorption based loadings are done on carbon based materials, high molecular weight polymerization and optimization of electrolytes by increasing their viscosities. Quinones are combined with several binders to give less solubility and to attain self-healing properties [26-28].

Quinones and their derivative including anthraquinone-2,6-dicarboxylic acid, 1,4-benzoquinone, 9,10-1nthraquinone, anthraquinone-2-carboxylic acid, 1,4-naphthoquinone, 2,6-diaminoanthraquinone and 2-aminoanthraquinone are studied by researchers with emphasizes on their redox properties, solvation energy, electron affinity, energy density and charge storage capacity [25, 29]. Schematic representation of working principle of benzoquinone is represented in Fig. 7.

Organic electrodes operational voltage can be modulate by electron-withdrawing functional groups. Halogens are strong electron-withdrawing groups with ability to lessen the energy of LUMO. Typically lowest unoccupied molecular orbital [LUMO] energy has direct effect on reduction voltage. Lesser LUMO energy will create more reduction voltage. Results ground on Density Functional Theory (DFT) shows redox voltage in following trend.

$$C_6F_4O_2 > C_6Cl_4O_2 > C_6Br_4O_2 > C_6H_4O_2$$

$C_6Cl_4O_2$ provides lower reduction potential than $C_6F_4O_2$ but later is irreversible during the charging process, forming NaF [28, 30].

Salt formation to enhance polarity of inorganic organic hybrid material can overcome organic electrode dissolution in aprotic electrolyte.

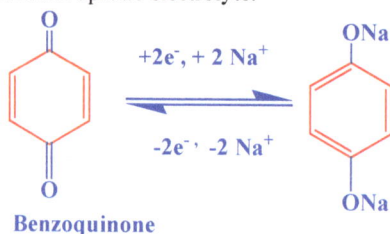

Figure 7. Schematic representation of working principle of benzoquinone.

2.1.4 Carboxylates

Sodium carboxylates are widely used electrode in NaIBs. Due to electron donating group (O-Na) directly bonded with carbonyl Na^+ insertion occurs at voltage below 1 V. Similarly, 4,4-biphenyldicarboxylate Na-salt ($Na_2C_{14}H_8O_4$) and disodium terephthalate salt ($Na_2C_8H_4O_4$) deliver 0.5 V discharge voltage with reversible capacity of 220mA h g⁻1 and Na insertion extraction mechanism was studied, respectively [31, 32]. Na^+ insertion extraction mechanism of disodium terephthalate is shown in Fig. 8.

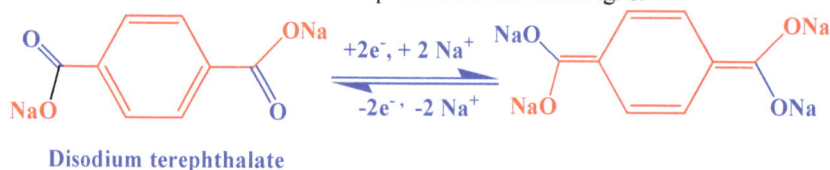

Figure 8. Schematic representation of working principle of disodium terephthalate.

Solubility issue is controlled by exchange of metallic cation in terephthalate salts. Calcium, zinc, potassium are also incorporated in terephthalate salt [33-35]. Sodium storage in disodium terephthalate was reported by Mahasin Alam Sk et.al with capacities of ~ 225mAh g^{-1}. At lower Na concentration carbonyl site take part in insertion while at high sodium concentration hexagonal ring part dominates [32].

2.1.5 Anhydrides

Extensively conjugated structures with tunable Na-ion insertion capacity are used as anhydride organic electrodes. PTCDA accommodates 2 Na-ions in between 1.0-3.0 V with 150mAhg-1. By lower discharge voltage more sodium ions can be inserted. At 0.01 V PTCDA results in 15 Na-ion insertion as shown in Fig.9 with 1017mAg-1 capacity.

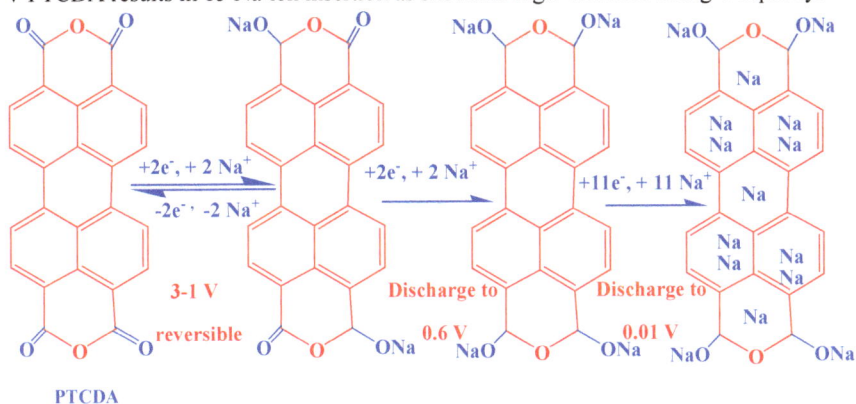

Figure 9. Schematic representation of working principles of PTCDA.

Extra capacity is due to generation of interface (solid electrolyte) and inserted Na-ion into aromatic rings [36]. NTCDA also gives extra capacity at 0.01 V discharge voltage with Na$^+$ insertion extraction mechanism as represented in Fig 10.

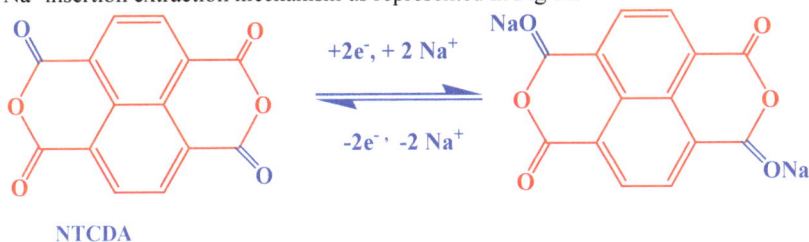

Figure 10. Schematic representation of working principles of NTCDA.

2.2   Organic electrodes based on doping reaction

2.2.1   Organic radical polymers

The main problem using organic radical polymer is self-discharging, low conductivity and highly soluble in organic electrolytes [37, 38]. Kim et al. worked on improving electrochemical performance of PTMA [39]. PTMA encapsulated in carbon nanotubes with 100-300 nm diameter overcome self-discharging issue [40].

Initially only p-doping reaction occurs with 130 mAhg^{-1} discharge voltage. During discharge both n and p-doping is observed with 220 mAhg^{-1} as shown in Fig 11. The fast kinetics for radical compounds can be endowed by oxidation induced structural changes with lesser electron rearrangement.

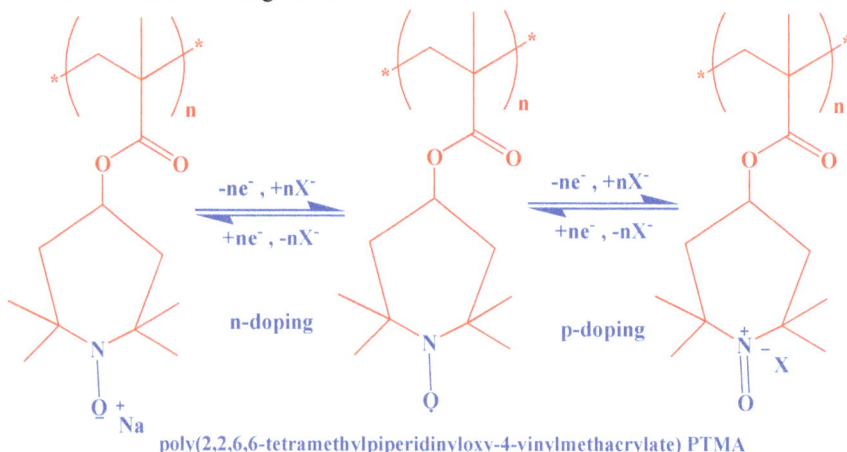

poly(2,2,6,6-tetramethylpiperidinyloxy-4-vinylmethacrylate) PTMA

Figure 11. Schematic representation of working principles of PTMA.

2.2.2   Conductive polymers

Conductive polymers both n and p doping reactions can be observed at different voltages. During early studies polyacetylene and polyphenylene were studied as conductive electrodes [15, 41]. Polybithiophene, n-type polymer retains 400 mAhg^{-1} at 1.0 V average discharge voltage. Oligopyrene, p-type polymer retains 121 mAhg^{-1} initial reversible capacity at 3.5 V discharge voltage [42]. Polypyrrole are the commonly used conductive polymers but deliver less capacity [43]. Their Na-ion insertion exertion mechanism is shown in Fig. 12. Copolymerization, doping anion and introducing nanostructured materials are effective strategies to overcome such issues. Furthermore, polypyrrole codoped with sodium p-toluenesulfonate (TsONa) and L-lactic acid (LA) through electropolymerization on Fe foil show high cathode performance [44].

Additionally polyaniline with nanostructured graphene/ polyaniline/polystyrene electrode and copolymer of o-nitro aniline /polyaniline showed enhanced capacity [12, 45].

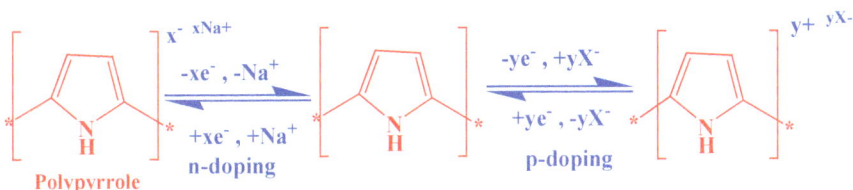

Figure 12. Schematic representation of working principles of polypyrrole.

2.2.3 Conjugated microporous polymers

The microporous conjugated organic polymers endowed several benefits, making them suitable for NaIBs [46]. Na-ion transport is facilitated due to high surface area and microporous structure. Dissolution problem is also overlooked as well as more redox-active sites gives high theoretical capacity. Triazine ring and benzene based porous, bipolar electrode material gives uniform 1.4 nm micropores [47]. Presently low conductivity is the main issue for microporous electrode batteries.

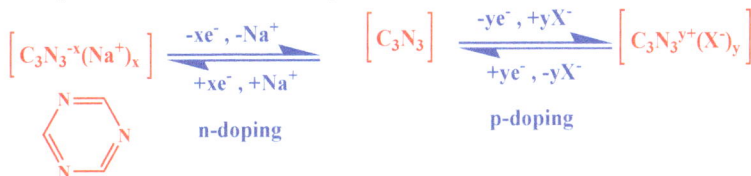

Figure 13. Schematic representation of working principles of conjugated microporous polymers.

2.2.4 Organometallic polymers

Organometallic like ferrocene electrodes have been used in electrochemical analysis due to their high air stability and ultrafast electrochemical response. Aprotic electrolytes solubilize simple organometallic electrodes so they are commonly employed in redox batteries. High molecular weight organometallics are commonly used cathodes for NaIBs. Poly(ferrocenyl-methylsilane) and its derivatives are also studied [48]. In ferrocene Fe exists in Fe^{+2} form. After p-doping reaction valency of Fe is in Fe^{+3} as shown in Fig. 14.

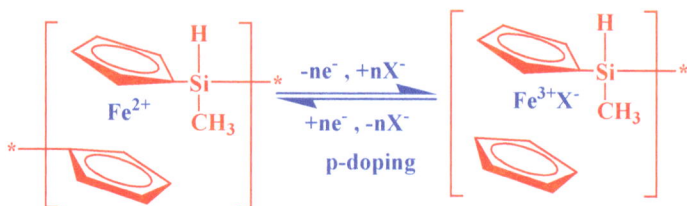

Figure 14. Schematic representation of working principles of organometallic polymers.

2.3   Organic electrode constituting of C=N based reaction

2.3.1 Schiff bases

The general repeating unit of Schiff base is (–N=C-Ar-C=N-), where Ar is aromatic rings having 10 πe^- satisfying Huckel rule. This repeating unit with conjugated double bonds is considered as active group while nonconjugated groups are termed as inactive centers. The aromatic linkers are more stable than aliphatic linkers. Aliphatic linker with 1 methyl group gives 0.9 Na-ion insertion while with aromatic linker 1.4 Na-ion insertion/extraction is observed as shown in Fig. 15. This phenomenon is due to planarity. By increasing methyl group as linker (up to 4) results in loss of planarity as well as drop capacity and shift reduction voltage downwards due to inductively electron-donating effect of methyl groups.

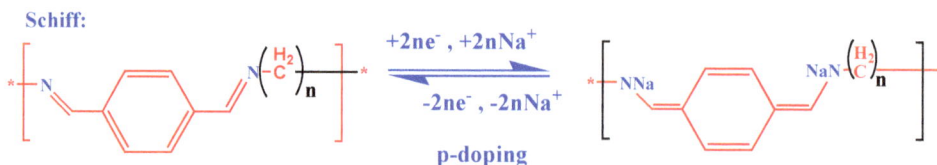

Figure 15. Schematic representation of working principles of Schiff bases.

2.3.2 Pteridine derivatives

Reversible tautomerism occurs in pteridine derivatives as shown in Fig.16. N is most commonly observed at the reactive site while O also assists the reaction. In addition to C=N, tetracyanoquinodimethane [49] and tetracyanoethylene having C≡N was also used [50].

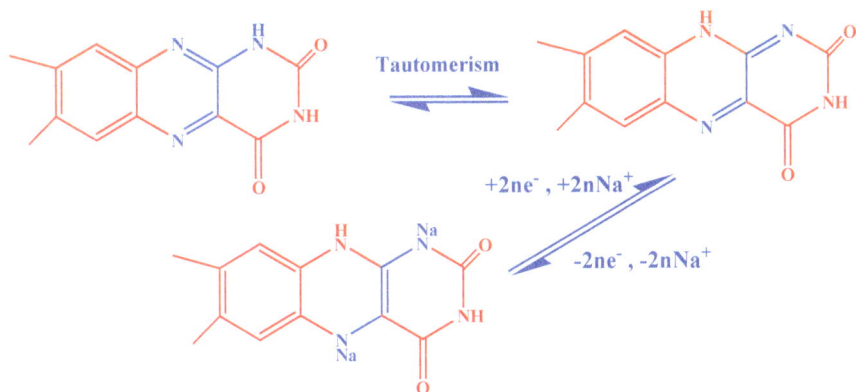

Figure 16. Schematic representation of working principles of Pteridine derivatives.

3.   Electrode design for sodium-ion batteries

Major issues dealing with sodium-ion batteries are

- ➤The inherently low conductivity of organic electrodes lower reaction kinetics,
- ➤Large volume change during phase transition results in pulverization causing capacity decay.
- ➤Organic electrode dissolution in organic electrolytes causes loss of active material.

Designing electrodes draws more attraction as compared to the search for new organic electrode materials [8, 9, 51]. Tunable statics adopted to modify organic electrodes in NaIBs are:

- ➤Molecular engineering.
- ➤Polymerization.
- ➤Carbon hybrid.
- ➤Modifying electrolyte.

With the help of the above the capacity, diffusion rate, conductivity, working voltage and solubility can be reasonably tuned resulting in enhanced cycle life, high rate performance and elevated capacity of NaIBs as schematically represented in Fig 17.

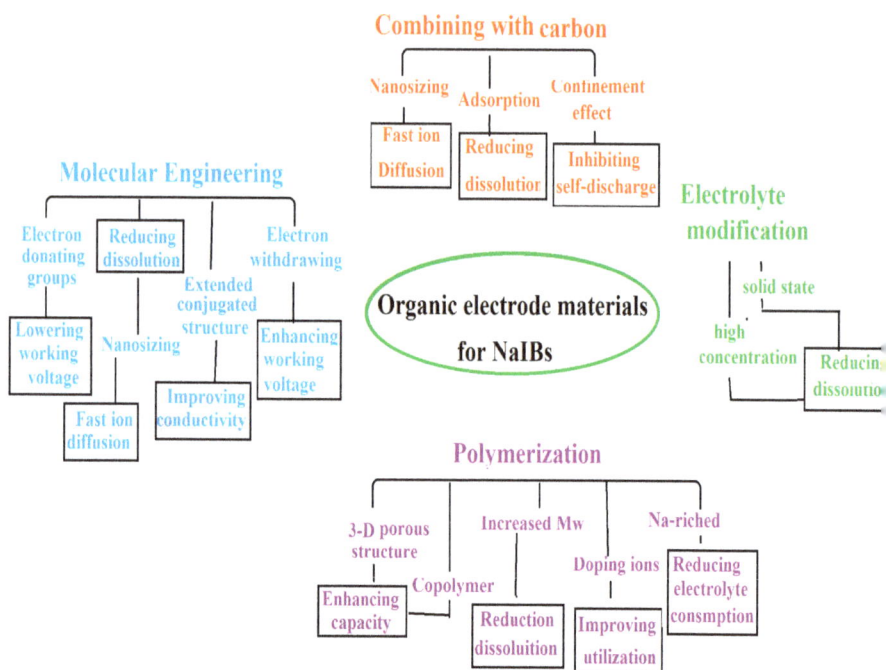

Figure 17. Electrode designs for Sodium ion batteries.

3.1 Molecular engineering

In molecular engineering the following strategies can be applied

➤Rate capacity of carbonyls can be improved by increasing the π-π aromatic conjugation of the core.

➤Rapid ion diffusion can be attained by organic nanostructured salt electrodes.

➤Dissolution of organic electrode can be overcome by salt formation.

➤LUMO energies can be increased or decreased by addition of electron-donating and electron-withdrawing groups resulting in reduced and enhanced working voltage, respectively.

3.2 Polymerization

Polymerization exhibits the following merits:

> Functionalizing conductive polymers with anions and cations can enhance their utilization and during the charging process lower electrolyte consumption of P-Type polymer is observed for Na based polymers.

> High molecular weight polymer gives lesser dissolution.

> Theoretical capacity can be enhanced by 3D porous polymers and copolymers.

3.3 Combining with carbon (carbon hybrid)

Combining with carbon material is endowed with the following merit:

> Incorporating carbon materials in organic materials can reduce dissolution with confinement and adsorption effects. Resultantly overcome with self-discharging issue of organic radical polymers.

The prepared material increases the active area of electrode and electrolyte giving high capacity and high rate capability.

3.4 Electrolyte modification

Concentrated electrolyte or solid electrolyte suppresses dissolution of soft active material, enhancing cycle performance.

4 Future challenges

With advances in portable electronics the rising need is smaller, flexible, wearable and thinner batteries. In comparison with inorganic materials, organic materials endow better film-forming ability, sublimability and solubility. Therefore, facile methods are used in designing organic sodium-ion batteries, flexible current collectors by printing, casting, vapor deposition, in situ polymerization and vacuum filtration. Moreover organic materials also endorse contributions in current collectors, binders and separators in flexible batteries.

Organic electrodes in aqueous electrolytes have been gaining the researcher's attention. Aqueous electrolytes come with several merits, economical, high ionic conductivity, nonflammable and facile battery assembling. Appropriate organic materials are advised for use in aqueous electrolytes with redox potential between H_2 and O_2 [52]. This limits the choice of polymeric electrode materials for sodium-ion batteries. To overcome organic electrode-based issues composite materials will be used to get better results like quinone with high theoretical capacity can be reinforced with polyimide having high cycle stability. Coating highly conductive polymers on lower conductive electrodes can be employed for enhanced electrochemical properties. Biomass-derived organic electrodes will be economical and of high demand.

References

[1] S. Jeong, B.H. Kim, Y.D. Park, C.Y. Lee, J. Mun, A. Tron, Artificially coated $NaFePO_4$ for aqueous rechargeable sodium-ion batteries, J. Alloys Compd. 784 (2019) 720-726. https://doi.org/10.1016/j.jallcom.2019.01.046

[2] J. Huang, Z. Wei, J. Liao, W. Ni, C. Wang, J. Ma, Molybdenum and tungsten chalcogenides for lithium/sodium-ion batteries: Beyond MoS_2, J. Energy Chem. 33 (2019) 100-124. https://doi.org/10.1016/j.jechem.2018.09.001

[3] Y. Zhang, Z. Gao, High performance anode material for sodium-ion batteries derived from covalent-organic frameworks, Electrochim. Acta 301 (2019) 23-28. https://doi.org/10.1016/j.electacta.2019.01.147

[4] F. Wu, C. Zhao, S. Chen, Y. Lu, Y. Hou, Y.-S. Hu, J. Maier, Y. Yu, Multi-electron reaction materials for sodium-based batteries, Mater. Today 21 (2018) 960-973. https://doi.org/10.1016/j.mattod.2018.03.004

[5] H. Gao, S. Xin, L. Xue, J.B. Goodenough, Stabilizing a high-energy-density rechargeable sodium battery with a solid electrolyte, Chem 4 (2018) 833-844. https://doi.org/10.1016/j.chempr.2018.01.007

[6] J.H. Kim, M.J. Jung, M.J. Kim, Y.S. Lee, Electrochemical performances of lithium and sodium ion batteries based on carbon materials, J. Industrial Eng. Chem. 61 (2018) 368-380. https://doi.org/10.1016/j.jiec.2017.12.036

[7] T. Liu, X. Cheng, H. Yu, H. Zhu, N. Peng, R. Zheng, J. Zhang, M. Shui, Y. Cui, J. Shu, An overview and future perspectives of aqueous rechargeable polyvalent ion batteries, Energy Storage Mater. 18 (2019) 68-91. https://doi.org/10.1016/j.ensm.2018.09.027

[8] S. Wang, L. Wang, K. Zhang, Z. Zhu, Z. Tao, J. Chen, Organic $Li_4C_8H_2O_6$ nanosheets for lithium-ion batteries, Nano Letters 13 (2013) 4404-4409. https://doi.org/10.1021/nl402239p

[9] C. Luo, R. Huang, R. Kevorkyants, M. Pavanello, H. He, C. Wang, Self-assembled organic nanowires for high power density lithium ion batteries, Nano Lett. 14 (2014) 1596-1602. https://doi.org/10.1021/nl500026j

[10] G. Li, R. Xue, L. Chen, The influence of polytetrafluorethylene reduction on the capacity loss of the carbon anode for lithium ion batteries, Solid State Ionics 90 (1996) 221-225. https://doi.org/10.1016/S0167-2738(96)00367-0

[11] C. Zhang, C. Lu, F. Zhang, F. Qiu, X. Zhuang, X. Feng, Two-dimensional organic cathode materials for alkali-metal-ion batteries, J. Energy Chem. 27 (2018) 86-98. https://doi.org/10.1016/j.jechem.2017.11.008

[12] J. Chen, Y. Liu, W. li, C. Wu, L. Xu, H. Yang, Nanostructured polystyrene/polyaniline/graphene hybrid materials for electrochemical supercapacitor and Na-ion battery applications, 50 (2015) 5466-5474. https://doi.org/10.1007/s10853-015-9092-z

[13] X. Dou, I. Hasa, D. Saurel, C. Vaalma, L. Wu, D. Buchholz, D. Bresser, S. Komaba, S. Passerini, Hard carbons for sodium-ion batteries: Structure, analysis, sustainability, and electrochemistry, Mater. Today 23 (2019) 87-104. https://doi.org/10.1016/j.mattod.2018.12.040

[14] N. Ingersoll, Z. Karimi, D. Patel, R. Underwood, R. Warren, Metal organic framework-derived carbon structures for sodium-ion battery anodes, Electrochim. Acta 297 (2019) 129-136. https://doi.org/10.1016/j.electacta.2018.11.140

[15] J. Wang, C. Lv, Y. Zhang, L. Deng, Z. Peng, Polyphenylene wrapped sulfur/multi-walled carbon nano-tubes via spontaneous grafting of diazonium salt for improved electrochemical performance of lithium-sulfur battery, Electrochim. Acta 165 (2015) 136-141. https://doi.org/10.1016/j.electacta.2015.03.013

[16] D. Wu, K. Luo, S. Du, X. Hu, A low-cost non-conjugated dicarboxylate coupled with reduced graphene oxide for stable sodium-organic batteries, J. Power Sources 398 (2018) 99-105. https://doi.org/10.1016/j.jpowsour.2018.07.067

[17] C. Luo, J. Wang, X. Fan, Y. Zhu, F. Han, L. Suo, C. Wang, Roll-to-roll fabrication of organic nanorod electrodes for sodium ion batteries, Nano Energy 13 (2015) 537-545. https://doi.org/10.1016/j.nanoen.2015.03.041

[18] X. Gao, G. Zhu, X. Zhang, T. Hu, Porous carbon materials derived from in situ construction of metal-organic frameworks for high-performance sodium ions batteries, Micropor. Mesopor. Mater. 273 (2019) 156-162. https://doi.org/10.1016/j.micromeso.2018.07.002

[19] D. Li, L. Chen, L. Chen, Q. Sun, M. Zhu, Y. Zhang, Y. Liu, Z. Liang, P. Si, J. Lou, J. Feng, L. Ci, Potassium gluconate-derived N/S Co-doped carbon nanosheets as superior electrode materials for supercapacitors and sodium-ion batteries, J. Power Sources 414 (2019) 308-316. https://doi.org/10.1016/j.jpowsour.2018.12.091

[20] J. Geng, J.-P. Bonnet, S. Renault, F. Dolhem, P. Poizot, Evaluation of polyketones with N-cyclic structure as electrode material for electrochemical energy storage: case of tetraketopiperazine unit, Energy Environ. Sci. 3 (2010) 1929-1933. https://doi.org/10.1039/c0ee00126k

[21] M. Armand, S. Grugeon, H. Vezin, S. Laruelle, P. Ribière, P. Poizot, J.M. Tarascon, Conjugated dicarboxylate anodes for Li-ion batteries, Nat. Mater. 8 (2009) 120. https://doi.org/10.1038/nmat2372

[22] H.-g. Wang, Y. Shuang, D.-l. Ma, X.-l. Huang, F. Meng, X.-b. Zhang, Tailored aromatic carbonyl derivative polyimides for high-power and long-cycle sodium-organic batteries, Adv. Energy Mater. 4 (2014) 1301651. https://doi.org/10.1002/aenm.201301651

[23] L. Chen, W. Li, Y. Wang, C. Wang, Y. Xia, Polyimide as anode electrode material for rechargeable sodium batteries, RSC Adv. 4 (2014) 25369-25373. https://doi.org/10.1039/C4RA03473B

[24] F. Xu, J. Xia, W. Shi, Anthraquinone-based polyimide cathodes for sodium secondary batteries, Electrochem. Commun. 60 (2015) 117-120. https://doi.org/10.1016/j.elecom.2015.08.027

[25] K.C. Kim, T. Liu, K.H. Jung, S.W. Lee, S.S. Jang, Unveiled correlations between electron affinity and solvation in redox potential of quinone-based sodium-ion batteries, Energy Storage Mater. 19 (2019) 242-250. https://doi.org/10.1016/j.ensm.2019.01.017

[26] X. Wang, Z. Shang, A. Yang, Q. Zhang, F. Cheng, D. Jia, J. Chen, Combining quinone cathode and ionic liquid electrolyte for organic sodium-ion Batteries, Chem 5 (2019) 364-375. https://doi.org/10.1016/j.chempr.2018.10.018

[27] C. Luo, X. Fan, Z. Ma, T. Gao, C. Wang, Self-healing chemistry between organic material and binder for stable sodium-ion batteries, Chem 3 (2017) 1050-1062. https://doi.org/10.1016/j.chempr.2017.09.004

[28] Y. Liang, P. Zhang, J. Chen, Function-oriented design of conjugated carbonyl compound electrodes for high energy lithium batteries, Chem. Sci. 4 (2013) 1330-1337. https://doi.org/10.1039/c3sc22093a

[29] L. Chen, S. Liu, Y. Wang, L. Zhao, Y. Zhao, 2, 3-Dicyano-5, 6-dichloro-1, 4-benzoquinone as a novel organic anode for sodium-ion batteries, J. Electroanal. Chem. 837 (2019) 226-229. https://doi.org/10.1016/j.jelechem.2019.02.029

[30] Z. Wang, A. Li, L. Gou, J. Ren, G. Zhai, Computational electrochemistry study of derivatives of anthraquinone and phenanthraquinone analogues: The substitution effect, RSC Adv. 6 (2016) 89827-89835. https://doi.org/10.1039/C6RA19128B

[31] A. Choi, Y.K. Kim, T.K. Kim, M.-S. Kwon, K.T. Lee, H.R. Moon, 4,4′-Biphenyldicarboxylate sodium coordination compounds as anodes for Na-ion batteries, J. Mater. Chem. A 2 (2014) 14986-14993. https://doi.org/10.1039/C4TA02424A

[32] M.A. Sk, S. Manzhos, Exploring the sodium storage mechanism in disodium terephthalate as anode for organic battery using density-functional theory calculations, J. Power Sources 324 (2016) 572-581. https://doi.org/10.1016/j.jpowsour.2016.05.101

[33] H. Zhu, J. Yin, X. Zhao, C. Wang, X. Yang, Humic acid as promising organic anodes for lithium/sodium ion batteries, Chem. Commun. 51 (2015) 14708-14711. https://doi.org/10.1039/C5CC04772B

[34] L. Wang, J. Zou, S. Chen, J. Yang, F. Qing, P. Gao, J. Li, Zinc terephthalates $ZnC_8H_4O_4$ as anodes for lithium ion batteries, Electrochim. Acta 235 (2017) 304-310. https://doi.org/10.1016/j.electacta.2017.03.095

[35] Q. Deng, C. Fan, L. Wang, B. Cao, Y. Jin, C.-M. Che, J. Li, Organic potassium terephthalate ($K_2C_8H_4O_4$) with stable lattice structure exhibits excellent cyclic and rate capability in li-ion batteries, Electrochim. Acta 222 (2016) 1086-1093. https://doi.org/10.1016/j.electacta.2016.11.079

[36] H.-g. Wang, S. Yuan, Z. Si, X.-b. Zhang, Multi-ring aromatic carbonyl compounds enabling high capacity and stable performance of sodium-organic batteries, Energy Environ. Sci. 8 (2015) 3160-3165. https://doi.org/10.1039/C5EE02589C

[37] J.T. Price, J.A. Paquette, C.S. Harrison, R. Bauld, G. Fanchini, J.B. Gilroy, 6-Oxoverdazyl radical polymers with tunable electrochemical properties, Polymer Chem. 5 (2014) 5223-5226. https://doi.org/10.1039/C4PY00829D

[38] K. Oyaizu, Y. Ando, H. Konishi, H. Nishide, Nernstian adsorbate-like bulk layer of organic radical polymers for high-density charge storage purposes, J. Am. Chem. Soc. 130 (2008) 14459-14461. https://doi.org/10.1021/ja803742b

[39] J.K. Kim, Y. Kim, S. Park, H. Ko, Y. Kim, Encapsulation of organic active materials in carbon nanotubes for application to high-electrochemical-performance sodium batteries, Energy Environ. Sci. 9 (2016) 1264-1269. https://doi.org/10.1039/C5EE02806J

[40] Q. Zhao, X. Hu, K. Zhang, N. Zhang, Y. Hu, J. Chen, Sulfur nanodots electrodeposited on ni foam as high-performance cathode for Li–S batteries, Nano Lett. 15 (2015) 721-726. https://doi.org/10.1021/nl504263m

[41] H. Shirakawa, E.J. Louis, A.G. MacDiarmid, C.K. Chiang, A.J. Heeger, Synthesis of electrically conducting organic polymers: Halogen derivatives of polyacetylene, Chem. Commun. (1977) 578-580. https://doi.org/10.1039/c39770000578

[42] Q. Zhao, A. Whittaker, X.S. Zhao, Polymer Electrode Materials for Sodium-ion Batteries, Mater. 17 (2018) E2567. https://doi.org/10.3390/ma11122567

[43] D. Su, J. Zhang, S. Dou, G. Wang, Polypyrrole hollow nanospheres: stable cathode materials for sodium-ion batteries, Chem. Commun. 51 (2015) 16092-16095. https://doi.org/10.1039/C5CC04229A

[44] Q. Liao, H. Hou, X. Liu, Y. Yao, Z. Dai, C. Yu, D. Li, l-lactic acid and sodium p-toluenesulfonate co-doped polypyrrole for high performance cathode in sodium ion battery, J. Phys. Chem. Solids 115 (2018) 233-237. https://doi.org/10.1016/j.jpcs.2017.12.015

[45] R. Zhao, L. Zhu, Y. Cao, X. Ai, H. X. Yang, An aniline-nitroaniline copolymer as a high capacity cathode for Na-ion batteries, Electrochem. Commun. 21 (2012) 36-38. https://doi.org/10.1016/j.elecom.2012.05.015

[46] S. Zhang, W. Huang, P. Hu, C. Huang, C. Shang, C. Zhang, R. Yang, G. Cui, Conjugated microporous polymers with excellent electrochemical performance for lithium and sodium storage, J. Mater. Chem A 3 (2014) 1896-1901. https://doi.org/10.1039/C4TA06058J

[47] J. Xie, P. Gu, Q. Zhang, Nanostructured conjugated polymers: Toward high-performance organic electrodes for rechargeable batteries, ACS Energy Lett. 2 (2017) 1985-1996. https://doi.org/10.1021/acsenergylett.7b00494

[48] H. Zhong, G. Wang, Z. Song, X. Li, H. Tang, Y. Zhou, H. Zhan, Organometallic polymer material for energy storage, Chem. Commun. 50 (2014) 6768-6770. https://doi.org/10.1039/C4CC01572J

[49] R. Precht, S. Stolz, E. Mankel, T. Mayer, W. Jaegermann, R. Hausbrand, Investigation of sodium insertion into tetracyanoquinodimethane (TCNQ): Results for a TCNQ thin film obtained by a surface science approach, Phys. Chem. Chem. Phys. 18 (2016) 3056-3064. https://doi.org/10.1039/C5CP06659J

[50] Y. Chen, S. Manzhos, A comparative computational study of lithium and sodium insertion into van der Waals and covalent tetracyanoethylene (TCNE)-based crystals as promising materials for organic lithium and sodium ion batteries, Phys. Chem. Chem. Phys. 18 (2016) 8874-8880. https://doi.org/10.1039/C5CP07474F

[51] J.K. Kim, J. Scheers, J.H. Ahn, P. Johansson, A. Matic, P. Jacobsson, Nano-fibrous polymer films for organic rechargeable batteries, J. Mater. Chem. A 1 (2013) 2426-2430. https://doi.org/10.1039/C2TA00743F

[52] H. Kim, J. Hong, K.Y. Park, H. Kim, S.-W. Kim, K. Kang, Aqueous rechargeable Li and Na ion batteries, Chem. Rev. 114 (2014) 11788-11827. https://doi.org/10.1021/cr500232y

Materials Research Forum LLC
https://doi.org/10.21741/9781644900833-4

Chapter 4

Alloys for Sodium-Ion Batteries

Vaishali Tomar[1], Ankita Dhillon[2], Kritika S. Sharma[3] and Dinesh Kumar[3,*]

[1]Formulation American Health Formulations New York 11788, USA

[2]Department of Chemistry, Banasthali Vidyapith, Rajastha304022, India

[3]School of Chemical Sciences, Central University of Gujarat, Gandhinagar 382030, India

*dsbchoudhary2002@gmail.com

Abstract

Sodium-ion batteries (SIBs) are developing as a substitution for lithium-ion batteries (LIBs). Sodium is in great abundance in the earth's crust. By the used of selective carbon as an anode, the expansion of sodium ion batteries anode has been accomplished. Due to the contribution of the carbon-based materials, the sodium ion batteries anode has been improved. Moreover, more investigation is still required to get the appropriate carbon materials that specify anode qualities. Alloy materials have shown a high capacity anode with the combination of carbon-based or other carbon-based materials for the development of SIBs. This chapter highlights the expansion of carbon-based materials and their complexes with alloy materials as well as their challenges and problems for sodium-ion batteries anodes.

Keywords

Sodium-Ion Batteries, Anode, Alloying, Electrolyte, Challenges

Contents

1.   Introduction

In the present primary area of energy storage devices, the rechargeable ion batteries have been slowly working. It is because of the benefits of excellent capabilities, long cycling life, larger capacity, etc. [1]. The green and safe energy sources are in demand nowadays. Due to the fast automation and demand for energy sources, environmental pollution and energy crunch have elevated the rising apprehensions across the world in current years. Green renewable energy sources should be demoralized and promoted to solve these issues. In present years, high-performance energy storing devices developed a current study matter [2]. Rechargeable ion batteries have been used with the advantage of excellent energy density and good cyclability to bring that responsibility.

Lithium-ion batteries (LIBs) developed the primary energy storing procedures. Meanwhile, Sony was commercialized them in 1991. They gained incredible achievement as authoritative bases for convenient electric devices and vehicles (EVs) [3]. But the scientific enhancement of LIBs cannot report the shortage of lithium reserves, and it gives a risk for electric vehicles to motorized by LIBs, which is no longer reasonable with their general procedure. It is challenging to change battery bits of knowledge based on an earth-abundant element other than LIBs. Na is a promising candidate who is next to Li in the periodic table [4]. The sixth most abundant element in the earth's crust is sodium. It shows many comparable alkali metal reactivity with Li. It is extensively dispersed. The model of SIBs is not novel because they were examined together with LIBs in the 1980s. So, in the primary 1990s, researchers have lost their interest because of the inferior energy density of Nickel-ion batteries (NIBs). Because of ambient temperature, NIBs have higher consideration for grid-level application.

SIBs contain more inferior electrochemical qualities than LIBs because of the better range of Na^+ (59 pm for Li^+ and 102 pm for Na^+) [5]. The big size Na^+ ion shows poor kinetic performance and significant volume variations. Lithium resources have been consumed heavily, and SIBs have fascinated the consideration due to large abundance and low price. Almost 6.8 million tons of lithium metal will be used up when 52% of gasoline operated vehicle is substituted by electric cars. The exclusivity of lithium resources increases the application of sodium ion batteries. The interest of researchers has been shifted towards SIBs. Figure 1 displays the charging and discharging mechanism of the sodium ion battery. During the mechanism, sodium ion from the cathode required to introduce and go through the electrolyte towards the anode. Through the external circuit, the charge evaluation electrons permit from cathode to anode at the same time. In the discharging process, different behavior occurs.

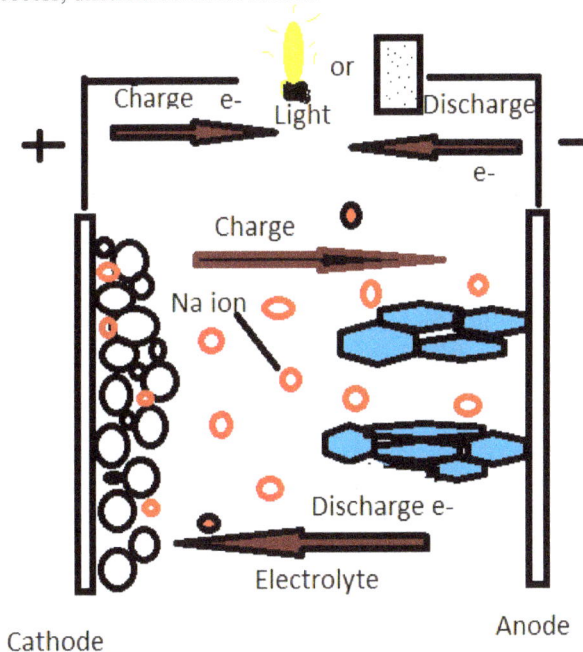

Figure 1. Diagrammatic representation of mechanism of sodium ion battery.

Sodium-Ion Batteries: Materials and Applications   Materials Research Forum LLC
Materials Research Foundations **76** (2020) 93--116   https://doi.org/10.21741/9781644900833-4

The exact energy density of SIBs is 18% inferior to LIBs. The anode constituents, graphite for LIBs do not work for sodium ion batteries with a low capacity [6]. It can be improved by solvent co-intercalation, but that method brings a set of challenges. Numerous efforts have been made to discover anodes for sodium ion batteries [7]. Sodium reacts with an extensive range of transition metal oxides.

Several review articles have been published related to sodium ion batteries anode. Carbon-based and alloy included anodes for sodium ion batteries are the current topic for researchers. Thus, this chapter highlights the high recital hard carbon anodes and sodium ion-storing mechanism in hard carbon. The alloying type anodes such as phosphorous to prove some of the maxima stated capacities are discussed. Lastly, metal oxide, sulfide, and organic based anodes are explored [8].

2. Sodium ion batteries anode materials

In the growth of sodium ion batteries, the anode material is expended with hard carbon. But their uses were restricted due to difficulties. Transition (d-block) metal oxides were used as an anode in LIBs because of the complexity of carbon resources. It was strained to sodium ion batteries as a substitute for carbon material [9]. Higher specific capacity is the factor in this substitution. So more considerable volume extension or reduction related to alkaline ion addition or removal to create remarkable loss to electrodes, which primary damage of electrical interaction with fast capacity is failing. Since these factors that delay metal-based anode considers the expansion of components that can be alloyed with sodium called alloy materials. For example, in the periodic table group IVA and VA such as Ge, Sn, Sb, P, etc. give alloy with sodium to make $NaGe$, $Na_{3.75}Sn$, Na_3Sb, and Na_3P, respectively [10-14]. Such type of alloying reaction was capable of reaching tremendous capabilities to a great extent, but difficulties were escorted with a massive increase in the volume of the host materials. It occurs because of the nonstop reduction of the electrode materials. It was studied by designing the novel nanostructures, defending the volume variation of the alloy anodes through sodium extension and reduction reaction. During the analysis of sodium ion battery anodes, carbon-based materials remain inspiring like lithium ion battery due to excellent conductivity and more abundance on earth [15].

3. Hard carbon

Experimentally and theoretically, various carbon-based resources were examined as an electrode material for energy storing. Cyclic efficiency, lifetime, and power of energy saving devices, supercapacitors, and lithium-ion batteries enhanced to improve specific capacity. In the 1980s, the electrochemical addition of sodium ion in carbon materials began initially [16] because of the arrangement of chemical and physical properties like price efficiency, high abundance, and high surface area. Hence, researchers began work on carbon-based electrode resources to recover the enactment of energy storing devices electrodes.

Some studies have been reported for the comparison of the functioning of carbon in Na and Li batteries. While the additional outcomes may construe for SIBs since the intercalation mechanism has appeared very close to that of lithium. The significant work has been reported for the electrochemical, microstructural properties of automated crushing on graphitized petroleum coke for SIBs and LIBs [17]. The storage capacity of sodium rises with elongated ball milling time. With three black carbon samples, the electrochemical reaction was reported for Na and Li. The lower nature of sodium for the productivity of the passivating sheet in sodium unit reduces irreversible capacity. A cost of a reversible function in carbon black SIBs was reported around 201 mAh/g [18]. This higher reversible capability was initiate to vary on the external surface of carbon black. This comparison was studied at different pyrolyzing temperatures. In hard carbon precursor, a high reversible capacity was observed around 210 mAh/g [19]. The structure of hard carbon easily allows sodium insertion with potential below 0.1V. PAN-based carbon fibers provided different carbon host samples. Solid coals possess excessive unalterable capabilities, almost 310 mAh/g. A saccharose coke carried out an extreme ability above 910 mAh/g and decreased with 202 mAh/g as a sample, which gives result in an unalterable capacity of 710 mAh/g [20]. It shows the high reversibility loss of hard carbons. This loss depends on a few factors like particle size, additives, electrolytes, and porosity capacities. The short opening volume and surface region of hard carbon substantial was studied to provide high reversible capacities around 345 mAh/g. Size of the stiff carbon offers good impression on capability [21].

4. Carbon nanostructures

In the fact of attainments, a growth of carbon anodes for sodium ion batteries (different means like developed methods, the calculation of electrolytes, etc.), the capability was barely cited. The energy storage devices with a suitable electrode have an attractive rate capability. The details of the size of the hard carbons as SIBs anodes are not sufficient to mention the rate abilities of the hard carbons [22]. Nano-sized materials are useful for the accomplishment of extreme rate capabilities, morphology, and size. They open opportunities to increase mass transportation and storage to recover the electrochemical performance of sodium. To regulate the morphology and size of electrode materials, the

Sodium-Ion Batteries: Materials and Applications
Materials Research Foundations **76** (2020) 93--116

Materials Research Forum LLC
https://doi.org/10.21741/9781644900833-4

synthesis of nanomaterials was a hopeful design. The development of electrochemical properties of electrode materials depends on different types of nanostructures [23].

Subsequently, the unique structure can increase the mass transference by proposing an excellent surface area and a small diffusion distance. And nanostructure design is a smart area of interest in LIBs. But nanostructure engineering has been prolonged toward sodium ion batteries and provides countless achievement. For the SIBs anodes, different types of nanostructured materials like hollow nanostructure, nanofibers, nanobubbles, nanosheets, and mesoporous carbon have been studied [24-26]. The rateability for hollow carbon nanospheres was established. The hollow carbon nanospheres distribute a discharge capacity of 100 mAh/g at the existing thickness of 2 A/g, which is 80% more than carbon spheres. The discharge ability of 120 mAh/g at density 100 A/g was also reported for carbon nanobubbles [27]. It was demonstrated by receiving the distinctive structure of peat moss leaves. A subsequent constituent is comprised of a 3D macroporous consistent network of carbon nanosheets. It provides an alterable capability of 299 mAh/g and 855 capability retention at 60 mAh/g over 211 cycles [28]. With 510 mA/g, the capability was observed 204 mAh/g. It has been studied that materials maintain a resonating nanowire arrangement with a plane surface which increases the particular area and connection area with electrolyte. The hollow carbon nanowire structure keeps the capability of 216 mAh/g after 330 rotations [29]. Retention was 82% because of the resonating structures. It gives a defending region for the discharge of mechanical stress by sodium ion addition or removal. The electrode was observed to the delivered alterable capacity of 150 mAh/g at density 500 mA/g [30]. The hollow carbon nanowire structure provides an increased rate of proficiency and long cycle immovability. This insertion property is accredited to small transmission space in resonating carbon nanowires, and it is interlaying space among the graphite areas. It provides excellent cycling immovability with rate ability of carbon nanomaterials on solid carbons.

5.   Carbon and alloy-based material composites

Electrode resources provide great capability and extensive steering life, but still, there is a need for excellent rate capability. It is hard to get all the purposes to a single-phase substantial, and dimensions to decrease simultaneously. The combination of separate nanostructures and perhaps synergistic effect has attracted the attention for energy storage devices as an electrode material [31]. Recently it has been studied in sodium ion batteries that metal oxides, metal Sulfides, elements of group IVA, VA, VIA (alloy-based constituents) and certain SIBs cathodes react with carbon to form compounds. Several alloy substantial components like Sn, Sb, Ge, Se, and P can alloy with sodium and produce the alloy phase like $Na_{3.75}Sn$ (857 mAh/g), Na_3Sb (670 mAh/g), Na_2Se (688 mAh/g), and Na_3P (2660 mAh/g) [32-37]. These materials provide appropriate implanting potential with high hypothetical specific capacities. So, the significant volume change of these materials delays their utility to SIBs. The development of composite alloys has been the only solution. Intermetallic or alloyed materials are being combined selenium with mesoporous carbon to give energy storage devices. After 380 rotations, the discharge capacity of 480 mAh/g to an ability of 340 mAh/g was delivered [38]. The selenium with mesoporous carbon showed a good result with capacity retention of about 35%. Some researchers showed the effect with phosphorous/carbon (P/C) composite materials for maximum Na ion-storing capability to 1765 mAh/g after 100 cycles [39]. The amorphous red p/c compound had released capability around 2400 mAh/g. With stable cycle performance, the high reversible capacity was recorded around 1880 mAh/g in some studies. The composites give excellent rate capabilities 1530 mAh/g at density 2.85 A/g [40]. Such rate capabilities are rarely present for some alloyed carbon complexes. The electrochemical implementation was accredited for carbon accumulation, which can play a role to increase the electrical conductivity of the conductor and provide support to expand the stability of the electrode. The elements of group IVA and VA like Sn/carbon and Sb/carbon compounds attract the attention towards SIB anode substantial [41]. The liberation capability of 547 mAh/g was studied for the composite of Sn and graphite. After the 100 cycles, there was less than 1% reduction in size. Sb nanomaterials are also reported with an ability of 420 mAh/g at the degree of 101 mA/g [42]. These materials are uniformly encapsulated in carbon fibers. After the 310 cycles, the capacity was 350 mAh/g. Such property provides the conductive way and intermediate carrier for the active discharge of mechanical pressure due to the addition or removal of Na ion. It avoids the aggregation of Sb nanoparticles [43].

Combinations of alloy constituents with further electrodes have been described for SIB anodes, which were named bimetallic. During sequential electrochemical reactions, the parallel electrode rich phase in the complex was designed, and it can serve to support each other. The composite SnSeSneC with sodium storage provides the capacity 660 mAh/g at 22 mA/g, and afterward, long cycling, the capability will retain 400 mAh/g [44]. During electrochemical cycles, the storage functioning of SnSeSneC compound can be connected to the self-supporting carrier. The sodium alloying reactions with SnSb/C was demonstrated as a connection for NIBs [45]. The charge or discharge behavior was studied due to the development and presence of Sn and Sb-rich stages during consecutive sodiation or desodiation. It provides the capability of around 720 mAh/g at 110 mA/g, which retain 85 % of original capability afterward, 55 rotations of sodium ion storage [46]. The alloy composition optimization is the factor for the notable functioning of the SnSb/C electrode. The alloy phase basic change occurs during the results and carbon

Materials Research Forum LLC
https://doi.org/10.21741/9781644900833-4

accumulation in the electrode. A multicomponent alloy reaction provides the track for the progress of extraordinary capability anode constituents for SIBs [47].

6. Alloying reactions-based anode materials

6.1 P-based materials

Phosphorous has the maximum theoretical capability of 2500 mAh/g and 0.4 V redox potential. Phosphorous is classified into three allotropes due to crystalline structure:

1. White phosphorous

2. Red phosphorous

3. Black phosphorous

White phosphorus contains more harmfulness and reactivity with air, so it is unsuitable to use as an anode material. Red phosphorous is used as an anode because of immovability. It has abundant properties. Black phosphorous has been studied because of extraordinary electrical conductivity (290 S/m) and encrusted arrangement [48]. The tool [49] of sodiation with red and black phosphorous is $P/Na_xP/Na_3P$ ($0 < x < 3$).

6.1.1 Red phosphorous

Red phosphorous contain chain-like structure, which is a byproduct of P_4 in which one P-P bond is wrecked, and extra bond is created with the neighbor tetrahedron. Due to the high hypothetical capacity and environmentally friendly nature make it more useable as anode substantial for Na ion batteries. So, exploitation of red P for sodium ion batteries is significantly delayed by great capacity variations and electrical conductivity (9^{-14} S cm^{-1}) [50-55]. In the reduction of active materials, big volumes causing the attention of the shell of active materials by thick SEI layer. Every part of these subjects has important influence like capability and cyclability on electrochemical performance. Different tactics have been observed to report these contests with building carbonaceous materials and decrease the element extend to the nanometer balance. Combination of red phosphorous with carbon-based constituents raise the electrical conductivity and reduce automated pressures in sodium ion approval or issue procedure. Different kinds of carbon constituents were verified like carbon nanotube, carbon black and graphene, etc. [56].

With uniform distribution between single-layer carbon nanotubes, the red phosphorous based electrode gives result in loose rate proficiency (~310 mAh/g at 210 mA/g) after 300 cycles the capability retain with 0.01% average [57] — a fitted interaction of active substances with electrolyte promoted by a huge quantity of interconnected and continuous pores. Unstructured red phosphorous in regular passages of mesoporous carbon medium CMK-3 (as P @ CMK-3) was confined in some studies [58]. The surface area for CMK-and P @ CMK-3 was observed 1138 and 16.02 m^2g^{-1} with pore volume 1.51 and 0.07 cm^3g^{-1}, respectively [59]. Due to the advantage of surface area, the rate proficiency with steady strength of P @ CMK-3 accomplished for the quantity of red phosphorous [60].

6.1.2 Black phosphorous

Black is maximum thermodynamically stable and layered structure among all three allotropes of phosphorous. It contains a similar structure like graphite. The interlayer space can be extended up to 4.18 Å in the case of black P to host Na ion [61]. By van der Waals forces, the layers are held together. Black P contains higher electrical conductivity than red P. All these things show that black P may be a good anode substantial for Na ion batteries. The solution process comprises of two phases: the intercalation development monitored by alloy kind results. Both intercalation and alloy type reaction is escorted by an alteration of the laminar structure [62]. There is no alteration in the black P crystal structure with no P-P bond breakage. The amorphous phase will be formed by the cleavage of the P-P bond. So, it is important in the diffusion of the sodium ion to not segment change from the covered construction to the amorphous phase. Phosphorous/graphene with inserted construction. The black phosphorous/multi carbon nanotubes make mix constituents. This hybrid material shows the particular capability of 2400 mAh/g at solidity 0.06 A/g [63]. Black phosphorous/multi-walled carbon nanotubes complexes are synthesized by extraordinary energy sphere with specific capacity of 1600 mAh/g after 110 rotations at 1.2 A/g density, and changeable capability was found 1010 mAh/g at 4 A/g density [64]. The application of carbon materials is here as automated strength with electrical artery. Various procedures for carbon incorporated composites provide information for black P such as anode for Na ion batteries. Black P like anode materials with Na ion batteries, should report the matter for high value, few layers phosphorene, large area, with a minor number of defects [65]. The most commonly used methods are automatic exfoliation and liquid segment peeling. The scotch tape technique was reported for a few layers' phosphorene, but the chemical residue is unavoidable in the scotch tape method. A long time ultrasonic exfoliated procedure is required for liquid-phase exfoliation in organic diluters. It is needed to discover an actual manner for the research of huge area phosphorene.

7.   Conversion based material

7.1   Metal oxides

Due to polyelectronic reactivity, metal oxides contain high reversible capacity. The reaction mechanism is divided into dual types: M is here a non-electrochemical dynamic component such as Fe, Co, Ni, and Cu. M is also active electrochemical components such as Sn and Sb. Metal constituent after the reaction with sodium oxide gives metal oxide. Hypothetically, metal oxide with transformation reactions has more particular capability than practical value due to poor conductivity, and it contains maximum volume expansion. The coulombic capacity is below 75% [66]. It can be accredited to the dilapidation of electrolytes, and it forms the compact electrolyte interphase level. To improve an electrochemical routine, nano crystallized, and nanostructured electrode material is the general pathway to restrict the volume change. Sn based oxides among various metallic oxides attract consideration due to maximum reversible capability, low cost, and abundance, and it is based on conversion reactions. In the alloying process, large volume change occurs in SnO_2, which is Sn to $Na_{15}Sn_4$ and accumulation of Sn Particles during the method of cycling. Several modifications in SnO_2 have been introducing to solve this issue with the use of new morphology and the introduction of carbon materials. Carbon cloth and graphene draped SnO_2 has been arranged to improve the electrochemical recital. These materials show the particular reversible capability of 500 mAh/g at 0.14 A/g with 440 mAh/g at 0.05 A/g, respectively [67-70]. The comparison between pristine $SnO_{2,\ a}$ complex of SnO_2 and carbon resources, particularly graphene, have exposed the electrochemical performance. It could be accredited the graphene qualities of electrical behavior, biochemical constancy with the maximum particular external zone. It was studied that the SnO_2 /N-doped graphene complex shows better performance than SnO_2 /bare graphene compounds. The investigation shows that the presence of electroactive site caused by N-doping is the chief motive. It increases the particular alterable capability from the increase in electron transmission proficiency. The amorphous SnO_2 showed more electrochemical implementation compared to crystal-like SnO_2 because of the change in volume from an amorphous state to crystalline oxide. To analysis the effect of SnO_2 crystalline on its electrochemical implementation, the experiments were designed. The alterable capacity was observed 370 mAh/g at the existing solidity of 55 mA/g over 110 rotations, but SnO_2 crystal-like provides one-third of the capacity. It is supplied by amorphous SnO_2 under equal circumstances [71, 72].

Due to the enhanced sodium ion distribution constant and the durable contact among amorphous SnO_2 with graphene, the electrochemical implementation of amorphous SnO_2

usually profited from the fundamental isotropic properties. Small amorphous SnO_2 particles give well release for volume strains [73].

7.2   Metal sulfides

These metallic sulfides are separated into two classes of configurations: coated and no coated structures. Most metallic sulfides like SnS_2, MoS_2, WS_2, and TiS_2 contain a characteristic layered structure. The atoms are firmly engaged with covalent bond where films were associated by frail van der Waals bonds. Through the process of adsorption, intercalation, and interruption without the damage of structure, some polar molecules can enter between the layers [74-79]. These kinds of materials display greater electrochemical implementation. A layered structure MoS_2 used as an anode for Na ion batteries. MoS_2 undergo two steps reactions: sodium ion intercalation in which two-phase effect exists from the trigonal to octahedral prismatic. During the process of intercalation, sodium ion 1.4 per formula was intercalated in MoS_2 [80-82].

WS_2 is a developing anode solid for Na ion batteries. It shows the alterable capability to 330 mAh/g on the concentration of 201 mA/g over 190 cycles. The diameter of the WS_2 nanowire is around 22 nm and can perform the long cycle life in the potential window between 0.6 to 4V. It is the lasting cycle life and firm routine for WS_2 in morphology modification. Particular non-layered metal sulfide was established to have motivating sodiation/desodiation qualities like FeS_2, Sb_2S_3, NiS_2, and Co_9S_8 [83-86].

This has been stiff to adjust the interlayer detachment of non-layered sulfides. It is limited by the essential kinds of stuff in comparison of coated metallic sulfides. Non-coated metallic sulfides display poorer rotation lifecycle activity. To improve the cycle performance of sulfides, modified electrolyte composition like $NaSO_3CF_3$ was added for solvent into the ether electrolyte. Conspicuously, it has been observed that below voltage 0.8V, FeS2 would undergo from a high capacity degeneration and low cycling implementation. Well, cycling performance and advanced energy density would be above 0.8 V [87-90].

8.   Graphene

Graphene-like carbonaceous material appeared as another energy storing substantial with possessions like small mass, chemically inactive, etc. Graphene contains big sp^2 attached microscopic carbon monolayer with numerous benefits such as exceptional optical, electrical, motorized, etc. qualities. It also contains favorable features of the huge external zone and conductivity for adsorption of ion [91-93]. It has been studied to make a complex electrode with graphene constituents for energy storing devices. In some papers, graphene has been described as anode for Na ion batteries. The oxide of Graphene (RGO) was discovered the first time like anode with a maximum capability of 140 mAh/g at 40 mA/g over 1000 cycles. The rate proficiency was also obtained, 95 mAh/g at a great amount of 1010 mA/g [94].

3D anode from nitrogen-doped graphene was discovered as high-performance Na ion battery. The nitrogen doped graphene shows the particular excessive capability of 2900 mAh/g paralleled to decrease graphene and simple nitrogen-based graphene towards a capability of beyond 2100 mAh/g at 0.2C [95]. The implementation of the nitrogen-doped graphene was observed at the 3D mesoporous structure. It consists of a huge external zone and distended matrix arrangement among sheets of the graphene. It has been defected by N-doping. It simplified a dispersion of Na ions and increased the storing of Na ions. It also minimizes the consequence of bulk expansions throughout the charge/discharge process. So, this was established that N-doping displays an encouraging result on carbon-based constituents. Recent work has been employed on graphene composite as anode for SIBs [96-102].

Metal oxides such as Fe_2O_3 can be uniformly attached on to graphene nanosheet (Fe_2O_3 @GNS). The capacity of Fe_2O_3 @GNS was observed 410 mAh/g over 220 cycles. The graphene functioned here as a conductive improver without any added reagents [103].

Conclusion and challenges

The utility of carbon-based materials undergoes enormous development among the variable SIB anode materials. Recently several studies have been dedicated to searching for a different carbon electrode, and review for these signs of progress have been concise in this review. It was observed that important developments had been done in making of carbon materials. Solid carbon is the furthermost extensively used carbon basis as sodium ion battery anodes and showed the electrochemical performance of SIBs. Various nanostructure materials have been evaluated. The composition of alloy constituents along with carbon, is graphene. It has been discussed & summarized. This complex shows well electrochemical implementation than pristine, consistent element and complexes. Constructed towards development was accomplished by the cathode constituents, carbon-based constituents with metal sulfides and cathode constituents were discussed in brief.

There are various challenges in the growth of SIBs for practical application. Firstly, the promising challenge is the additives into the electrolyte of sodium ion battery [104]. It was described that additives might progress the Na storing implementation of carbon-related constituents with complexes. Occasionally, additives with low coulombic efficiency could encourage the expansion of the SEI layer. Research should focus on the aim of why they encourage sheet creation. The additive was added in the electrolyte, selecting a suitable additive is another task. So, the extra investigation should be carried out. Another task is that if additive was not added in the electrolyte how to accomplish sodium ion battery anodes with Na storing implementation.

The second challenge is the rateability of hard carbon should more enhance [105]. Numerous statements for stiff carbon are established. The rateability is very small related to carbonize based carbon. Furthermost methods have been used to improve the rateability of solid carbon; it is the calculation of additives. To some extent, the process can recover the rate ability.

With all these factors, carbon will donate to further growth of sodium ion battery as different generations of the battery. Extraordinary improvement in SIBs should be considered so that the lithium-ion battery can be substituted.

The third challenge is the rate capability of another carbon-based electrode, and it embraces important challenges. Carbon materials in nanostructure need to be produced more. Recently, there are a lot of developments regarding the synthesis and manufacturing of carbon substantial nanostructure but still need to develop further.

Moreover, the process accepted in manufacturing the complexes of carbon-related constituents with alloy ingredients requires further development. The reports regarding the collaborating of dynamic carbon with further constituents increase quickly.

Sodium ion battery cathode constituents have fascinated great consideration in the last few years, but newly they were less measured. This is because of the little capability and modest rate capabilities of cathode materials. Such a problem can be concentrated by carbon crust on the external of cathode materials or graphene. The coated layers of carbonaceous were discovered to assist as a conductive system path for cathode constituents in sodium ion battery. There are still a lot of research areas requiring advancement. Much attention is required on carbon related materials in terminologies of storing capability improvement.

Acknowledgments

Dinesh Kumar is also thankful DST, New Delhi for financial support to this work (sanctioned vide project Sanction Order F. No. DST/TM/WTI/WIC/2K17/124(C).

References

[1] J.M. Tarascon, M. Armand, Issues and challenges facing rechargeable lithium batteries, Nature 414 (6861) (2001) 359-367. https://doi.org/10.1038/35104644

[2] M.S. Balogun, C. Li, Y. Zeng, M. Yu, Q. Wu, M. Wu, Titanium dioxide@ titanium nitride nanowires on carbon cloth with remarkable rate capability for flexible lithium-ion batteries, J. Power Sources 272 (2014) 946-953. https://doi.org/10.1016/j.jpowsour.2014.09.034

[3] M.S. Balogun, W. Qiu, W. Wang, P. Fang, X. Lu, Y. Tong, Recent advances in metal nitrides as high-performance electrode materials for energy storage devices, J. Mater. Chem. A 3 (4) (2015) 1364-1387. https://doi.org/10.1039/C4TA05565A

[4] M.S. Balogun, M. Yu, Y. Huang, C. Li, P. Fang, Y. Liu, Binder-free Fe_2N nanoparticles on carbon textile with high power density as novel anode for high-performance flexible lithium-ion batteries, Nano Energy 11 (2015) 348-355. https://doi.org/10.1016/j.nanoen.2014.11.019

[5] M.S. Balogun, M. Yu, C. Li, T. Zhai, Y. Liu, X. Lu, Facile synthesis of titanium nitride nanowires on carbon fabric for flexible and high-rate lithium-ion batteries, J. Mater. Chem. A 2 (28) (2014) 10825-10829. https://doi.org/10.1039/C4TA00987H

[6] M.S. Balogun, W. Qiu, Y. Luo, Y. Huang, H. Yang, M. Li, Improving the lithium-storage properties of self-grown nickel oxide: A back up from TiO_2 nanoparticles, ChemElectroChem 2 (9) (2015) 1243-1248. https://doi.org/10.1002/celc.201500146

[7] G.N. Zhu, Y.G. Wang, Y.Y. Xia, Ti-based compounds as anode materials for Li-ion batteries, Energy Environ. Sci. 5 (5) (2012) 6652-6667. https://doi.org/10.1039/c2ee03410g

[8] M.D. Slater, D. Kim, E. Lee, C.S. Johnson, Sodium-Ion Batteries, Adv. Funct. Mater. 23 (8) (2013) 947-958. https://doi.org/10.1002/adfm.201200691

[9] B. Dunn, H. Kamath, J.-M. Tarascon, Electrical energy storage for the grid: A battery of choices, Science 334 (6058) (2011) 928-935. https://doi.org/10.1126/science.1212741

[10] V. Palomares, P. Serras, I. Villaluenga, K.B. Hueso, J. Carretero-Gonzalez, T. Rojo, Na-ion batteries, recent advances and present challenges to become low cost energy storage systems, Energy Environ. Sci. 5 (3) (2012) 5884-5901. https://doi.org/10.1039/c2ee02781j

[11] C. Nithya, S. Gopukumar, Sodium ion batteries: a newer electrochemical storage, Wiley Interdiscip. Rev. Energy Environ. 4 (2014) 253-278. https://doi.org/10.1002/wene.136

Materials Research Forum LLC
https://doi.org/10.21741/9781644900833-4

[12]   J. Molenda, C. Delmas, P. Hagenmuller, Electronic and electrochemical properties of Na_xCoO_{2-y} cathode, Solid State Ionics 9 (1983) 431-435. https://doi.org/10.1016/0167-2738(83)90271-0

[13]   A.S. Nagelberg, W.L. Worrell, A thermodynamic study of sodium-intercalated TaS_2 and TiS_2, J. Solid State Chem. 29 (3) (1979) 345-354. https://doi.org/10.1016/0022-4596(79)90191-9

[14]   J. Tarascon, G. Hull, Sodium intercalation into the layer oxides $Na_xMo_2O_4$, Solid State Ionics 22 (1) (1986) 85-96. https://doi.org/10.1016/0167-2738(86)90062-7

[15]   V. Chevrier, G. Ceder, Challenges for Na-ion negative electrodes, J. Electrochem. Soc. 158 (9) (2011) A1011-A1014. https://doi.org/10.1149/1.3607983

[16]   J. Qian, X. Wu, Y. Cao, X. Ai, H. Yang, High capacity and rate capability of amorphous phosphorus for sodium ion batteries, Angew. Chem. 125 (17) (2013) 4731-4734. https://doi.org/10.1002/ange.201209689

[17]   M.S. Whittingham, Chemistry of intercalation compounds: metal guests in chalcogenide hosts, Prog. Solid State Chem. 12 (1) (1978) 41-99. https://doi.org/10.1016/0079-6786(78)90003-1

[18]   J.S. Kim, H.J. Ahn, H.S. Ryu, D.J. Kim, G.B. Cho, K.W. Kim, The discharge properties of Na/Ni_3S_2 cell at ambient temperature, J. Power Sources 178 (2) (2008) 852-856. https://doi.org/10.1016/j.jpowsour.2007.09.067

[19]   M. Reynaud, P. Barpanda, G. Rousse, J.-N. Chotard, B.C. Melot, N. Recham, et al., Synthesis and crystal chemistry of the $NaMSO_4F$ family ($M_{1/4}$ Mg, Fe, Co, Cu, Zn), Solid State Sci. 14 (1) (2012) 15-20. https://doi.org/10.1016/j.solidstatesciences.2011.09.004

[20]   B. Ellis, W. Makahnouk, Y. Makimura, K. Toghill, L. Nazar, A multifunctional 3.5 V iron-based phosphate cathode for rechargeable batteries, Nat. Mater. 6 (10) (2007) 749-753. https://doi.org/10.1038/nmat2007

[21]   K.T. Lee, T. Ramesh, F. Nan, G. Botton, L.F. Nazar, Topochemical synthesis of sodium metal phosphate olivines for sodium-ion batteries, Chem. Mater. 23 (16) (2011) 3593-3600. https://doi.org/10.1021/cm200450y

[22]   Y. Yamada, T. Doi, I. Tanaka, S. Okada, J. Yamaki, Liquid-phase synthesis of highly dispersed $NaFeF_3$ particles and their electrochemical properties for sodium-ion batteries, J. Power Sources 196 (10) (2011) 4837-4841. https://doi.org/10.1016/j.jpowsour.2011.01.060

[23]   Y. Park, D.S. Shin, S.H. Woo, N.S. Choi, K.H. Shin, S.M. Oh, Sodium terephthalate as an organic anode material for sodium ion batteries, Adv. Mater. 24 (26) (2012) 3562-3567. https://doi.org/10.1002/adma.201201205

[24]   H. Kim, I. Park, D.-H. Seo, S. Lee, S.-W. Kim, W.J. Kwon, New iron-based mixed-polyanion cathodes for lithium and sodium rechargeable batteries: combined first principles calculations and experimental study, J. Am. Chem. Soc. 134 (25) (2012) 10369-10372. https://doi.org/10.1021/ja3038646

[25]   R. Berthelot, D. Carlier, C. Delmas, Electrochemical investigation of the P2-Na_xCoO_2 phase diagram, Nat. Mater. 10 (1) (2011) 74-80. https://doi.org/10.1038/nmat2920

[26]   Y. Lu, L. Wang, J. Cheng, J.B. Goodenough, Prussian blue: A new framework of electrode materials for sodium batteries, Chem. Commun. 48 (52) (2012) 6544-6546. https://doi.org/10.1039/c2cc31777j

[27]   M. Zhou, L. Zhu, Y. Cao, R. Zhao, J. Qian, X. Ai, et al., $Fe(CN)_6$ doped polypyrrole: A high-capacity and high-rate cathode material for sodium-ion batteries, RSC Adv. 2 (13) (2012) 5495-5498. https://doi.org/10.1039/c2ra20666h

[28]   M. Zhou, Y. Xiong, Y. Cao, X. Ai, H. Yang, Electroactive organic anion-doped polypyrrole as a low cost and renewable cathode for sodium-ion batteries, J. Polym. Sci. Pol. Phys. 51 (2) (2013) 114-118. https://doi.org/10.1002/polb.23184

[29]   M.M. Doeff, S.J. Visco, M. Yanping, M. Peng, D. Lei, L.C. De Jonghe, Thin film solid state sodium batteries for electric vehicles, Electrochim. Acta 40 (13) (1995) 2205-2210. https://doi.org/10.1016/0013-4686(95)00164-A

[30]   M.M. Doeff, Y. Ma, S.J. Visco, L.C. De Jonghe, Electrochemical insertion of sodium into carbon, J. Electrochem. Soc. 140 (12) (1993) L169-L170. https://doi.org/10.1149/1.2221153

[31]   H. Xiong, M.D. Slater, M. Balasubramanian, C.S. Johnson, T. Rajh, Amorphous TiO_2 nanotube anode for rechargeable sodium-ion batteries, J. Phys. Chem. Lett. 2 (20) (2011) 2560-2565. https://doi.org/10.1021/jz2012066

[32]   Y. Jiang, M. Hu, D. Zhang, T. Yuan, W. Sun, B. Xu, Transition metal oxides for high-performance sodium ion battery anodes, Nano Energy 5 (2014) 60-66. https://doi.org/10.1016/j.nanoen.2014.02.002

Materials Research Forum LLC
https://doi.org/10.21741/9781644900833-4

[33]   H. Yu, Y. Ren, D. Xiao, S. Guo, Y. Zhu, Y. Qian, An ultrastable anode for long-life room-temperature sodium-ion batteries, Angew. Chem. Int. Ed. 53 (34) (2014) 8963-8969. https://doi.org/10.1002/anie.201404549

[34]   Y. Kim, Y. Park, A. Choi, N.S. Choi, J. Kim, J. Lee, An amorphous red phosphorus/carbon composite as a promising anode material for sodium ion batteries, Adv. Mater. 25 (22) (2013) 3045-3049. https://doi.org/10.1002/adma.201204877

[35]   L. Wu, X. Hu, J. Qian, F. Pei, F. Wu, R. Mao, Sb-C nanofibers with long cycle life as an anode material for high-performance sodium-ion batteries, Energy Environ. Sci. 7 (1) (2014) 323-328. https://doi.org/10.1039/C3EE42944J

[36]   P. Thomas, J. Ghanbaja, D. Billaud, Electrochemical insertion of sodium in pitch-based carbon fibers in comparison with graphite in $NaClO_4$ ethylene carbonate electrolyte, Electrochim. Acta 45 (3) (1999) 423-430. https://doi.org/10.1016/S0013-4686(99)00276-5

[37]   B. Jache, P. Adelhelm, Use of graphite as a highly reversible electrode with superior cycle life for sodium-ion batteries by making use of co-intercalation phenomena, Angew. Chem. Int. Ed. 53 (38) (2014) 10169-10173. https://doi.org/10.1002/anie.201403734

[38]   B.L. Ellis, L.F. Nazar, Sodium and sodium-ion energy storage batteries, Curr. Opin. Solid State Mater. Sci. 16 (4) (2012) 168-177. https://doi.org/10.1016/j.cossms.2012.04.002

[39]   S.Y. Hong, Y. Kim, Y. Park, A. Choi, N.-S. Choi, K.T. Lee, Charge carriers in rechargeable batteries: Na ions vs. Li ions, Energy Environ. Sci. 6 (7) (2013) 2067-2081. https://doi.org/10.1039/c3ee40811f

[40]   V. Palomares, M. Casas-Cabanas, E. Castillo-Martinez, M.H. Han, T. Rojo, Update on Na-based battery materials. A growing research path, Energy Environ. Sci. 6 (8) (2013) 2312-2337. https://doi.org/10.1039/c3ee41031e

[41]   N. Yabuuchi, K. Kubota, M. Dahbi, S. Komaba, Research development on sodium-ion batteries, Chem. Rev. 114 (23) (2014) 11636-11682. https://doi.org/10.1021/cr500192f

[42]   D. Kundu, E. Talaie, V. Duffort, L.F. Nazar, The emerging chemistry of sodium ion batteries for electrochemical energy storage, Angew. Chem. Int. Ed. 54 (11) (2015) 3431-3448. https://doi.org/10.1002/anie.201410376

[43]   L.P. Wang, L. Yu, X. Wang, M. Srinivasan, Z.J. Xu, Recent developments in electrode materials for sodium-ion batteries, J. Mater. Chem. A 3 (18) (2015) 9353-9378. https://doi.org/10.1039/C4TA06467D

[44]   C. Liang, S. Huang, W. Zhao, W. Liu, J. Chen, H. Liu, Polyhedral Fe_3O_4 nanoparticles for lithium-ion storage, New J. Chem. 39 (4) (2015) 2651-2656. https://doi.org/10.1039/C4NJ02032D

[45]   C. Liang, T. Zhai, W. Wang, J. Chen, W. Zhao, X. Lu, Fe_3O_4/reduced graphene oxide with enhanced electrochemical performance towards lithium storage, J. Mater. Chem. A 2 (20) (2014) 7214-7220. https://doi.org/10.1039/C3TA15426B

[46]   Y. Luo, M.-S. Balogun, W. Qiu, R. Zhao, P. Liu, Y. Tong, Sulfurization of FeOOH nanorods on a carbon cloth and their conversion into Fe_2O_3/Fe_3O_4-S coreshell nanorods for lithium storage, Chem. Commun. 51 (2015) 13016-13019. https://doi.org/10.1039/C5CC04700E

[47]   J.-W. Wen, D.-W. Zhang, Y. Zang, X. Sun, B. Cheng, C.-X. Ding, Li and Na storage behavior of bowl-like hollow Co_3O_4 microspheres as an anode material for lithium-ion and sodium-ion batteries, Electrochim. Acta 132 (2014) 193-199. https://doi.org/10.1016/j.electacta.2014.03.139

[48]   Z. Yan, L. Liu, J. Tan, Q. Zhou, Z. Huang, D. Xia, One-pot synthesis of bicrystalline titanium dioxide spheres with a coreeshell structure as anode materials for lithium and sodium ion batteries, J. Power Sources 269 (2014) 37-45. https://doi.org/10.1016/j.jpowsour.2014.06.150

[49]   K.T. Kim, G. Ali, K.Y. Chung, C.S. Yoon, H. Yashiro, Y.K. Sun, Anatase titania nanorods as an intercalation anode material for rechargeable sodium batteries, Nano Lett. 14 (2) (2014) 416-422. https://doi.org/10.1021/nl402747x

[50]   J. Park, J.W. Park, J.H. Han, S.W. Lee, K.Y. Lee, H.S. Ryu, Charge-discharge properties of tin dioxide for sodium-ion battery, Mater. Res. Bull. 58 (2014) 186-189. https://doi.org/10.1016/j.materresbull.2014.04.051

[51]   G. Qin, X. Zhang, C. Wang, Design of nitrogen doped graphene grafted TiO_2 hollow nanostructures with enhanced sodium storage performance, J. Mater. Chem. A 2 (2014) 12449-12458. https://doi.org/10.1039/C4TA01789G

[52]   P.R. Abel, M.G. Fields, A. Heller, C.B. Mullins, Tine germanium alloys as anode materials for sodium-ion batteries, ACS Appl. Mater. Interfaces 6 (18) (2014) 15860-15867. https://doi.org/10.1021/am503365k

Materials Research Forum LLC
https://doi.org/10.21741/9781644900833-4

[53] B. Farbod, K. Cui, W.P. Kalisvaart, M. Kupsta, B. Zahiri, A. Kohandehghan, et al., Anodes for sodium ion batteries based on TineGermaniumeAntimony alloys, ACS Nano 8 (5) (2014) 4415-4429. https://doi.org/10.1021/nn4063598

[54] M. He, K. Kravchyk, M. Walter, M.V. Kovalenko, Monodisperse antimony nanocrystals for high-rate Li-ion and Na-ion battery anodes: nano versus bulk, Nano Lett. 14 (3) (2014) 1255-1262. https://doi.org/10.1021/nl404165c

[55] L. Wu, D. Bresser, D. Buchholz, G.A. Giffin, C.R. Castro, A. Ochel, et al., Unfolding the mechanism of sodium insertion in anatase TiO_2 nanoparticles, Adv. Energy Mater. 5 (2) (2015) 1401142-1401146. https://doi.org/10.1002/aenm.201401142

[56] J. Qian, Y. Xiong, Y. Cao, X. Ai, H. Yang, Synergistic Na-storage reactions in Sn4P3 as a high-capacity, cycle-stable anode of Na-Ion batteries, Nano Lett. 14 (4) (2014) 1865-1869. https://doi.org/10.1021/nl404637q

[57] J. Ding, H. Wang, Z. Li, A. Kohandehghan, K. Cui, Z. Xu, et al., Carbon nanosheet frameworks derived from peat moss as high performance sodium ion battery anodes, ACS Nano 7 (12) (2013) 11004-11015. https://doi.org/10.1021/nn404640c

[58] K. Tang, L. Fu, R.J. White, L. Yu, M.M. Titirici, M. Antonietti, Hollow carbon nanospheres with superior rate capability for sodium-based batteries, Adv. Energy Mater. 2 (7) (2012) 873-877. https://doi.org/10.1002/aenm.201100691

[59] L. Jia, Y. Tian, Q. Liu, C. Xia, J. Yu, Z. Wang, et al., A direct carbon fuel cell with (molten carbonate)/(doped ceria) composite electrolyte, J. Power Sources 195 (17) (2010) 5581-5586. https://doi.org/10.1016/j.jpowsour.2010.03.016

[60] Z.Z. Jiang, Z.B. Wang, Y.-Y. Chu, D.M. Gu, G.P. Yin, Ultrahigh stable carbon riveted Pt/TiO_2C catalyst prepared by in situ carbonized glucose for proton exchange membrane fuel cell, Energy Environ. Sci. 4 (3) (2011) 728-735. https://doi.org/10.1039/C0EE00475H

[61] W. Li, C. Liang, J. Qiu, W. Zhou, H. Han, Z. Wei, et al., Carbon nanotubes as support for cathode catalyst of a direct methanol fuel cell, Carbon 40 (5) (2002) 791-794. https://doi.org/10.1016/S0008-6223(02)00039-8

[62] J.E. Mink, J.P. Rojas, B.E. Logan, M.M. Hussain, Vertically grown multiwalled carbon nanotube anode and nickel silicide integrated high performance microsized (1.25 ml) microbial fuel cell, Nano Lett. 12 (2) (2012) 791-795. https://doi.org/10.1021/nl203801h

Materials Research Forum LLC
https://doi.org/10.21741/9781644900833-4

[63]   X. Xie, M. Ye, L. Hu, N. Liu, J.R. McDonough, W. Chen, Carbon nanotubecoated macroporous sponge for microbial fuel cell electrodes, Energy Environ. Sci. 5 (1) (2012) 5265-5270. https://doi.org/10.1039/C1EE02122B

[64]   J. Chmiola, C. Largeot, P.-L. Taberna, P. Simon, Y. Gogotsi, Monolithic carbidederived carbon films for micro-supercapacitors, Science 328 (5977) (2010) 480-483. https://doi.org/10.1126/science.1184126

[65]   Z. Fan, J. Yan, L. Zhi, Q. Zhang, T. Wei, J. Feng, A three-dimensional carbon nanotube/graphene sandwich and its application as electrode in supercapacitors, Adv. Mater. 22 (33) (2010) 3723-3728. https://doi.org/10.1002/adma.201001029

[66]   D. Pech, M. Brunet, H. Durou, P. Huang, V. Mochalin, Y. Gogotsi, Ultrahighpower micrometre-sized supercapacitors based on onion-like carbon, Nat. Nanotechnol. 5 (9) (2010) 651-654. https://doi.org/10.1038/nnano.2010.162

[67]   G. Wang, H. Wang, X. Lu, Y. Ling, M. Yu, T. Zhai, Solid-state supercapacitor based on activated carbon cloths exhibits excellent rate capability, Adv. Mater. 26 (17) (2014) 2676-2682. https://doi.org/10.1002/adma.201304756

[68]   L. Yuan, X.-H. Lu, X. Xiao, T. Zhai, J. Dai, F. Zhang, Flexible solid-state supercapacitors based on carbon nanoparticles/MnO_2 nanorods hybrid structure, Acs Nano 6 (1) (2011) 656-661. https://doi.org/10.1021/nn2041279

[69]   Y. Zhu, S. Murali, M.D. Stoller, K. Ganesh, W. Cai, P.J. Ferreira, Carbon based supercapacitors produced by activation of graphene, Science 332 (6037) (2011) 1537-1541. https://doi.org/10.1126/science.1200770

[70]   M. Endo, C. Kim, K. Nishimura, T. Fujino, K. Miyashita, Recent development of carbon materials for Li ion batteries, Carbon 38 (2) (2000) 183-197. https://doi.org/10.1016/S0008-6223(99)00141-4

[71]   S. Flandrois, B. Simon, Carbon materials for lithium-ion rechargeable batteries, Carbon 37 (2) (1999) 165-180. https://doi.org/10.1016/S0008-6223(98)00290-5

[72]   B.J. Landi, M.J. Ganter, C.D. Cress, R.A. DiLeo, R.P. Raffaelle, Carbon nanotubes for lithium ion batteries, Energy Environ. Sci. 2 (6) (2009) 638-654. https://doi.org/10.1039/b904116h

[73]   Q. Wang, H. Li, L. Chen, X. Huang, Novel spherical microporous carbon as anode material for Li-ion batteries, Solid State Ionics 152 (2002) 43-50. https://doi.org/10.1016/S0167-2738(02)00687-2

[74]   Y. Wu, E. Rahm, R. Holze, Carbon anode materials for lithium ion batteries, J. Power Sources 114 (2) (2003) 228-236. https://doi.org/10.1016/S0378-7753(02)00596-7

[75]   S. Talapatra, S. Kar, S. Pal, R. Vajtai, L. Ci, P. Victor, Direct growth of aligned carbon nanotubes on bulk metals, Nat. Nanotechnol. 1 (2) (2006) 112-116. https://doi.org/10.1038/nnano.2006.56

[76]   S. Bose, T. Kuila, A.K. Mishra, R. Rajasekar, N.H. Kim, J.H. Lee, Carbon-based nanostructured materials and their composites as supercapacitor electrodes, J. Mater. Chem. 22 (3) (2012) 767-784. https://doi.org/10.1039/C1JM14468E

[77]   M.D. Levi, G. Salitra, N. Levy, D. Aurbach, J. Maier, Application of a quartzcrystal microbalance to measure ionic fluxes in microporous carbons for energy storage, Nat. Mater. 8 (11) (2009) 872-875. https://doi.org/10.1038/nmat2559

[78]   R. Alcantara, P. Lavela, G.F. Ortiz, J.L. Tirado, R. Menendez, R. Santamarı a, et al., Electrochemical, textural and microstructural effects of mechanical grinding on graphitized petroleum coke for lithium and sodium batteries, Carbon 41 (15) (2003) 3003-3013. https://doi.org/10.1016/S0008-6223(03)00432-9

[79]   D. Stevens, J. Dahn, High capacity anode materials for rechargeable sodiumion batteries, J. Electrochem. Soc. 147 (4) (2000) 1271-1273. https://doi.org/10.1149/1.1393348

[80]   P. Thomas, D. Billaud, Electrochemical insertion of sodium into hard carbons, Electrochim. Acta 47 (20) (2002) 3303-3307. https://doi.org/10.1016/S0013-4686(02)00250-5

[81]   Y. Li, S. Xu, X. Wu, J. Yu, Y. Wang, Y.S. Hu, Amorphous monodispersed hard carbon micro-spherules derived from biomass as a high-performance negative electrode material for sodium-ion batteries, J. Mater. Chem. A 3 (1) (2015) 71-77. https://doi.org/10.1039/C4TA05451B

[82]   S.J.R. Prabakar, J. Jeong, M. Pyo, Nanoporous hard carbon anodes for improved electrochemical performance in sodium ion batteries, Electrochim. Acta 161 (0) (2015) 23-31. https://doi.org/10.1016/j.electacta.2015.02.086

[83]   W. Luo, C. Bommier, Z. Jian, X. Li, R. Carter, S. Vail, Low-surface-area hard carbon anode for Na-ion batteries via graphene oxide as a dehydration agent, ACS Appl. Mater. Interfaces 7 (4) (2015) 2626-2631. https://doi.org/10.1021/am507679x

[84]   S.W. Kim, D.H. Seo, X. Ma, G. Ceder, K. Kang, Electrode materials for rechargeable sodium-ion batteries: potential alternatives to current lithiumion batteries, Adv. Energy Mater. 2 (7) (2012) 710-721. https://doi.org/10.1002/aenm.201200026

[85]   R. Alcantara, J.M. Jimenez-Mateos, P. Lavela, J.L. Tirado, Carbon black: a promising electrode material for sodium-ion batteries, Electrochem. Commun. 3 (11) (2001) 639-642. https://doi.org/10.1016/S1388-2481(01)00244-2

[86]   A. Naji, P. Thomas, J. Ghanbaja, D. Billaud, Identification by TEM and EELS of the products formed at the surface of a carbon electrode during its reduction in $MClO_4$eEC and MBF_4-EC electrolytes (M= Li, Na), Micron 31 (4) (2000) 401-409. https://doi.org/10.1016/S0968-4328(99)00118-3

[87]   C. Bommier, W. Luo, W.-Y. Gao, A. Greaney, S. Ma, X. Ji, Predicting capacity of hard carbon anodes in sodium-ion batteries using porosity measurements, Carbon 76 (2014) 165-174. https://doi.org/10.1016/j.carbon.2014.04.064

[88]   P. Thomas, D. Billaud, Effect of mechanical grinding of pitch-based carbon fibers and graphite on their electrochemical sodium insertion properties, Electrochim. Acta 46 (1) (2000) 39-47. https://doi.org/10.1016/S0013-4686(00)00542-9

[89]   P. Thomas, D. Billaud, Sodium electrochemical insertion mechanisms in various carbon fibres, Electrochim. Acta 46 (22) (2001) 3359-3366. https://doi.org/10.1016/S0013-4686(01)00536-9

[90]   X. Cao, Y. Li, X. Li, J. Zheng, J. Gao, Y. Gao, Novel phosphamide additive to improve thermal stability of solid electrolyte interphase on graphite anode in lithium-ion batteries, ACS Appl. Mater. Interfaces 5 (22) (2013) 11494-11497. https://doi.org/10.1021/am4024884

[91]   S. Komaba, T. Itabashi, B. Kaplan, H. Groult, N. Kumagai, Enhancement of Li ion battery performance of graphite anode by sodium ion as an electrolyte additive, Electrochem. Commun. 5 (11) (2003) 962-966. https://doi.org/10.1016/j.elecom.2003.09.003

[92]   R. Alcantara, P. Lavela, G.F. Ortiz, J.L. Tirado, Carbon microspheres obtained from resorcinol-formaldehyde as high-capacity electrodes for sodium-ion batteries, Electrochem. Solid-State Lett. 8 (4) (2005) A222-A225. https://doi.org/10.1149/1.1870612

[93]   S. Komaba, T. Ishikawa, N. Yabuuchi, W. Murata, A. Ito, Y. Ohsawa, Fluorinated ethylene carbonate as electrolyte additive for rechargeable Na batteries, ACS Appl. Mater. Interfaces 3 (11) (2011) 4165-4168. https://doi.org/10.1021/am200973k

[94]   Ponrouch, A. Go~Ni, M.R. Palacín, High capacity hard carbon anodes for sodium ion batteries in additive free electrolyte, Electrochem. Commun. 27 (2013) 85-88. https://doi.org/10.1016/j.elecom.2012.10.038

[95]   H. Kim, J. Hong, Y.-U. Park, J. Kim, I. Hwang, K. Kang, Sodium storage behavior in natural graphite using ether-based electrolyte systems, Adv. Funct. Mater. 25 (4) (2015) 534-541. https://doi.org/10.1002/adfm.201402984

[96]   S. Komaba, W. Murata, T. Ishikawa, N. Yabuuchi, T. Ozeki, T. Nakayama, et al., Electrochemical Na insertion and solid electrolyte interphase for hard-carbon electrodes and application to Na-ion batteries, Adv. Funct. Mater. 21 (20) (2011) 3859-3867. https://doi.org/10.1002/adfm.201100854

[97]   J. Zhao, L. Zhao, K. Chihara, S. Okada, J-i Yamaki, S. Matsumoto, et al., Electrochemical and thermal properties of hard carbon-type anodes for Na-ion batteries, J. Power Sources 244 (2013) 752-757. https://doi.org/10.1016/j.jpowsour.2013.06.109

[98]   F. Yang, Z. Zhang, K. Du, X. Zhao, W. Chen, Y. Lai, Dopamine derived nitrogen-doped carbon sheets as anode materials for high-performance sodium ion batteries, Carbon 91 (2015) 88-95. https://doi.org/10.1016/j.carbon.2015.04.049

[99]   V.G. Pol, E. Lee, D. Zhou, F. Dogan, J.M. Calderon-Moreno, C.S. Johnson, Spherical carbon as a new high-rate anode for sodium-ion batteries, Electrochim. Acta 127 (2014) 61-67. https://doi.org/10.1016/j.electacta.2014.01.132

[100] S. Wenzel, T. Hara, J. Janek, P. Adelhelm, Room-temperature sodium-ion batteries: Improving the rate capability of carbon anode materials by templating strategies, Energy Environ. Sci. 4 (9) (2011) 3342-3345. https://doi.org/10.1039/c1ee01744f

[101] A.S. Aricò, P. Bruce, B. Scrosati, J.-M. Tarascon, W. Van Schalkwijk, Nanostructured materials for advanced energy conversion and storage devices, Nat. Mater. 4 (5) (2005) 366-377. https://doi.org/10.1038/nmat1368

[102] P.G. Bruce, B. Scrosati, J.M. Tarascon, Nanomaterials for rechargeable lithium batteries, Angew. Chem. Int. Ed. 47 (16) (2008) 2930-2946. https://doi.org/10.1002/anie.200702505

[103] Y.G. Guo, J.S. Hu, L.J. Wan, Nanostructured materials for electrochemical energy conversion and storage devices, Adv. Mater. 20 (15) (2008) 2878-2887. https://doi.org/10.1002/adma.200800627

[104] B. Koo, H. Xiong, M.D. Slater, V.B. Prakapenka, M. Balasubramanian, P. Podsiadlo, et al., Hollow iron oxide nanoparticles for application in lithium ion batteries, Nano Lett. 12 (5) (2012) 2429-2435. https://doi.org/10.1021/nl3004286

[105] W. Li, L. Zeng, Z. Yang, L. Gu, J. Wang, X. Liu, et al., Free-standing and binderfree sodium-ion electrodes with ultralong cycle life and high rate performance based on porous carbon nanofibers, Nanoscale 6 (2) (2014) 693-698. https://doi.org/10.1039/C3NR05022J

Sodium-Ion Batteries: Materials and Applications
Materials Research Foundations **76** (2020) 117-134

Materials Research Forum LLC
https://doi.org/10.21741/9781644900833-5

Chapter 5

Mn-Based Materials for Sodium-Ion Batteries

N. Suresh Kumar[1], R. Padma Suvarna[1*], S. Ramesh[2], D. Baba Basha[3], K. Srinivas[2],
K. Chandra Babu Naidu[2*]

[1]Department of Physics, JNTUA, Anantapuramu-515002, A.P, India

[2]Department of Physics, GITAM Deemed to be University, Bangalore-562163, India

[3]College of Computer and Information Sciences, Majmaah University Al'Majmaah, K.S.A-11952, India

*chandrababu954@gmail.com, padmajntua@gmail.com

Abstract

In this chapter, we discussed thoroughly about the Na-ion batteries along with Mn-based sodium-ion batteries. In addition, electrochemical parameters like potential, current density, and reversible capacity were described for Mn-based sodium-ion batteries applications. Using this comparison, we pointed out the high-performance Mn-based materials for sodium-ion batteries.

Keywords

Sodium-Ion Batteries, Reversible Capacity, Mn-Based Materials

Contents

1. Introduction

Energy is a key concern for people in all walks of life. These concerns will continue to increase owing to the deterioration in the availability of fossil fuels like oil and gas which are non-renewable [1]. Also, an impending conversion from fossil fuel based on economy to one based on renewable energy sources, leads to an ever-increasing need for storage solution [2-6]. By storing energy, the energy generated from renewable resources like solar and wind power in reliable, powerful batteries, 24 x 7 power supply can be available [7, 8]. Moreover, the demand for advanced storage system is gradually increasing all over the world due to increasing population and energy consumption. Therefore, a large-scale energy storage system plays a pivotal role in integrating these renewable energies into the electric grid [9]. When flexibility, simple maintenance and high energy conversion efficiency is considered, electrochemical secondary storage is the favorable candidate for the energy storage in large scale [10].

2. History

Battery is a device with one or more cells in which energy in chemical form is transformed into electrical form and used as a power source, so batteries have become ubiquitous today. The term battery was first used by American scientist Benjamin Franklin in 1749 when he was doing experiments with electricity using capacitors [11]. Italian physicist Alexandro Volta invented the first true battery in 1800. The fixed discs of copper and zinc separated by cloth soaked in saltwater wires associated to both ends of the stack produces constant current. 0.76 Volts will be produced by a set of Cu and Zn disc (each cell). By stacking the cells together multiple of 0.76 V can be obtained [12]. The oldest rechargeable battery is the Lead (Pb)-acid battery, the most durable one invented in 1859. Till today, to start most of the internal combustion engine cars, lead-acid batteries are being used. Nowadays, batteries are available in different sizes from Megawatt sizes used to store energy from solar farms or substations to tiny batteries used in electronic watches. Based on different chemistries, batteries are available in the range 1.0 to 3.6 V and voltage can be enhanced by joining the cells in series and current can be enhanced by connecting them in parallel. By stacking in this way required voltages and currents can be obtained up to Megawatt sizes [13]. Any battery mainly consists of two metals or compounds having different chemical potentials separated by a porous insulator. The energy stored in the chemicals (atoms or bonds) is then transmitted to the moving electrons connected to an external device. Electrolyte like salt and water is used to transfer the ions from one metal to another during the reaction. During discharge, the metal which loses electrons is an anode and which accepts the electrons is the cathode [14].

3.   Types

Based on flow of electrons, batteries are divided into two types: primary and secondary batteries.

Primary battery

This type of battery is designed to be used only once and cannot be recharged or reused. Also, the electrochemical reaction or flow of electrons cannot be reversible. Example for this type is Zinc-Carbon battery [15, 16]. Fig.1 indicates the schematic representation of primary battery. The problem with these batteries is disposing them after using. Now the challenge before us is to recharge these primary batteries as it is not commercially feasible to replace the larger batteries.

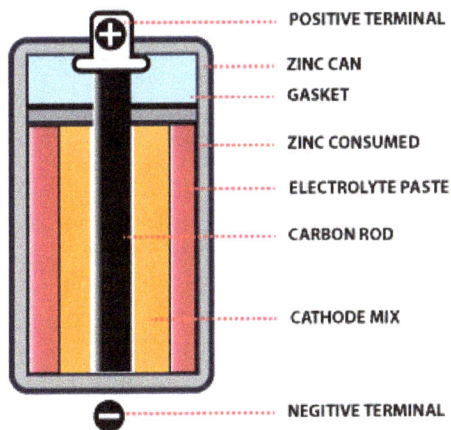

POSITIVE TERMINAL

ZINC CAN

GASKET

ZINC CONSUMED

ELECTROLYTE PASTE

CARBON ROD

CATHODE MIX

NEGITIVE TERMINAL

Figure 1. Primary battery.

Secondary battery

In this type of batteries, the electrochemical reactions or flow of electrons can be reversible i.e., these are rechargeable batteries. Hence secondary batteries act as both source and energy storage systems [17] Example for this type of batteries is Ni-Cd battery [18]. The earliest rechargeable batteries used alkali as an electrolyte. With alkali electrolyte, it was observed that the lifetime is less. NiMH (nickel-metal hydrogen) batteries established in 1989 had a large lifetime than Ni-Cd batteries [19], but the problem with these batteries is overcharging and overheating while charging. As larger applications of batteries are expected, the safety has become a significant consideration. Fig.2 shows the secondary battery.

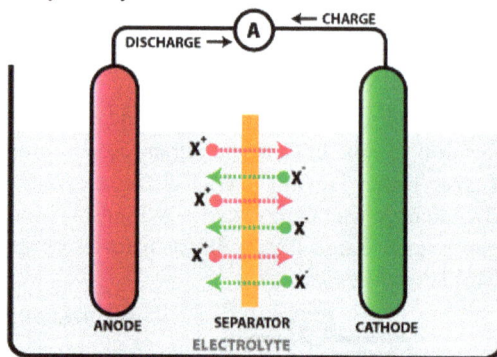

Figure 2. Secondary battery.

Quite often, new technologies demand more compact, safe, rechargeable and large capacity batteries. American physicist Prof. John Good enough invented lithium battery in 1980[20-22]. Lithium being the lightest element in the periodic table and having the longest electrochemical potential produces larger voltage in the high dense and light volumes. This is the root for the Li-ion battery. In this battery, Li is amalgamated with transition metals like Co, Ni, Mn or iron and oxygen to form the positive electrode (cathode). While recharging when a potential is applied, the cathode transports the positively charged Li-ion to the anode (graphite) and becomes Li metal [23, 24]. Since Li is having a durable electrochemical driving force to be oxidized, it transports back to the cathode and becomes Li^+ ion again and release electrons. This electrons movement in the circuit generates current which can be used for many applications in day-to-day life.

4.   Sodium-ion batteries

Rechargeable lithium (Li-ion) batteries are utmost extensive rechargeable in electronic and portable device markets owing to their remarkable features such as high density, long lifetime, etc. [25, 26]. The cost of lithium is gradually increasing following the commercialization and also the availability of lithium is only 20 ppm on earth's crust [27, 28]. The production of Li-ion batteries is difficult due to uneven distribution of lithium sources. Because of all these reasons Li-ion batteries alone cannot satisfy the market demands [29-31]. So, researchers concentrated on searching for alternative and suggested that, a metal which is similar to lithium can also be used to build rechargeable sodium-ion (Na-ion) batteries [32-42]. There is a substantial similarity between the chemistry of Na and Li.

Some of cathodic and anodic materials has been reported which shows good cycle ability with high capacity Na-storage performance, mainly owing to the bigger radii of sodium (Na) ions (1.06 Å) than lithium (Li) ions (0.76 Å) [43]. Moreover, the relative abundance of sodium (2.64 % on Earth) is very high when compared to lithium and also Na is extensively available in various forms such as rock salts, sea salts etc. Due to this sodium-ion batteries are contemplated as most favorable substitutes for lithium-ion batteries for next-generation rechargeable batteries [44-49].

5. Mn-based sodium-ion batteries

Even though there are lots of similarities between Na and Li ions, the battery characteristics are relatively distinct. For example, a layered chromium dioxide is electrochemically inactive in Li-ions but active in Na-ions [50]. Meanwhile the size variation between the ions causes each ion to have unique stability within the same structure [51]. Hence, it is necessary to investigate high potential electrode materials for Na-ion batteries [52]. Various electrodes have been introduced for example tunnel-type oxides [53], layer oxides [54], organic compounds [55] and polyanionic compounds [56]. Here most attention has been focused on layered oxides with a common formula Na_xMO_2 where M indicates metal like Fe, Cu, Mn, Cr, Co, Ni etc. This chapter had been focused on Mn-based materials for Na-ion batteries. Manganese oxide is the material which exhibits electrochemically inactive property in macro scale, but in micro-level it experiences large voltage polarization in between charge and discharge. In addition, it exhibits high initial irreversible capacity which makes the manganese oxide a suitable candidate (cathode or anode) for battery applications [57]. The Table.1 shows some of the Mn-based electrodes (cathode or anode) materials for sodium-ion batteries.

Abouimrane et al. [58] prepared $Na_{0.75}Mn_{0.7}Ni_{0.23}O_2$ cathode materials with Na_2BDC as a negative electrode (anode) for Na-ion batteries. They reported that these materials exhibit reversible specific capacity of 238 mAh g^{-1} after 50 cycles at an operating voltage of 3.6 V & temperature of 55 °C. Also reported is that these materials are favorable materials for rechargeable Na-ion batteries and by using these compounds as anodes and cathodes, the sodium ion batteries become cheap when compared to Li-ion batteries. Kim et al. [59] tested the sodium-ion battery performance using carbon anode coupled with $Na(Ni_{1/3}Fe_{1/3}Mn_{1/3})O_2$ cathode material and organic ester carbonate as an electrolyte. They reported that these cathode and anode materials exhibit high stability and long cyclic capacity (they retained 100 mAh/g after 150 cycles). Weng et al. [60] used ultra-

Sodium-Ion Batteries: Materials and Applications
Materials Research Foundations **76** (2020) 117-134

Materials Research Forum LLC
https://doi.org/10.21741/9781644900833-5

fine manganese oxide (MnO) as a negative electrode for sodium-ion batteries. They synthesized this negative electrode material through hypothermal route. The electrochemical studies revealed that UF-MnO_2 with average crystallite of size 4 nm shows a reversible specific capacity of around 567 mAhg^{-1} and also exhibits high life cycles i.e., almost 70% after 500 cycles. In contrast, same material with average crystallite 14 nm shows negligible behavior. So, some materials which are electrochemically in active in bulk form exhibits excellent electrochemical properties at nano level. Kim et al. [61] developed a new cathode (Na$_{0.44}$MnO$_2$) for Na-ion batteries synthesized via modified Pechini process, which can exhibit excellent cyclic performance and large capacity around 130 mAh/g. They reported that the synthesized material is isostructural. Also investigated were the Na insertion and deinsertion mechanisms through a calculation and found that the outcomes are well-matched with the investigational values.

Investigations on different electrode (cathode or anode) materials for Na-ion batteries have been increased. Barpanda et al. [62] discovered a new manganese-based cathode material for Na-ion batteries. This Mn-based cathode material (β-Na$_2$MnP$_2$O$_7$) was optimized in triclinic structure. Fig. 3 in ref. [62] it is clear that the material shows electrochemically active behavior at a functioning voltage of 36mV and also exhibits a reversible capacity of 80 mAhg^{-1}. All these results established a fact that β-Na$_2$MnP$_2$O$_7$ can serve as a cathode material for Mn-based Na-ion batteries. Continuing the searching for new electrode materials for Na-batteries, Chan Sun Park et al. [63] reported new cathode material for Na-ion batteries which is Mn-based pyrophosphate, Na$_2$MnP$_2$O$_7$. Compared to other Mn-based cathode materials this new material exhibits excellent electrochemical property at 3.8 V. In addition, it shows reversible capacity at around 90 mAh/g at room temperature. Furthermore, the cycling capacity is also very high for this new compound i.e., they retain almost 96% of the capacity after 90 cycles. Fig.2 in ref. [63] represents the charge-discharge cycles of prepared cathode material. So, Mn-based polymorphic structures may exhibit good electrochemical properties which may be useful for next-generation Na-ion batteries. Buchholz et al. [64] synthesized a new P$_2$-layered Na$_{0.45}$Ni$_{0.22}$Co$_{0.11}$Mn$_{0.66}$O$_2$ cathode material and compared the electrochemical performance of the prepared cathode material for both Na-based and Li-based systems. Fig. 3 and Fig.4 in ref. [64] shows the comparison of electrochemical properties of both Na-based and Li-based systems. From the results it is clear that Ni-based system dominated Li-based system which is due to redox reactions of Mn^{2+}/Mn^{4+} ions. Using this prepared material as a cathode Na based system retains 82% of the capacity after 100 cycles which is very high when compared to Li based system which retains 61% of the capacity after 100 cycles. Moreover, Columbia efficiency also dominates in Na based

(99.5%) system over Li based (98.5%) system. These results suggest that Mn-based cathode materials can serve as a prominent candidate for Na-ion batteries.

Table:-1. Various parameters of Mn-based materials for sodium ion batteries

S. No.	Material (Cathode/Anode)	Potential (V)	Current density (mA/g)	Reversible Capacity (mAh/g)	Reference
1	$Na_{0.75}Mn_{0.7}Ni_{0.23}O_2/$ Na_2BDC	2.0-4.2	20 (-C/13)	238	58
2.	$Na (Ni_{1/3}Fe_{1/3}Mn_{1/3}) O_2/$ Carbon	1.5-4.0	C/2	100	59
3.	$UF-MnO_x$(anode)	0.001-2.8	15	567	60
4.	$\beta-Na_2MnP_2O_7$	3.6	-	80	62
5.	$Na_2MnP_2O_7$	3.8	10	90	63
6.	$Na_{0.67} [Mn_{0.65}Co_{0.2}Ni_{0.15})_{0.9}]$ O_2	2.3-4.25	20	141	65
7.	$Na_2Mn_3O_5$	1.5-4.7	5C	219	66
8.	$Na_{0.5} [Ni_{0.23}Fe_{0.12}Mn_{0.63}] O_2$	1.5-4.6	C/10	180	67
9.	$Na_{0.67}Mn_{0.65}Fe_{0.2}Ni_{0.15}O_2$	1.5-4.3	1C	208	68
10.	$Na_{0.6}Ni_{0.22}Fe_{0.11}Mn_{0.66}O_2/$ Sb-C	0.7-4.1	30	180	69
11.	$Na_xNi_{0.22}Co_{0.11}Mn_{0.66}O_2/$ hard carbon	0.5-4.0	1C	220	70
12.	$Na [Cu_{0.1}(Fe_{1/3}Mn_{2/3})_{0.9}]C_2O_4/$ P-TiP_2 -C	1.2-3.4	12 (C/10)	135	71
13.	$Na_{0.76} Mn_{0.5}Ni_{0.3}Fe_{0.1} Mg_{0.1}$ $O_2/$ hard carban	2.0-4.2	12	131	72
14.	$Na_{0.67}[Fe_{0.5}Mn_{0.5}] O_2$-$NaN_3/$ hard carbon	1.0-4.4	15	76	73
15.	$NaMnO_2/$ FeSb alloy ribbons	1.1-3.5	50	300	74
16.	$Na_xNi_{0.2}Mn_{0.6}Co_{0.2}O_2/$ hard carbon	0.5-4.0	C/50	150	75
17.	$Na_{2/3}Fe_{1/3}Mn_{2/3}O_2/$ hard carbon	1.5-4.1	C/20	79	76

Yuan et al. [65] prepared pure $Na_{0.67} [Mn_{0.65}Co_{0.2}Ni_{0.15})_{0.9}] O_2$ micro flakes by sol-gel route for the applications in Na-ion batteries. These materials serve as good cathode materials. From the electrochemical studies the reversible capacity was observed to be 141 mAh/g and cycling capacity is also very high i.e., they exhibit 125 $mAhg^{-1}$ after 50 cycles. Moreover, they recommended that the substitution of cobalt and aluminium in the above-prepared cathode material enhances the lifetime as well as reversible capacity. Fig.4 in ref. [65] evidenced that aluminium substituted compound retains almost 95.4%

Materials Research Forum LLC
https://doi.org/10.21741/9781644900833-5

of initial capacity after 50 cycles. So, these aluminium substituted materials can serve as high capacity cathodes for Na-ion batteries.

Guo et al. [66] prepared a layered sodium manganese oxide ($Na_2Mn_3O_5$) material using redox reaction method followed by hydrothermal process. They reported that the prepared material crystallized in Birnessite structure with lamellar stacking of the synthetic nanosheets. In addition, the studies on electrochemical behavior revealed that the material exhibit high reversible capacity of 219 mAh g^{-1} at an operating voltage of 1.5-4.7 V. Also, the prepared material exhibits large density of energy around 605 Wh/kg. This energy density is very large associated to other transition metal oxides. Hasa et al. [67] used co-precipitation technique to prepare layered $Na_{0.5}[Ni_{0.23}Fe_{0.12}Mn_{0.63}]O_2$positive electrode material for Na-ion batteries. The prepared sample exhibit P_2 layered structure as confirmed by using XRD and FESEM. Fig. 1 in ref. [67] represents the XRD pattern and FESEM images of the prepared cathode material. Furthermore, the investigations of electrochemical studies revealed that this layered cathode material exhibit high reversible specific capacity of around 180 mAhg^{-1} and cycling life. Owing to environmentally friendly, low cost and high performance, this P_2 layered $Na_{0.5}$ $[Ni_{0.23}Fe_{0.12}Mn_{0.63}]O_2$ is a promising electrode material for Na-ion batteries. Using sol-gel technique, Yuan et al. [68] prepared pure P_2-layered $Na_{0.67}Mn_{0.65}Fe_{0.2}Ni_{0.15}O_2$ cathode material for Na-ion batteries. X-ray diffraction analysis showed that during sodium insertion and extraction there is a reversible two-phase transition. The substitution of Ni causes smoother phase transformation. Fig. 1 in ref. [68] displays the X-ray diffraction pattern of the synthesized material. In addition, the electrochemical studies revealed that $Na_{0.67}Mn_{0.65}Fe_{0.35}O_2$ shows reversible capacity of 136 mAh/g but $Na_{0.67}Mn_{0.65}Fe_{0.2}Ni_{0.15}O_2$ exhibit the reversible specific capacity of 208 mAhg^{-1}and also Ni substituted material has high cycling life which shows 71% retention after 50 cycles. Therefore, Ni- substitution causes smoother phase transition and improves the cycling capacity. These types of Ni-substituted Mn-based layered oxide materials are also the prominent candidates for Na-ion batteries which improves cycling life and capacity.

Hasa et al. [69] presented new P_2 type layered cathode and nanostructured anode (Sb-C) materials for the applications in sodium-ion batteries. Their studies revealed that with the above said positive and negative electrode materials the Na-ion cells deliver stable capacity about 120 mAhg^{-1} at a functioning voltage of about 27 mV. Fig. 3 reference [69] represents the electrochemical properties of Na-ion cells prepared with the above-mentioned cathode and anode materials. Wu et al. [70] synthesized hard carbon naturally from apple bio waste. They reported that the prepared hard carbon is suitable for the preparation of cheap and powerful carbon-based anode materials for sodium-ion batteries. These anode materials exhibit good cycling stability (112 mAh/g around 1000 cycles) and also showed the reversible capacity of about 250 mAhg^{-1}around 80 cycles. Also, the above said anode materials coupled with cathode material (Na$_x$Ni$_{0.22}$Co$_{0.11}$Mn$_{0.66}$O$_2$) exhibits excellent electrochemical performance (220 mAh/g after 100 cycles). Finally, the usage of biowaste is the suitable approach for the advancement of long life, high-performance active materials for the applications in Na-ion batteries. Seung-Min Oh [71] prepared a low-cost Na-ion cell using O$_3$ incrusted Na [Cu$_{0.1}$(Fe$_{1/3}$Mn$_{2/3}$)$_{0.9}$]C$_2$O$_4$ as a positive electrode prepared by oxalate coprecipitation route and P-TiP$_2$ -C as anode. They reported that the cell exhibits 130 mAh g^{-1} reversible capacities after 100 cycles. The presence of copper causes the suppression of phase transition to P$_3$ phase from O$_3$ phase. Keller et al. [72] proposed a new cathode material Na$_{0.76}$ Mn$_{0.5}$Ni$_{0.3}$Fe$_{0.1}$ Mg$_{0.1}$ O$_2$ for Na-ion batteries and prepared a sodium ion cell with above-said material as a positive electrode and starch-derived hard carbon as a negative electrode. The results revealed that these materials show high performance and good cyclic stability i.e., they show almost 80% of retention after 700 cycles. So, these materials are well suitable for preparation of high-performance batteries with long lifetime. Martinez et al. [73] studied the electrochemical properties of Na-ion cell with Na$_{0.67}$[Fe$_{0.5}$Mn$_{0.5}$] O$_2$ as positive electrode and hard carbon as negative electrode. They reported that the deficiency of sodium in cathode material is compensated by adding NaN$_3$and also observed an increase in reversible capacity with respect to NaN$_3$ concentration. Fig. 3 and Fig. 4 in reference [73] show the electrochemical behavior of Na-ion cell with varying NaN$_3$. The cell shows the reversible specific capacity of 76 mAh g^{-1} around 40 cycles. Fig. 8 in ref. [73] shows NMR spectra of cathode material with varying NaN$_3$ concentration. From NMR studies it is clear that sodium chemical shift occurred as a function of NaN$_3$ content in cathode material i.e., there is increase in Na ions which causes the improving electrochemical performance of Na-ion batteries.

New type of alloys as cathode and anode materials have gained much attention for Na-ion batteries, Edison et al. [74] synthesized a novel type of negative electrode (Fe-Sb alloy) for Na-ion cells by using melt spinning process. These anode materials exhibit high cyclic stability (466 mAh/g). With cathode (NaMnO$_2$) combination these materials exhibit reversible capacity of around 300 mAh/g. These new type of anode materials plays a vital role in development of next-generation Na-ion batteries with less cost. Jose et al. [77] used emulsion-based method to synthesize P$_2$ layered Na$_{2/3}$Fe$_{1/3}$Mn$_{2/3}$O$_2$ as cathode material for Na-ion batteries. In general, emulsion-based synthesis technique affords a homogeneous high pure sample. The particles exhibit plate-like or elongated nano-rod structures confirmed by electron microscopy (Fig. 2 in ref. [77]). The results evidenced that nanosize particles exhibits large cyclic capacity than micron size particles.

Fig. 12 in ref. [77] shows the electrochemical properties of cathode material. From the results it is clear that the nano rode like oxides are promising electrodes (cathode) for Na-ion batteries. Nayak et al. [78] prepared MnO_2 thin films through pulsed laser deposition and investigated electrochemical behavior & phase formation of different polymorphs of MnO_2. XRD and XPS studies disclosed that the polymorphs of MnO_2 depends on substrate temperature. The films exhibit different mixed phases (like γ and ß, ß and R, ß and R MnO_2) at different temperatures but especially at 550 and 650 °C the films exhibit a mixture of R-$MnO_2$and a-Mn_2O_3 phase. From Fig.5 in ref. [78] and electrochemical analysis, the films containing ß and R- MnO_2 phases at 450 °C exhibits optimum electrochemical behavior against Na-electrode. All the results suggest that these new type of materials serves as good negative electrodes for sodium-ion batteries. Zhang et al. [79] prepared $Na_{0.833}[Li_{0.25}Mn_{0.75}]$ O_2 (NLMO) and $(Na_{0.833}Zn_{0.075})$ $[Li_{0.25}Mn_{0.675}]O_2$ (NLMO-Zn) through sol-gel technique continued by heat treatment. Pure P_2-type structure was observed using XRD also investigated the electronic structures using XPS. Fig. 2 in ref. [79] shows the phase stability of NLMO and Zn doped NLMO. Also, reported the electrochemical behavior (Fig. 7 in ref. [79]) of the prepared samples, Zn-doped NLMO shows high cyclic capacity than undoped NLMO. After 100 cycles NLMO shows the reversible capacity 0.074 Ah g^{-1} but Zn-doped NLMO retains 0.161 Ah g^{-1} reversible capacity after 200 cycles which is almost twice compared to NLMO. Owing to high theoretical capacities and sodium ion conductivities, manganese (Mn) based layered oxides gained considerable attraction as positive electrodes for Na-ion batteries. Nevertheless, Jahn–Teller distortion arising from Mn-centers weakens the host structure and reduces the cycling life of these Mn-based materials. However, the performance of these layered cathode materials has increased by doping. Now research is going on to develop good electrode materials for next generation Na-ion batteries. In this connection Mn-based layered oxide materials have attracted much attention and will play a vital role in developing new electrode materials.

References

[1] D. Larcher, J. M. Tarascon, Towards greener and more sustainable batteries for electrical energy storage, Nat. Chem. 7 (2015) 19-29. https://doi.org/10.1038/nchem.2085

[2] P. G. Bruce, S. A. Freunberger, L.J. Hardwick, and J.M. Tarascon, Li-O and Li-S batteries with high energy storage, Nat. Mater. 11 (2012) 19–29. https://doi.org/10.1038/nmat3191

[3] B. Dunn, H. Kamath, J. M. Tarascon, Electrical energy storage for the grid: A battery of choices, Sci. 334 (2011) 928–935. https://doi.org/10.1126/science.1212741

[4] C. Liu, F. Li, L.P. Ma, H.M. Cheng, Advanced materials for energy storage, Adv. Mater. 22 (2010) 28–62. https://doi.org/10.1002/adma.200903328

[5] J. Yang, S. Muhammad, M. R. Jo, H. Kim, K. Song, D. A. Agyeman, Y. Kim, W. Yoon and Y. Kang, In situ analyses for ion storage materials, Chem. Soc. Rev. **45** (2016) 5717-5770. https://doi.org/10.1039/C5CS00734H

[6] K. Zhang, X. Han, Z. Hu, X. Zhang, Z. Tao and J. Chen, Nanostructured Mn-based oxides for electrochemical energy storage and conversion, **Chem. Soc. Rev. 44** (2015) 699-728. https://doi.org/10.1039/C4CS00218K

[7] J. M. Carrasco, L. G. Franquelo, J. T. Bialasiewicz, E. Galvan, R. C. Portillo Guisado, M. A. M. Prats, J. I. Leon, N. Moreno Alfonso, Power-electronic systems for the grid integration of renewable energy sources: A Survey, IEEE T.. Ind. Electron. 53 (2006) 1002-1016. https://doi.org/10.1109/TIE.2006.878356

[8] S. Hameer, J. L. Van Niekerk, A review of large-scale electrical energy storage, Int. J. Energy Res. 39 (2015) 1179-1195. https://doi.org/10.1002/er.3294

[9] B. Dunn, H. Kamath, J. M. Tarascon, Electrical energy storage for the grid: A battery of choices, Sci. 334 (2011) 928-935. https://doi.org/10.1126/science.1212741

[10] K. C. DivyaK, J. Østergaard, Battery energy storage technology for power systems-An overview, *Electr. Power Syst. Res.* 79 (2009) 511-520. https://doi.org/10.1016/j.epsr.2008.09.017

[11] W. Leonard, W. Labaree, The Papers of Benjamin Franklin (New Haven, Connecticut: Yale University Press, 1961), Benjamin Franklin et al., 3 (1961) 352: Letter to Peter Collinson, April 29, 1749. Paragraph 18. Franklinpapers.org. Retrieved 2012-08-29.

[12] *S. Bernard Finn,* Origin of Electrical Power, *National Museum of American History. (September 2002).* Retrieved 2012-08-29.

[13] Gaston Planté, Corrosion Doctors, Retrieved 2012-08-29.

[14] J. Desilvestro, O. Haas, Metal Oxide Cathode Materials for Electrochemical Energy Storage: A Review, J. Electrochem. Soc. 137 (1990) 5-22. https://doi.org/10.1149/1.2086438

[15] D. Linden, Handbook of Batteries and Fuel cells, Mc Gran- Hill Book Company, New York, 1984.

[16] T. R Crompton, Battery Reference Book, Third Edition. Newnes, Oxford, 2010

[17] T. Takamura, Trends in advanced batteries and key materials in the new century, Solid State Ion. 152-153 (2002) 19-34. https://doi.org/10.1016/S0167-2738(02)00325-9

[18] B. Scrosati, The present status of battery technology, Renew. Energy. 5 (1994) 285-294. https://doi.org/10.1016/0960-1481(94)90385-9

[19] F. Yu, L. Zhang, Y. Li, Y. An, M. Zhu and B. Dai, Mechanism Studies of $LiFePO_4$ Cathode Material: Lithiation/delithiation process, electrochemical modification and synthetic   reaction,   RSC   Adv. 4   (2014)   54576-54602. https://doi.org/10.1039/C4RA10899J

[20] S. Miyazaki, S. Kikkawa, M. Koizumi, Chemical and electrochemical deintercalactions of layered compounds, $LiCrO_2$, $LiCoO_2$ and $NaCrO_2$, $NaFeO_2$, $NaCoO_2$ and $NaNiO$. Synth. Met. 6 (1983) 211–217.2. https://doi.org/10.1016/0379-6779(83)90156-X

[21] M. Thomas, P.G. Bruce, J.B. Goodenough, AC impedance analysis of polycrystalline insertion electrodes – application to $Li_{1-x}CoO_2$, J. Electrochem. Soc. 132 (1986) 1521–1528. https://doi.org/10.1149/1.2114158

[22] K. Mizushima, P. C. Jones, P. J. Wiseman, John B Good enough, Li_xCoO_2 (0<x<=1): a new cathode material for batteries of high energy density, Solid State Ion. 3–4 (1981) 171–174. https://doi.org/10.1016/0167-2738(81)90077-1

[23] G. E. Blomgren, The development and future of lithium ion batteries, J. Electrochem. Soc. 164 (2017) 5019–5025. https://doi.org/10.1149/2.0251701jes

[24] J. M. Tarascon, The Li-ion battery: 25 years of exciting and enriching experiences. Electrochem. Soc. Interface 25 (2016) 78–82. https://doi.org/10.1149/2.F08163if

[25] M. Armand, J. M. Tarascon, Building better batteries, Nature 451 (2008) 652–657. https://doi.org/10.1038/451652a

[26] B. Kang, G. Ceder, Battery materials for ultrafast charging and discharging, Nat. 458 (2009) 190 –193. https://doi.org/10.1038/nature07853

[27] H. M. Vikström, S. Davidsson, M. Höök, Lithium availability and future production   outlooks,   Appl.   Energy   110   (2013)   252-266. https://doi.org/10.1016/j.apenergy.2013.04.005

[28] P. W. Gruber, P. A. Medina, G. A. Keoleian, S. E. Kesler, M. P. Everson, T. J. Wallington, Global Lithium Availability: A Constraint for Electric Vehicles?, J. Ind. Ecol. 15 (2011) 760-775. https://doi.org/10.1111/j.1530-9290.2011.00359.x

[29] S.W. Kim, D.H. Seo, X.H. Ma, G. Ceder, K. Kang, Electrode Materials for Rechargeable Sodium-Ion Batteries: Potential Alternatives to Current Lithium-Ion Batteries,   Adv.   Energy   Mater.   2   (2012)   710   –721. https://doi.org/10.1002/aenm.201200026

[30] C. S. Park, H. Kim, R. A. Shakoor, E. Yang, S. Y. Lim, R. Kahraman, Y. Jung, J. W. Choi, Anomalous manganese activation of a pyrophosphate cathode in sodium ion batteries: a combined experimental and theoretical study, J. Am. Chem. Soc. 135 (2013) 2787 – 2792. https://doi.org/10.1021/ja312044k

[31] Y.S. Wang, X.Q. Yu, S.Y. Xu, J.M. Bai, R.J. Xiao, Y.S. Hu, H. Li, X.Q. Yang, L.Q. Chen, X.J. Huang, A zero-strain layered metal oxide as the negative electrode for long-life sodium-ion batteries, Nat. Commun. 4 (2013) 2365. https://doi.org/10.1038/ncomms3858

[32] Y. You, A. Manthiram, Progress in high-voltage cathode materials for rechargeable sodium-ion batteries. Adv. Energy Mater. 8 (2018) 1701785. https://doi.org/10.1002/aenm.201701785

[33] P. K. Nayak, L.T. Yang, W. Brehm, P. Adelhelm, From lithium-ion to sodium-ion batteries: advantages, challenges, and surprises. Angew. Chem. Int. Ed. 57 (2018) 102–120. https://doi.org/10.1002/anie.201703772

[34] M. M Lao, Y. Zhang, W. B. Luo, Q. Y. Yan, W. P. Sun, S. X. Dou, Alloy-based anode materials toward advanced sodium-ion batteries, Adv. Mater. 29 (2017) 1700622. https://doi.org/10.1002/adma.201700622

[35] S. Q. Chen, C. Wu, L. F. Shen, Challenges and perspectives for NASICON-type electrode materials for advanced sodium-ion batteries. Adv. Mater. 29 (2017)1700431. https://doi.org/10.1002/adma.201700431

[36] H. S. Hou, X. Q. Qiu, W. F. Wei, Y. Zhang, X. B. Ji, Carbon anode materials for advanced sodium-ion batteries, Adv. Energy Mater. 7 (2017) 1602898. https://doi.org/10.1002/aenm.201602898

[37] I. Hasa, J. Hassoun, S. Passerini, Nanostructured Na-ion and Li-ion anodes for battery application: a comparative overview, Nano Res. 10 (2017) 3942–3969. https://doi.org/10.1007/s12274-017-1513-7

[38] S.P. Guo, J.C. Li, Q.T. Xu, Z. Ma, H.G. Xue, Recent achievements on polyanion-type compounds for sodium-ion batteries: syntheses, crystal chemistry and electrochemical performance, J. Power Sources 361 (2017) 285–299. https://doi.org/10.1016/j.jpowsour.2017.07.002

[39]   A. Paolella, C. Faure, V. Timoshevskii, S. Marras, G. Bertoni, A. Guerfi, A. Vijh, M. Armand and K. Zaghib, A review on hexacyanoferrate-based materials for energy storage and smart windows: challenges and perspectives, J. Mater. Chem. A. 5 (2017)18919–18932. https://doi.org/10.1039/C7TA05121B

[40]   M. A. Munoz-Marquez, D. Saurel, J.L. Gomez-Camer, M. Casas-Cabanas, E. Castillo-Martinez, T. Rojo, Na-ion batteries for large scale applications: a review on anode materials and solid electrolyte inter phase formation, Adv. Energy Mater. 7 (2017) 1700463. https://doi.org/10.1002/aenm.201700463

[41]   D. Kundu, E. Talaie, V. Duffort, L. F. Nazar, The emerging chemistry of sodium ion batteries for electrochemical energy storage, Angew. Chem. Int. Ed. 54 (2015) 3431–3448. https://doi.org/10.1002/anie.201410376

[42]   R.J. Clement, P. G. Bruce, C. P Grey, Review-manganese-based P2-type transition metal oxides as sodium-ion battery cathode materials, J. Electrochem. Soc. 162 (2015) 2589–2604. https://doi.org/10.1149/2.0201514jes

[43]   A. F. Hollemann, E. Wiberg, N. Wiberg, Lehrbuch der Anorganischen Chemie (Book), de Gruyter, Berlin/NewYork (1985)

[44]   K. B. Hueso, M. Armand and T. Rojo, High temperature sodium batteries: status, challenges and future trends, Energy Environ. Sci. 6 (2013) 734–749. https://doi.org/10.1039/c3ee24086j

[45]   H. Pan, Y. S. Hu and L. Chen, Room-temperature stationary sodium-ion batteries for large-scale electric energy storage, Energy Environ. Sci. 6 (2013) 2338–2360. https://doi.org/10.1039/c3ee40847g

[46]   K. Kubota, N. Yabuuchi, H. Yoshida, M. Dahbi and S. Komaba, Layered oxides as positive electrode materials for Na-ion batteries, MRS Bull. 39 (2014) 416–422. https://doi.org/10.1557/mrs.2014.85

[47]   Z. L. Jian, L. Zhao, H. L. Pan, Y. S. Hu, H. Li, W. Chen, L. Q. Chen, Spinel lithium titanate ($Li_4Ti_5O_{12}$) as novel anode material for room-temperature sodium-ion battery, Electrochem. Commun.14 (2012) 86– 89. https://doi.org/10.1016/j.elecom.2011.11.009

[48]   J. F. Qian, M. Zhou, Y. L. Cao, X. P. Ai, H. X. Yang, Nanosized $Na_4Fe(CN)_6$/C Composite as a Low-Cost and High-Rate Cathode Material for Sodium-Ion Batteries, Adv. Energy Mater. 2 (2012) 410 –414. https://doi.org/10.1002/aenm.201100655

[49]   H. Pan, X. Lu, X. Yu, Y. S. Hu, H. Li, X. Q. Yang, L. Chen, Sodium Storage and Transport Properties in Layered $Na_2Ti_3O_7$ for Room-Temperature Sodium-Ion

Batteries, Adv. Energy Mater. 3 (2013) 1186– 1194. https://doi.org/10.1002/aenm.201300139

[50] S. Komaba, T, Nakayama, A. Ogata, T. Shimizu, C. Takei, S. Takada, A. Hokura, A. Nakai, High-performance sodium-ion full cell with a layered oxide cathode and a phosphorous-based composite anode, I. ECS Trans. 16 (2009) 43.

[51] L. Pauling, The principles determining the structure of complex ionic crystals, J. Am. Chem. Soc. 51 (1929) 1010. https://doi.org/10.1021/ja01379a006

[52] S. P. Ong, V. L. Chevrier, G. Hautier, A. Jain, C. Moore, S. Kim, X. Ma, G. Ceder, Voltage, stability and diffusion barrier differences between sodium-ion and lithium-ion intercalation materials, Energy Environ. Science 4 (2011) 3680. https://doi.org/10.1039/c1ee01782a

[53] S. Komaba, C. Takei, T. Nakayama, A. Ogata, N. Yabuuchi, Electrochemical intercalation activity of layered $NaCrO_2$ vs. $LiCrO_2$, Electrochem. Commun. 12 (2010) 355-358. https://doi.org/10.1016/j.elecom.2009.12.033

[54] K. Kubota, N. Yabuuchi, H. Yoshida, M. Dahbi, S. Komaba, Layered oxides as positive electrode materials for Na-ion batteries, Mater. Res. Bull. 39 (2014) 416-422. https://doi.org/10.1557/mrs.2014.85

[55] L. Zhao, L. Zhao, J.M. Zhao, Y.S. Hu, H. Li, Z. Zhou, M. Armand, Q. Chen, Disodium terephthalate ($Na_2C_8H_4O_4$) as high performance anode material for low-cost room-temperature sodium-ion battery, Adv. Energy Mater.2 (2012) 962. https://doi.org/10.1002/aenm.201200166

[56] P. Barpanda, S.I. Nishimura, A. Yamada, High-voltage pyrophosphate cathodes, Adv. Energy Mater.2 (2012) 841. https://doi.org/10.1002/aenm.201100772

[57] F. Jiao, P.G. Bruce, Mesoporous crystalline ß-MnO_2- A reversible positive electrode for rechargeable lithium batteries, Adv. Mater. 19 (2017) 657–660. https://doi.org/10.1002/adma.200602499

[58] A. Abouimrane, W. Weng, H. Eltayeb, Y. Cui, J. Niklas, O. Poluektov and K. Amine, Sodium insertion in carboxylate based materials and their application in 3.6 V full sodium cells, Energy Environ. Sci. 5 (2012) 9632. https://doi.org/10.1039/c2ee22864e

[59] D. Kim, E. Lee, M. Slater, W. Lu, S. Rood, C. S. Johnson, Layered $Na(Ni_{1/3}Fe_{1/3}Mn_{1/3})O_2$ cathodes for Na-ion battery application. Electrochem. Commun. 18 (2012) 66–69. https://doi.org/10.1016/j.elecom.2012.02.020

[60] Y. Weng, T. Huang, C. Lim, P. Shao, S. Hy, C. Kuo, J. Cheng, B. Hwang, J. Lee and N. Wu, Unexpected Large Capacity of Ultrafine Manganese Oxide as Sodium-Ion Battery Anode, Nanoscale 7 (2015) 20075. https://doi.org/10.1039/C5NR07100C

[61] H. Kim, D. Jun Kim, Dong-Hwa Seo, M. Yeom, K. Kang, D. Kim, and Y. Jung, Ab Initio study of the sodium intercalation and intermediate phases in $na_{0.44}mno_2$ for sodium-ion battery, Chem. Mater. 24 (2012) 1205-1211. https://doi.org/10.1021/cm300065y

[62] P. Barpanda, T. Ye, M. Avdeev, S. Chung and A. Yamada, A new polymorph of $Na_2MnP_2O_7$ as a 3.6 V cathode material for sodium-ion batteries, J. Mater. Chem. A. 1 (2013) 4194-4197. https://doi.org/10.1039/c3ta10210f

[63] C. Park, H. Kim, A.Shakoor, E. Yang, S. Lim, R. Kahraman, Y. Jung, and J. Wook Choi, Anomalous manganese activation of a pyrophosphate cathode in sodium ion batteries: a combined experimental and theoretical study, *J. Am. Chem. Soc.* 135 (2013) 2787–2792. https://doi.org/10.1021/ja312044k

[64] D. Buchholz, L. G. Chagas, M. Winter, S. Passerini, Layered P_2-$Na_{0.45}Ni_{0.22}Co_{0.11}Mn_{0.66}O_2$ as intercalation host material for lithium and sodium batteries, Electrochim. Acta 110 (2013) 208-213. https://doi.org/10.1016/j.electacta.2013.02.109

[65] D. Yuan, W. He, F. Pei, F. Wu, Y. Wu, J. Qian, Y. Cao, X. Ai and H. Yang, Synthesis and electrochemical behaviors of layered $Na_{0.67}[Mn_{0.65}Co_{0.2}Ni_{0.15}]O_2$ microflakes as a stable cathode material for sodium-ion batteries, J. Mater. Chem. A 1 (2013) 3895-3899. https://doi.org/10.1039/c3ta01430d

[66] S. Guo, H. Yu, Zelang Jian, P. Liu, Y. Zhu, X. Guo, A High-capacity, low-cost layered sodium manganese oxide material as cathode for sodium-ion batteries, Chem. Sus. Chem. 7 (2014) 2115-2119. https://doi.org/10.1002/cssc.201402138

[67] I. Hasa, D. Buchholz, S. Passerini, B. Scrosati, J. Hassoun, High performance $Na_{0.5}[Ni_{0.23}Fe_{0.13}Mn_{0.63}]O_2$ cathode for sodium-ion batteries, Adv. Energy mater. 4 (2014) 15. https://doi.org/10.1002/aenm.201400083

[68] D. Yuan, X. Hu, J. Qian, F. Pei, F. Wu, X. Ai, H. Yang, Y. Cao, R. Mao, P_2-Type $Na_{0.67}Mn_{0.65}Fe_{0.20}Ni_{0.15}O_2$ microspheres as a positive electrode material with a promising electrochemical performance for Na-ion batteries, J. Electrochem. Soc. 164 (2017) 2176-2182. https://doi.org/10.1149/2.1301713jes

[69] I. Hasa, S. Passerini, J. Hassoun, A rechargeable sodium-ion battery using a nanostructured Sb–C anode and P_2-type layered $Na_{0.6}Ni_{0.22}Fe_{0.11}Mn_{0.66}O_2$ cathode. RSC Adv. 5 (2015) 48928–48934. https://doi.org/10.1039/C5RA06336A

[70] L. Wu, D. Buchholz, C. Vaalma, G.A. Giffin, S. Passerini, Apple-biowaste-derived hard carbon as a powerful anode material for Na-ion batteries, Chem. Electro. Chem. 3 (2016) 292–298. https://doi.org/10.1002/celc.201500437

[71] S-M. Oh, P. Oh, S.O. Kim, A. Manthiram, A high-performance sodium ion full cell with a layered oxide cathode and a phosphorous-based composite anode, J. Electrochem. Soc.164 (2017) 321–326. https://doi.org/10.1149/2.0931702jes

[72] M. Keller, C. Vaalma, D. Buchholz, S. Passerini, Development and characterization of high-performance sodium-ion cells based on layered oxide and hard carbon, ChemElectroChem. 3 (2016) 124–1132. https://doi.org/10.1002/celc.201600329

[73] M.J Ilarduya, L. Otaegui, J.M. López del Amo, M. Armand, G. Singh, NaN_3 addition, a strategy to overcome the problem of sodium deficiency in P2-$Na_{0.67}[Fe_{0.5}Mn_{0.5}]O_2$ cathode for sodium-ion battery, J. Power Sources 337 (2017)197–203. https://doi.org/10.1016/j.jpowsour.2016.10.084

[74] E. Edison, S. Sreejith, S. Madhavi, Melt-spun Fe-Sb intermetallic alloy anode for performance enhanced sodium-ion batteries, ACS Appl. Mater. Interfaces 9 (2017) 39399–39406. https://doi.org/10.1021/acsami.7b13096

[75] M. Sathiya, J. Thomas, D. Batuk, V. Pimenta, R. Gopalan, J. M. Tarascon, Dual stabilization and sacrificial effect of Na_2CO_3 for increasing capacities of Na-ion cells based on P2- Na_XMO_2 electrodes, Chem. Mater. 29 (2017) 5948–5956. https://doi.org/10.1021/acs.chemmater.7b01542

[76] M.J. Aragón, P. Lavela, G. Ortiz, R. Alcántara, J.L. Tirado, Nanometric P2-$Na_{2/3}Fe_{1/3}Mn_{2/3}O_2$ with controlled morphology as cathode for sodium-ion batteries, J. Alloy. Compd. 724 (2017) 465–473. https://doi.org/10.1016/j.jallcom.2017.07.044

[77] M. Aragón, P. Lavela, G. Ortiz, R. Alcántara, J. Tirado, Nanometric P2-$Na_{2/3}Fe_{1/3}Mn_{2/3}O_2$ with controlled morphology as cathode for sodium-ion batteries, J. Alloy. Compd. 724 (2017) 465-473. https://doi.org/10.1016/j.jallcom.2017.07.044

[78] D. Nayak, S. Ghosh, A. Venimadhav, Thin film manganese oxide polymorphs as anode for sodium-ion batteries: An electrochemical and DFT based study, Mater. Chem. Phys. 217 (2018) 82-89. https://doi.org/10.1016/j.matchemphys.2018.06.065

Sodium-Ion Batteries: Materials and Applications
Materials Research Forum LLC
Materials Research Foundations **76** (2020) 117-134
https://doi.org/10.21741/9781644900833-5

[79] K. Zhang, D. Kim, Z. Hu, M. Park, G. Noh, Y. Yang, J. Zhang, V. Wing-hei Lau, S. Chou, M. Cho, S. Choi and Y. Kang, Manganese based layered oxides with modulated electronic and thermodynamic properties for sodium ion batteries, Nat. Commun. 10 (2019) 5203. https://doi.org/10.1038/s41467-018-07646-4

Sodium-Ion Batteries: Materials and Applications
Materials Research Foundations **76** (2020) 135-158

Materials Research Forum LLC
https://doi.org/10.21741/9781644900833-6

Chapter 6

Tin-Based Materials for Sodium-Ion Batteries

Bhawana Jain[a], Ajaya K. Singh[a*], Md. Abu Bin Hasan Susan[b]

[a]Department of Chemistry, Govt. V. Y. T. PG. Autonomous, College, Durg, Chhattisgarh, 491001, India

[a]Department of Chemistry, University of Dhaka, Dhaka 1000, Bangladesh

*ajayaksingh_au@yahoo.co.in; bhawanajain123@gmail.com

Abstract

Due to intermittent behavior, renewable energy sources can be used for storage of sustainable electrical energy in stationary devices. Sodium ion batteries have the bright prospect for energy storage from economic point of view because of its high abundance in nature. But batteries of this kind are associated with lower energy density, which moves scientist back to Li-ion batteries. However, sodium ion batteries (NaIBs) with Sn as advanced anode material may be more suitable for energy storage with high cycling capacity and negligible capacity loss. Anode material of NaIBs highly affects the basic characteristics of such devices such as cycling effect, capacity etc. In this account, we deliberate recent developments in Sn based anode material of NaIBs. We have highlighted the role of Sn as anode along with the mechanism. We focused and discussed in detail Sn-alloys which offer highest reported capacities albeit some challenges and solution.

Keywords

Na-Ion Batteries, Sn-Anode, Types of Sn-Based Anode, Mechanism, Performance

Contents

Materials Research Forum LLC
https://doi.org/10.21741/9781644900833-6

1.   Introduction

NaIBs, due inter alia to higher natural abundance, greater safety and widespread distribution of Na along with economic efficiency and reduction in political tensions, show necessary promise as a substitute to lithium ion batteries (LiIBs) for storage of electrochemical energy [1]. Lithium ion batteries cannot fulfill ever increasing demand because of limited lithium resources especially demand for hybrid electric vehicle from industries and storage of stationary energy at a large-scale. Alloy types anodes of NaIBs have high capacities, which attract researcher around the globe. But, on electrochemical cycling larger volume change occurs which creates pulverization of electrode, loss of capacity of anodes, and disentanglement with the current collector [2]. Graphite has promises as anode material for Li-ion batteries although it cannot easily accommodate with sodium ion due to its large radius. Thus sodium ion batteries is limited to nongraphitic carbon based anode with low specific capacities (< 300 mA h g^{-1}) [3-4]. But metal alloy anodes made of carbonaceous materials for instance, hard carbon have low theoretical specific capacities compared to partial Group IVA and VA, which is based on the reaction mechanism of alloying exhibit superior theoretical specific capacities for storage of sodium. It demonstrates immense potential of NaIBs for high-energy such as antimony (Sb), tin (Sn), phosphorus (P), lead (Pb), and germanium (Ge) [5]. Organic complexes are formed by carbon anodes for storage of Na^+ ions, while inorganic complexes are provided by alloyed anodes with Na^+ ions, such as Na_3Sn, Na_3Sb, and Na_3P [6]. Therefore, alloy anodes have greater theoretical capacity than carbon. Capacity between 300-400 mAh g^{-1} is reported for amorphous carbon, while a theoretical capacity of 2596 mAh g^{-1} is noted for Na_3P [7]. However, a large volume of as much as 400% is necessary for the alloying process. This causes the formation of fractures and displacement of the alloying material to bring about passivation and 'dead weight' and ultimately render it incapable of accepting sodium ions. Cycle life is reduced by these volume changes. It is therefore not surprising that mitigation of volume changes and their negative impacts have been the focus of research in this particular area. However, huge volume change associated with Na^+ extraction is a common phenomenon for alloyed anodes to form cracks. This causes pulverization of the particles and electric contact between the active materials, thus current collector is lost. There have so far been considerable efforts to solve these issues [8-9].

Recently, there have been numerous attempts to develop open electrode materials having extended durability and high specific capacities. Elements of partial group IVA and VA such as Sn, P, and Sb have high theoretical specific capacities compared to carbonaceous anode for alloying based sodium storage. Upon alloying, final capacity of charge carriers increases. This gives new insights into preparation of high capacity sodium ion anodes. Literature review implies that group of IVA and VA metals and their compounds such as sulphide, selenides, and oxides offer high capacity by making sodium-rich intermetallic compounds by alloying reaction mechanism such as $Na_{15}Sn_4$, Na_3Sb [10-11]. Among all types of anode material for NaIBs, Sn is most intensively studied. Sn has very high theoretical specific capacity 847 mAh g^{-1} with low reaction potential and quick pulverization. Formation of $Na_{15}Sn_4$ grasps 420% volume expansion [12]. Metallic Sn has received considerable attention for use as a NaIBs anode material owing to good sodium storage capacity of 847 mA h g^{-1}[13-15]. First of all, Sn combines with Na to form Na_xSn phase, which later on converts into $Na_{15}Sn_4$ with 60% and 420% volume expansion (Fig.1) [16].

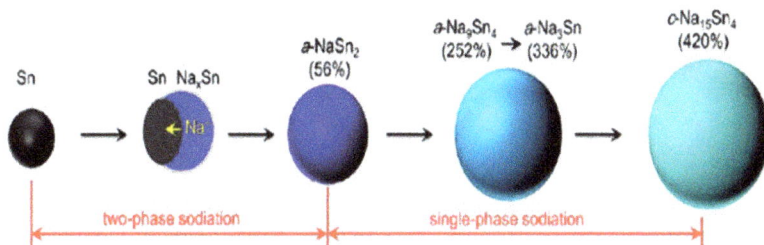

Figure 1. The structure evolution of Sn-anode upon sodiation [16]. (Reprint with permission, Copyright American Chemical Society 2012).

To prevail over the problem associated with larger volume expansion and enhanced cycles, composites of Sn or Sn covered over other metals, such as nickel, carbon etc. and some optimizing binders (i.e. Ge) even hierarchical wood fiber substrate have proved useful [17-19]. Even surface modification remarkably enhances the cycling performance [20]. Due to quick pulverization it loses electrical contact with current collector (Fig. 2). Researcher needs to find more focuses on NaIBs that will replace lithium ion batteries for capacity, life cycle, charging speed, power density and energy.

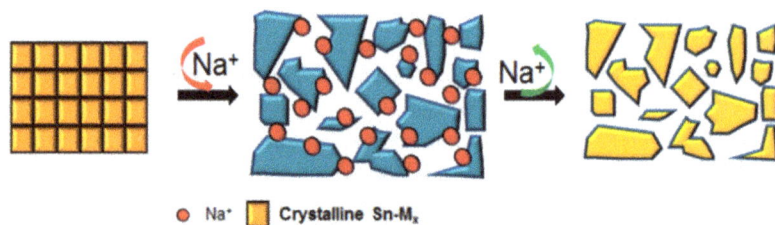

○ Na⁺ ▢ Crystalline Sn-M$_x$

Figure 2. Issue of pulverization in crystalline alloys. (Reprint with permission [21]. Reprint with permission, Copyright American Chemical Society 2015).

Despite the fact that high capacity Sn-based materials have experienced thorough investigation and developed as materials for anode, their thermal stability still remains an issue. Furthermore, for large-scale applications of NaIBs anode materials, safety issues have major technological challenges.

In this review, we focused mainly on the developments of anode materials based on Sn. Specific efforts and approaches to increase efficiency of NaIBs by introduction of Sn as anodic material are also summarized with emphasis on challenges and opportunities.

2.   Types of Sn-based anodes

Metallic tin: At ambient temperature different phases, $NaSn_2$, $NaSn_3$, $NaSn_4$, $NaSn_6$, Na_3Sn, Na_9Sn_4, and $Na_{15}Sn_4$ are present at equilibrium Sn−Na phase diagram [22]. Except Na_9Sn_4 (orthorhombic) and $Na_{15}Sn_4$ (cubic), phases are not fully understood. Transmission electron micrscopy (TEM) images reveal that $Na_{15}Sn_4$ has spherical shapes and other alloy rich compounds have flakes like structure [18]. Pure Sn and Sn rich alloy has four plateaus and it becomes distinct on recycling [23-24], Table1, shows characteristics properties of various metallic Sn and Sn-composite anodes.

Table 1. Sn based anodes for NaIBs

Modification	Potential (V)	Current density (mA g^{-1})	Reversible Capacity (mAhg^{-1})	Cycle no	Capacity retention (%)	Ref.
Porous Sn	0.001-1	847	674	500	77	[25]
8-Sn@C	0.01-2	500	447	200	98.4	[26]
Sn@2rGO + G	0.01-2	423.5	381.3	50	84	[27]
Sn-nanofibers	0.001-0.65	84.7	816.4	100	95.1	[28]
F-G/Sn@C	0.01-2	100	413	100	99	[29]
L-type nanoparticles	0.001-0.65	50	618.6	40	98.2	[30]

Tin oxide: Low toxicity associated with high storage and higher availability in nature have also made tin oxides (SnO and SnO$_2$) as strong candidates for anodic materials [31-32]. Initially conversion of tin oxide into tin is occurred which is followed by alloying of Sn with Na (Eqs. 1-3). During conversion, Na$_2$O is produced which in effect accommodate with volume change in Sn and thus agglomeration is inhibited in the alloying process [33].

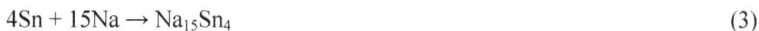

$$SnO_{2 +} 4Na^+ + 4e^- \leftrightarrow Sn + Na_2O \tag{1}$$

$$SnO + 2Na^+ + 2e^- \rightarrow Sn + Na_2O \tag{2}$$

$$4Sn + 15Na \rightarrow Na_{15}Sn_4 \tag{3}$$

If we compare lithiation and sodiation process with tin oxide, it may be concluded that lithiation process is more effective with high storage capacity along with cycling performance. This is due to the fact that sodiation process does not take place directly with tin oxide [34]. Table 2 shows cycling capacity and performance of various kinds of tin oxide anode for batteries.

Table 2. SnO/SnO$_2$ based anodes for NaIBs

Modification	Potential (V)	Current density (mA g^{-1})	Reversible Capacity (mAhg^{-1})	Cycle no.	Capacity retention (%)	Ref.
SnO thick film	0.05-2	50	580	50	43	[35]
Mesoporous SnO/SnO$_2$	0.05-2	50	525.8	50	70.8	[36]
Flower like SnO	0.05-2	50	440	40	120	[37]
SnO$_2$@graphene	0.01-3	20	505	100	82	[38]
SnO$_2$@graphene	0.01-3	20	741	100	86.1	[39]
SnO$_2$ on mesoporous carbon	0.005-2	50	525.8	50	70.8	[40]

Tin sulfides: Tin sulfides are superior to pristine Sn in terms higher theoretical capacity. Even hexagonal SnS_2 shows unusually high theoretical capacity (1137 $mAhg^{-1}$). Initially, SnS_2 converts into Sn by conversion reaction with Na. After that alloying reaction takes place.

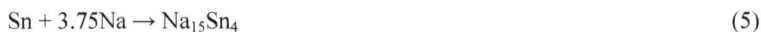

$$SnS_2 + 2Na \rightarrow Sn + 2Na_2S \tag{4}$$

$$Sn + 3.75Na \rightarrow Na_{15}Sn_4 \tag{5}$$

While orthorhombic SnS also has high theoretical capacity (1022 $mAhg^{-1}$), it undergoes following reaction

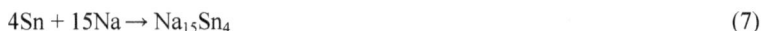

$$SnS + 2Na \rightarrow Na_2S \tag{6}$$

$$4Sn + 15Na \rightarrow Na_{15}Sn_4 \tag{7}$$

In tin sulphide, lower sodium diffusion was generally observed, which makes it more promising candidate for NaIBs than tin oxide along with few drawbacks. for instances, low rate capability and cycling stability. Further this drawback can be overcome by using 2D nano structure since it can buffer volume expansion [41-43]. Tin sulfide based anodes show good agreement for cycling performance (Table 3).

Other tin compounds: The tin−phosphide system is another promising alloying material. For instance, Sn_4P_3 has theoretical capacity of 1132 $mAhg^{-1}$. The system is very advantageous since it offers the majority of the capacity in the practical voltage range of below 0.5 V vs. Na/Na^+ [53-54].

Sodiation and desodiation of tin sufide are as follows:

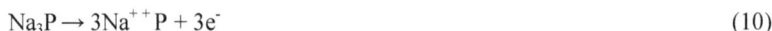

$$Sn_3P_4 + 24Na^+ + 24e^- \rightarrow Na_{15}Sn_4 + 3Na_3P \tag{8}$$

$$Na_{15}Sn_4 \leftrightarrow 4Sn + 15Na^+ + 15e^- \tag{9}$$

$$Na_3P \rightarrow 3Na^{++} P + 3e^- \tag{10}$$

Sn-carbonaceous composites are promising candidates to prevent aggregation of the nanoparticles of Sn over cycling and unsteadiness along with desodiation. Carbon can effectively balance the volume expansion and increase charge transfer. Even porous carbon nanostructure has proved to be useful and efficient for providing sufficient ionic diffusion enough pathways [55-57]. Tin composite shows marvelous changes in batteries (Table 4).

Table 3. SnS/SnS$_2$ based anodes for NaIBs

Modification	Potential (V)	Current density (mA g^{-1})	Reversible Capacity (mAhg^{-1})	Cycle no	Capacity retention (%)	Ref.
SnS	0.1–2.5	125	500	30	74	[44]
SnS-3D flower	0.01-2	150	434	-	-	[45]
SnS$_2$ nanosheets	0.001-3	100	733	50	88.3	[46]
SnS$_2$ nanoarrays	0.01-3	200	910	200	89	[47]
SnS@rGO	0.01-2	100	410	100	94	[48]
SnS@graphene	0.01-3	30	1037	50	92.4	[49]
SnS@C	0.001-3	100	916	-	-	[50]
SnS@CNT	0.01-2.5	100	758	300	87	[51]
SnS$_2$-GO-300	0-2	50	610	-	-	[52]

Table 4. SnSe/Sn$_4$P$_3$ based anodes for NaIBs

Modification	Potential (V)	Current density (mA g^{-1})	Reversible Capacity (mAhg^{-1})	Cycle (no.)	Capacity retention (%)	Ref.
SnSe nanoplates	0.001-3	300	558	50	100	[58]
SnSe@carbon	0.01-2	500	748.5	200	72.5	[59]
SnSe-rGO	0.01-2	50	590	-	-	[60]
SnSe$_2$-rGO	0.005-2.5	100	515	100	78	[61]
Sn$_4$P$_3$/rGO	0.01-3	100	775	100	84.6	[62]

3.   Electrochemical performance

Tables 1-4 provide general information for Sn based NaIBs. Almost all highly efficient NaIBs have nanostructure rather than microstructure anode. Micro scale SnO$_2$ shows limited cycling rate [63]. Nanostructures, nanowires or nanoparticles show magical changes in cycling limitation [64-66]. Upon alloying, pure Sn expands to the terminal Na$_{15}$Sn$_4$ to 420%, which is much higher for the formation of L$_{22}$Sn$_5$ that is 260%. In the case of Sn-based alloys, such an expansion may also be very high [67-69]. In the absence of a secondary matrix, the Sn electrode will instantly break and during cycling electrical contact may get disconnected. The dual ion transportation enables in NaIBs because of

Sodium-Ion Batteries: Materials and Applications   Materials Research Forum LLC
Materials Research Foundations **76** (2020) 135-158   https://doi.org/10.21741/9781644900833-6

porous nature of the Sn substrate. On doping or surface modification Sn nanoparticles do not separate from basic substrate on charging and cycling ability increases effectively.

4.   Structure and design

Practicable structure design of NaIBs will ensure void space to incorporate the volume changes of Sn. The unique structural premises also modify both kinetics and stability of the electrode materials in terms of structure. Thus, design of the structure is a very common strategy to effectively handle the issues related to Sn-based anode materials. In the present review, we have addressed the typical and advanced design of structures of Sn-based composites encompassing 0D to 3D. Table 2 categorizes these Sn-based composites with 0D to 3D dimension.

0D materials possess exceptionally high specific surface area to give large numbers of reactive sites. Therefore as compared to bulk materials, the electrode materials shows improved electrochemical activity for 0D materials [70-71]. Aerosol assisted method is best to fabricate 0D material [72-74].

Great internal space in 1D material can facilitate the relief of the stress and thereby restrain the accumulation of the stress in the radial direction. Generally, 1D nanofiber is fabricated by electrospinning method [75-79].

After the discovery of 2D graphene, researchers focused on 2D materials since they reduce the volume expansion along with more numbers of reactive sites [80-81]. Graphene is highly used along with Sn electrode due to its high flexibility, strength, conductivity, surface area etc. [82-84]. 2D/2D composite has been a representative of modification in Sn electrode to increase cycling capacity of NAIBs. Carbon acts as protective layer to Sn to lead to novel structure and graphene and Sn nanosheets exhibit surface-to-surface integration [85].

3D Sn-based structure generally includes the benefits of lower dimension materials, i.e. high efficiency, surface area, transport speed for ions and electrons etc. 3D nanomaterial always have unique properties, they conquer the pulverization issue due to huge volume changes. Various kinds of 3D anode structures are available in literature such as 3D nano, 3D porous and 3D network structures etc [86-91]. 3D structures have high resistance to volume expansion, and beneficial for kinetics and they showed enhance electrochemical performance in batteries [92].

5.   Performance

Sodium ion cells have been reported to maintain 115 mAh/kg after 50 cycles, with voltage of 3.6 volts and are capable to balance a cathode-specific energy of approximately 400 Wh/kg. The non-aqueous Na-ion batteries don't vie well with commercial Li-ion cells because of inferior cycling performance. Sn shows potential for anode material with its maximum capacity of 847 mAh/kg, which is twice to that of graphite anode used in LiIBs [93]. Sn is highly stable, non-toxic, and highly abundant in nature and has high charge storage capacities. Nano electrodes show higher performance than micro electrodes. Even nano scale Sn-anodes increases cycles of NaIBs than micro anodes. SnO_2 based nanowires without secondary supporting phase shows high cycle [94-96]. Expansion of Sn-alloy in $Na_{15}Sn_4$ is 420%. The colossal expansion is much higher than associated with $Li_{22}Sn_5$ (260%). In the absence of secondary matrix, pure Sn electrode is instantly decrepitated and loose electrical contact. However, during sodiation and desodiation, the electrode would continuously and reversibly breath and increase the cycle. Addition of viral nanoforests, graphene, 3D graphene aerogel, wood fiber, increases the performance of Sn base anodes [97-98]. Secondary matrix highly increases the efficiency of NaIBs. An interconnected array would reduce the electric loses during contact time and ultimately raises the rate efficiency. Secondary phase will encapsulate the Sn and stabilize it to give better performance by increasing cycling nature. Nano range of Sn particles is also important and key factor for the performance of NaIBs [10].

6.   Thermal stability

The susceptibility to thermal reactions related to charged anodes is considered to be higher in SIBs than in LIBs since sodium metal has high reactivity. Furthermore, it is envisaged that the stable solid electrolyte interface (SEI) does not form with ease on the anode surface in Na-ion batteries. The chemical structure as well as morphology of SEI layer can critically influence the safety of the batteries; this also increases the thermal risk for the cells [99-102]. Thus, study of the thermal properties for sodiated Sn electrodes is very crucial. Generally fully charged anode without electrolyte are thermally stable. Thus Sn electrodes alone, without electrolyte in sodiated state are not exothermic in nature [103-104].

7.   Mechanism

Sn-NaIBs have almost same mechanism as LiIBs, which give high reusability and long cycling life. Mechanism of the reaction demonstrates great potential for storage of high energy. Mechanism of the Sn-based NaIBs can be explained on the basis of phase

Materials Research Forum LLC
https://doi.org/10.21741/9781644900833-6

diagram. In Na-Sn binary system except $Na_{15}Sn_4$, other intermediate phases are either unstable or highly reactive. The Na-Sn phases at equilibrium consists of Na-deficient equilibrium stages corresponding to $NaSn_2$, $NaSn_3$, Na_3Sn_5, $NaSn_5$ and amorphous NaSn and Na-rich equilibrium stages correspond to Na_9Sn_4, Na_3Sn alloys and $Na_{15}Sn_4$ (it is the only crystalline phase). During mechanism success and failure of sodiation and desodiation are clearly explained. Proper development of Sn based anode materials for sodium ion batteries is restricted due to the lack of knowledge of sodiation/de-sodiation mechanisms inside the active material. This results in damage of electrode and renders the Sn electrode surface highly porous during sodiation with absence of a crack [105].

In Fig. 3, scanning electron microscopic (SEM) observations explains that as the de-sodiation commences and attains the next de-sodiation plateau (0.28 V), Sn film shows natural cracks with porous islands of active material. As the de-sodiation potential attains its maximum value (2.0 V), these islands of agglomerated particles split into smaller islands. Importantly, the experimental value of sodium diffusion coefficient inside Sn could for the first time be measured (3.9×10^{-14} cm^2 s^{-1}).

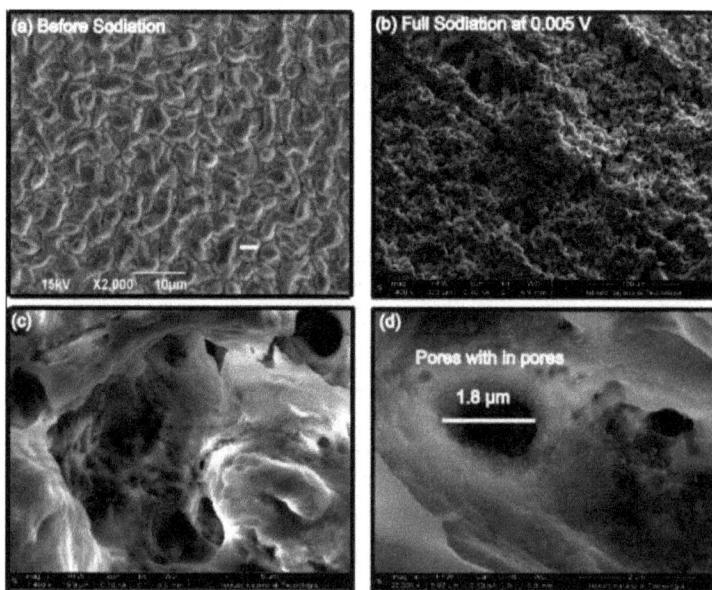

Figure 3. SEM images of thin film of Sn (a) before (b) after sodiation. c & d are magnifying images of a & b. Reprint with permission [105]. Copyright 2019, American Chemical Society.

144

8.   Drawbacks

As noted before, higher availability of sodium resources make sodium ion batteries best substitute to lithium ion batteries. But finite energy density, average cycling life and less developed technology for manufacturing of sodium ion battery hinder their practical applications. The radius of Na^+ (102 pm) is larger than that of Li^+ (76 pm). It will slow down the reaction and undergo large volume change during reactions to give cycling stability and specific capacity. It is the major concern to develop anodic material with high specific capacity with prolonged cycling life. Slow reaction kinetics of Sn based anodic materials generates unstable solid electrolyte interphase film during sodiation and desodiation is also a big challenge. Upon recycling, it undergoes capacity fading. Practical application of Sn-based SIBs is still a big challenge. Sn based electrodes ever suffer large volume expansion with sever mechanical stress cause loss in electrodes efficiency [106].

Aggregation and pulverization affect loss of nanoscale diffusion distance; they ultimately show slower kinetics and large variation in volume (larger radius of Na^+ gives rise to larger volume expansion) and the active materials ultimately experience loss of electrical contact [107-111]. Efforts have been made to remove these drawbacks [112-113].

9.   Factors affecting the capacity of Sn based sodium ion batteries

On the basis of above studies we concluded that various factors are responsible for efficiency and recycling capacity of NaIBs. Some of them are:

Size control

On size reduction to nanoscale, volume expansion of individual nanoparticle is also reduced and mitigates the inner strain and ultimately inhibits the particle pulverization. In case, of nanomaterials fast transportation of carrier ions is caused by promoting surface diffusion with a large surface area. Even rapid charging is possible for batteries composed of nanomaterials. But size control alone cannot improve properties of Sn electrode [114].

Doping and coating

Doping and coating of anode material increases the energy efficiency and performance of NaIBs by increasing the electrical conductivity of anode [115-116]. Complete penetration of carrier ions is increased by this technique in the anode.

Replacement of anode material

Alternative means to alter battery performance is to change the anode material itself. Anodic materials which have larger bond distance and high electrical conductivity show better performance since to disconnect the atomic bond requires less energy and carrier ions can move rapidly.

Electrolyte, additives and binder

On the basis of literature, it was observed that electrolyte, additives and binders are very important to make Sn-NaIBs effective. Protective layer is created by an electrolyte and binders at both the cathode and anode, the surface layer (SL) and the solid electrolyte interfaces (SEI), respectively [118].

Conclusion

Sn based materials for use as anode in sodium ion batteries have superior capacity. Practical utilization of Sn anodes is predominantly inhibited by stern attenuation of capacity resulting from the drastic fluctuations of volume. By the surface modification and changes in structural design of Sn based electrodes, we can overcome drawback of particle pulverization and low coulombic efficiency. Reduction in particle size in nano range effectively increases the cycling life of Sn anodes. Modification in Sn by alloying with other materials i.e. metal and non metal enhances the efficiency. Atomic layer depositions on Sn anodes open a new way to enhance capacity of NaIBs. Size reduction alone cannot increase cycling stability and rate capability; composite with Sn anode material can magically increase the theoretical capacity and cycling stability. Significant changes have been observed in NaIBs capacities in the last few years. But still it is in infant stage, we need to focus more on addressing these challenging issues associated with the commercial application. A bright future with these is ahead in the world of batteries.

Conflicts of interest

The authors declare no competing financial interest.

Acknowledgement

Bhawana Jain (Ph. D.) acknowledges the support obtained from University Grant Commission, New Delhi, for Women Post doctoral fellowship (Award no. F.15-1/2013-14/PDFWM-2013-14-GE-CHH-18784 (SA-II)).

References

[1]   J. Deng, W. B. Luo, S. L. Chou, H. K. Liu, S. X. Dou, Sodium ion batteries: From academic research to practical commercialization, Adv. Energy Mater. 8(5) (2018) 1701428. https://doi.org/10.1002/aenm.201701428

[2]   V. L. Chevrier, G. Ceder, Challenges for Na-ion negative electrodes, J. Electrochem. Soc. 158 (2011) A1011-A1014. https://doi.org/10.1149/1.3607983

[3]   C. Bommier, X. L. Ji, Recent development on anodes for Na-ion batteries, Isr. J. Chem. 55 (2015) 486-507. https://doi.org/10.1002/ijch.201400118

[4]   M. Dhabi, N. Yabuuchi, K. Kubota, K. Tokiwa, S. Komaba, Negative electrodes for Na-ion batteries, Phys. Chem. Chem. Phys. 16 (2014) 15007-15028. https://doi.org/10.1039/c4cp00826j

[5]   L. Li, Y. Zheng, S. Zhang, J. Yang, Z. Shao, Z. Guo, Recent progress on sodium ion batteries:Potential high performance anode, Energy Environ. Sci. 11 (2018) 2310-2340. https://doi.org/10.1039/C8EE01023D

[6]   M. Lao, Y. Zhang, W. Luo, Q. Yan, W. Sun, S. X. Dou, Alloy based anode materials toward advanced sodium batteries, Adv. Mater. 29 (2017) 1700622 https://doi.org/10.1002/adma.201700622

[7]   M. S. Balogun, Y. Luo, W. Qiu, P. Liu, Y. Tong, A review of carbon materials and their composites with alloy metals for sodium ion battery anodes, Carbon 98 (2016) 162-178. https://doi.org/10.1016/j.carbon.2015.09.091

[8]   G. J. Jung, Y. Lee, Y. S. Mun, H. Kim, J. Hur, T. Y. Kim, K. S. Suh, J. H. Kim, D. Lee, W. Choi, I. T. Kim, Sb-AlC$_{0.75}$-composite anodes for high performance sodium ion batteries, J. Power Sources 340 (2017) 393-400. https://doi.org/10.1016/j.jpowsour.2016.11.086

[9]   L. Wu, H. Lu, L. Xiao, X. Ai, H. Yang and Y. Cao, Electrochemical properties and morphological evolution of pitaya like Sb@C microspheres as high performance anode for sodium ion batteries, J. Mater. Chem. A 3 (2015) 5708-5713. https://doi.org/10.1039/C4TA06086E

[10]   L. Baggetto, P. Ganesh, R. P. Meisner, R. R. Unocic, J. C. Jumas, C. A. Bridges, G. M. Veith, Characterization of sodium ion electrochemical reaction with tin anodes:Experiment and theory, J. Power Sources 234 (2013) 48-59. https://doi.org/10.1016/j.jpowsour.2013.01.083

[11] N. Yabuuchi, K. Kubota, M. Dhabi, S. Komaba, Research development on sodium ion batteries, Chem. Rev. 11 (2014) 11636-11682. https://doi.org/10.1021/cr500192f

[12] Z. Li, J. Ding, D. Mitlin, Tin and tin compounds for sodium ion battery anodes:Phase transformations and performance, ACC. Chem. Res. 48 (2015) 1657-1665. https://doi.org/10.1021/acs.accounts.5b00114

[13] M. D. Slater, D. Kim, E. Lee and C. S. Johnson, Sodium ion batteries, Adv. Funct. Mater. 23 (2013) 947-958. https://doi.org/10.1002/adfm.201200691

[14] Y. Kim, K. H. Ha, S. M. Oh, K. T. Lee, High capacity anode materials for sodium ion batteries, Chem. Eur. J. 20 (2014) 11980. https://doi.org/10.1002/chem.201402511

[15] H. Pan, Y. S. Hu, L. Chen, Room temperature stationary sodium ion batteries for large scale electric energy storage, Energy Environ. Sci. 6 (2013) 2338-2360. https://doi.org/10.1039/c3ee40847g

[16] J. W. Wang, X. H. Liu, S. X. Mao, J. Y. Huang, Microstructural evolution of tin nanoparticles during insitu sodium insertion and extraction, Nano Lett. 12 (2012) 5897-5902. https://doi.org/10.1021/nl303305c

[17] Y. Liu, Y. Xu, Y. Zhu, J. N. Culver C.A. Lundgren, K. Xu, C. Wang, Tin coated viral nanoforests as sodium ion battery anodes, ACS Nano 7 (2013) 3627-3634. https://doi.org/10.1021/nn400601y

[18] B. Farbod, K. Cui, W.P. Kalisvaart, M. Kupsta, B. Zahiri, A. Kohandehghan, E.M. Lotfabad, Z. Li, E. J. Luber, D. Mitlin, Anodes for sodium ion batteries based on tin-germanium-antimony alloys. ACS Nano 8 (2014) 4415-4429. https://doi.org/10.1021/nn4063598

[19] J. Liu, Y. Wen, P. A. van Aken, J. Maier, Y. Yu, Facile synthesis of highly porous Ni−Sn intermetallic microcages with excellent electrochemical performance for lithium and sodium storage, Nano Lett. 14 (2014) 6387−6392. https://doi.org/10.1021/nl5028606

[20] X. Xie, K. Kretschmer, J. Zhang, B. Sun, D. Su, G. Wang, Sn@CNT nanopillars grown perpendicularly on carbon paper: A novel free standing anode for sodium ion batteries, Nano Energy 13 (2015) 208-217. https://doi.org/10.1016/j.nanoen.2015.02.022

[21] J. Zhu, D. Deng, Amorphous bimetallic Co_3Sn_2 nanoalloys are better than crystalline counterparts for sodium storage, J. Phys. Chem. C 119 (2015) 21323-21328. https://doi.org/10.1021/acs.jpcc.5b05232

[22] L. D. Ellis, T. D. Hatchard, M. N. Obrovac, Reversible insertion of sodium in tin, J. Electrochem. Soc. 159 (2012) A1801−A1805. https://doi.org/10.1149/2.037211jes

Sodium-Ion Batteries: Materials and Applications
Materials Research Foundations **76** (2020) 135-158

Materials Research Forum LLC
https://doi.org/10.21741/9781644900833-6

[23]   Y. M. Lin, P.R. Abel, A. Gupta, J.B. Goodenough, A. Heller, C.B. Mullins, Sn−Cu nanocomposite anodes for rechargeable sodium-ion batteries, ACS Appl. Mater. Interfaces 5 (2013) 8273−8277. https://doi.org/10.1021/am4023994

[24]   L. Baggetto, J.C. Jumas, J. Gorka, C.A. Bridges, G.M. Veith, Predictions of particle size and lattice diffusion pathway requirements for sodium-ion anodes using [small eta]-Cu_6Sn_5 thin films as a model system, Phys. Chem. Chem. Phys. 15 (2013) 10885−10894. https://doi.org/10.1039/c3cp51657a

[25]   C. Kim, K.Y. Lee, I. Kim, J. Park, G. Cho, K. W. Kim, J. H. Ahn, H.J. Ahn, Long term cycling stability of porous Sn anode for sodium ion batteries, J. Power Sources 317 (2016) 153-158. https://doi.org/10.1016/j.jpowsour.2016.03.060

[26]   Y. Liu, N. Zhang, L. Jiao, Z. Tao, J. Chen, Ultra small Sn nanoparticles embedded in carbon as high performance anode for sodium ion batteries, Adv. Funct. Mater. 25 (2015) 214-220. https://doi.org/10.1002/adfm.201402943

[27]   Y. Jeon, X. Han, K. Fu, J. Dai, J. H. Kim, L. Hu, T. Song, U. Paik, Flash induced reduced graphene oxide as a Sn anode host for high performance sodium ion batteries, J. Mater. Chem. A 4 (2016) 18306-18313. https://doi.org/10.1039/C6TA07582G

[28]   D. H. Nam, T. H. Kim, K. S. Hong, H. S. Kwon, Template free electrochemical synthesis of Sn nanofiber as high performance anode materials for Na-ion batteries, ACS Nano 8 (2014) 11824-11835. https://doi.org/10.1021/nn505536t

[29]   B. Luo, T. Qiu, D. Ye, L. Wang, L. Zhi, Tin nanoparticles encapsulated in graphene backbone carbonaceous foams as high performance anodes for lithium ion and sodium ion storage, Nano Energy 22 (2016) 232-240. https://doi.org/10.1016/j.nanoen.2016.02.024

[30]   D. H. Nam, K. S. Hong, S. J. Lim, T. H. Kim, H. S. Kwon, Electrochemical properties of electrodeposited Sn anodes for Na-ion batteries, J. Phys. Chem. C 118 (2014) 20086. https://doi.org/10.1021/jp504055j

[31]   M. Dirican, Y. Lu, Y. Ge, O. Yildiz, X. Zhang, Carbon confined SnO_2 electrodeposited porous carbon nanofiber composite as high capacity sodium ion battery anode material, ACS Appl. Meter. Interfaces 7 (2015) 18387-18396. https://doi.org/10.1021/acsami.5b04338

[32]   N. Oehl, P. Michalowski, M. Knipper, J. Kolny-Olesiask, T. Plaggenborg, J. Parisi, Size dependent strain of Sn/SnOx core/shell nanoparticles, J. Phys. Chem C 118 (2014) 30238-30243. https://doi.org/10.1021/jp5096147

[33]   J. Ding, Z. Li, H. Wang, K. Cui, A. Kohandehghan, X. Tan, D. Karpuzov, D. Mitlin, Peanut shell hybrid sodium ion capacitor with extreme energy power rivals lithium ion capacitor, Energy Environ. Sci. 8 (2015) 941-955. https://doi.org/10.1039/C4EE02986K

[34]   M. Gu, A. Kushima, Y. Shao, J. G. Zhang, J. Liu, N.D. Browning, J. Li, C. Wang, Probing the failure mechanism of SnO_2 nanowires for sodium ion batteries, Nano Lett. 13 (2013) 5203-5211. https://doi.org/10.1021/nl402633n

[35]   Y. Zhang, J. Xie, S. Zhang, P. Zhu, G. Cao, X. Zhao, Ultrafine tin oxide on reduced graphene oxide as high performance anode for sodium ion batteries, Electrochim. Acta 151 (2015) 8-15. https://doi.org/10.1016/j.electacta.2014.11.009

[36]   D. Tang, Q. Huang, R. Yi, F. Dai, M. L. Gordin, S. Hu, S. Chen, Z. Yu, H. Sohn, J. Song, D. Wang, Room temperature synthesis of mesoporous Sn/SnO_2 composite as anode for sodium ion batteries, Eur. J. Inorg. Chem. 2016 (2016) 1950-1954. https://doi.org/10.1002/ejic.201501441

[37]   Y. C. Lu, C. Ma, J. Alvarado, T. Kidera, N. Dimov, Y. S. Meng, S. Okada, Electrochemical properties of tin oxide anodes for sodium ion batteries, J. Power Sources 284 (2015) 287-295. https://doi.org/10.1016/j.jpowsour.2015.03.042

[38]   J. Patra, H. C. Chen, C. H. Yang, C. T. Hsieh, C.Y. Su, J. K. Chang, High dispersion of 1 nm SnO_2 nanoparticles between graphene nanosheet constructed using supercritical CO_2 fluid for sodium ion battery anode, Nano Energy 28 (2016) 124-134. https://doi.org/10.1016/j.nanoen.2016.08.044

[39]   D. Su, H. J. Ahn, G. Wang, SnO_2@graphene nanocomposites as anode materials for Na-ion batteries with superior electrochemical performance, Chem. Commun. 49 (2013) 3131-3133. https://doi.org/10.1039/c3cc40448j

[40]   A. Jahel, C. M. Ghimbeu, A. Darwiche, L. Vidal, S. H. Garreau, C. V. Guterl, L. Monconduit, Exceptionally high performing Na-ion battery anode using crystalline SnO_2 nanoparticles confined in mesoporous carbon, J. Mater. Chem. A 3 (2015) 11960-11969. https://doi.org/10.1039/C5TA01963J

[41]   P. K. Dutta, U. K. Sen, S. Mitra, Excellent electrochemical performance of tinmonosulfide (SnS) as a sodium battery anode, RSC Adv. 4 (2014) 43155-43159. https://doi.org/10.1039/C4RA05851H

[42]   W. Sun, X. Rui, D. Yang, Z. Sun, B. Li, W. Zhang, Y. Zong, S. Madhavi, S. X. Dou, Q. Yan, Two dimensional tin disulfide nanosheets for enhanced sodium storage, ACS Nano 9 (2015) 11371-11381. https://doi.org/10.1021/acsnano.5b05229

Materials Research Forum LLC
https://doi.org/10.21741/9781644900833-6

[43]   G. Li, R. Su, J. Rao, J. Wu, P. Rudolf, G. R. Blake, R. A. de Groot, F. Besenbacher, T. T. M. Palstra, Band gap narrowing of SnS_2 superstructures with improved hydrogen production, J. Mater. Chem. A 4 (2016) 209-216. https://doi.org/10.1039/C5TA07283B

[44]   D. Chao, P. Liang, Z. Chen, L. Bai, H. Shen, X. Liu, X. Xia, Y. Zhao, S. V. Savilov, J. Lin, Z. X. Shen, Psuedocapacitive Na ion storage boosts high rate and areal capacity of self branched 2D layer metal chalcogenide nanoarrays, ACS Nano 10 (2016) 10211-10219. https://doi.org/10.1021/acsnano.6b05566

[45]   E. Cho, K. Song, M. H. Park, K. W. Nam, Y. M. Kang, SnS 3D flower with superb kinetic properties for anodic use in next generation sodium rechargeable batteries, Small 12 (2016) 2510-2517. https://doi.org/10.1002/smll.201503168

[46]   L. Shi, D. Li, P. Yao, J. Yu, C. Li, B. Yang, C. Zhu, J. Xu, Sodium ion batteries:SnS2 nanosheets coating on nanohollow cubic CoS_2/C for ultralong life and high rate capability half/full sodium ion batteries, Small 14 (2018) 1870187. https://doi.org/10.1002/smll.201870187

[47]   D. Chao, C. Zhu, P. Yang, X. Xia, J. Liu, J. Wang, X. Fan, S. V. Savilov, J. Lin, H. J. Fan, Z. X. Shen, Array of nanosheets render ultrafast and high capacity Na-ion storage by tunable psuedocapacitance, Nature Commun. 7 (2016) 12122. https://doi.org/10.1038/ncomms12122

[48]   L. Wu, H. Lu, L. Xiao, X. Ai, H. Yang, Y. Cao, Improved sodium storage performance of stannous sulfide@reduced graphene oxide composite as high capacity anodes for sodium ion batteries, J. Power Sources 293 (2015) 784-789. https://doi.org/10.1016/j.jpowsour.2015.06.015

[49]   T. Zhou, W. Pang, C. Zhang, J. Yang, Z. Chen, H. Liu, Z. Guo, Enhanced sodium ion battery performance by structural phase transition from two dimensional hexagonal SnS2 to orthorhombic SnS, ACS Nano 8 (2014) 8323-8333. https://doi.org/10.1021/nn503582c

[50]   C. Zhu, P. Kopold, W. Li, P. A. van Aken, J. Maier, Y. Yu, A general stratergy to fabricate carbon coated 3D porous interconnected metal sulfides:Case study of SnS/C nanocomposite for high performance lithium and sodium ion batteries, Adv. Sci. 2 (2015) 1500200. https://doi.org/10.1002/advs.201500200

[51]   ᵕ H. Li, M. Zhou, W. Li, K. Wang, S. Cheng, K. Jiang, Layered SnS_2 cross linked by carbon nanotubes as high performance anode for sodium ion batteries, RSC Adv. 6 (2016) 35197-35202. https://doi.org/10.1039/C6RA04941A

[52]   P. V. Prikhodchenko, D. Y. W. Yu, S. K. Batabyal, V. Uvarov, J. Gun, S. Sladkevich, A. A. Mikhaylov, A. G. Medvedev, O. Lev, Nanocrystalline tin disulfide coating of reduced graphene oxide produced by the peroxostannate deposition route for the sodium battery anodes, J. Mater. Chem. A 2 (2014) 8431-8437. https://doi.org/10.1039/c3ta15248k

[53]   Y. Kim, Y. Kim, A. Choi, S. Woo, D. Mok, N. S. Choi, Y. S. Jung, J. H. Ryu, S. M. Oh, K. T. Lee, Tin phosphide as a promising anode material for Na-ion batteries, Adv. Mater. 26 (2014) 4139−4144. https://doi.org/10.1002/adma.201305638

[54]   W. Li, S. L. Chou, J.Z. Wang, J. H. Kim, H. K. Liu, S. X. Dou, $Sn_{4+x}P_3$ @ amorphous Sn-P composites as anodes for sodium-ion batteries with low cost, high capacity, long life, and superior rate capability. Adv. Mater. 26 (2014) 4037−4042. https://doi.org/10.1002/adma.201400794

[55]   Y. Liu, N. Zhang, L. Jiao, Z. Tao, J. Chen, Ultrasmall Sn nanoparticles embedded in carbon as high performance anodes for sodium ion batteries, Adv. Funct. Mater. 25 (2015) 214-220. https://doi.org/10.1002/adfm.201402943

[56]   T. Jin, Y. Liu, Y. Li, K. Cao, X. Wang, L. Jiao, Electrospun NaVPO4F/C nanofibers as self standing cathode material for ultralong cycle life Na-ion batteries, Adv. Energy Mater. 7 (2015) 1700087. https://doi.org/10.1002/aenm.201700087

[57]   Z. Li, W. Lv, C. Zhang, J. Qin, W. Wei, J. J. Shao, D. W. Wang, B. Li, F. Kang, Q. H. Yang, Nanospace confined formation of flattened Sn sheets in pre seeded graphene for lithium ion batteries, Nanoscale 16 (2014) 9554-9558. https://doi.org/10.1039/C4NR01924E

[58]   G. D. Park, J. H. Lee, Y. C. Kang, Superior Na-ion storage properties of high aspect ratio SnSe nanoplates prepared by spray pyrolysis process, Nanoscale 8 (2016) 11889-11896. https://doi.org/10.1039/C6NR02983C

[59]   Z. Zhang, X. Zhao, J. Li, SnSe/carbon naocomposite synthesized by high energy ball milling as an anode materials for sodium ion and lithium ion batteries, Electrochim. Acta 176 (2015) 1296-1301. https://doi.org/10.1016/j.electacta.2015.07.140

[60]   X. Yang, R. Zhang, N. Chen, X. Meng, P. Yang, C. Wang, Y. Zhang, Y. Wei, G. Chen, F. Du, Assembly of SnSe nanoparticles confined in graphene for enhanced sodium ion storage performance, Chemistry 22 (2016) 1445-1451. https://doi.org/10.1002/chem.201504074

[61] F. Zhang, C. Xia, J. Zhu, B. Ahmed, H. Liang, D.B. Velusamy, U. Schwingenschlögl, H.N. Alshareef, SnSe$_2$ 2D anodes for advanced sodium ion batteries, Adv. Energy Mater. 6 (2016) 1601188. https://doi.org/10.1002/aenm.201601188

[62] Q. Li, Z. Li, Z. Zhang, C. Li, J. Ma, C. Wang, X. Ge, S. Dong, L. Yin, Low temperature solution based phophorization reaction route to Sn$_3$P$_4$/reduced graphene oxide nanohybrid as anodes for sodium ion batteries, Adv. Energy Mater. 6 (2016) 1600376. https://doi.org/10.1002/aenm.201600376

[63] Y. Xu, Y. Zhu, Y. Liu, C. Wang, Electrochemical performance of porous carbon/tin composite anodes for sodium-ion and lithium-ion batteries, Adv. Energy Mater. 3 (2013) 128−133. https://doi.org/10.1002/aenm.201200346

[64] Y. Wang, D. Su, C. Wang, G. Wang, SnO$_2$@MWCNT nanocomposite as a high capacity anode material for sodium-ion batteries, Electrochem. Commun. 29 (2013) 8−11. https://doi.org/10.1016/j.elecom.2013.01.001

[65] X. Han, Y. Liu, Z. Jia, Y. C. Chen, J. Wan, N. Weadock, K.J. Gaskell, T. Li, L. Hu, Atomic-layer-deposition oxide nanoglue for sodium ion batteries, Nano Lett. 14 (2013) 139−147. https://doi.org/10.1021/nl4035626

[66] M. K. Datta, R. Epur, P. Saha, K. Kadakia, S. K. Park, P. N. Kumta, Tin and graphite based nanocomposites: Potential anode for sodium ion batteries, J. Power Sources 225 (2013) 316−322. https://doi.org/10.1016/j.jpowsour.2012.10.014

[67] J. Ding, Z. Li, H. L. Wang, K. Cui, A. Kohandehghan, X. H. Tan, D. Karpuzov, D. Mitlin, Sodiation vs. lithiation phase trans- formations in a high rate-high stability SnO$_2$ in carbon nanocomposite, J. Mater. Chem. A 3 (2015) 7100−7111. https://doi.org/10.1039/C5TA00399G

[68] K. Dai, H. Zhao, Z. Wang, X. Song, V. Battaglia, G. Liu, Toward high specific capacity and high cycling stability of pure tin nanoparticles with conductive polymer binder for sodium ion batteries, J. Power Sources 263 (2014) 276−279. https://doi.org/10.1016/j.jpowsour.2014.04.012

[69] Y. Zhang, P. Zhu, L. Huang, J. Xie, S. Zhang, G. Cao, X. Zhao, Few-layered SnS$_2$ on few-layered reduced graphene oxide as Na-ion battery anode with ultralong cycle life and superior rate capability, Adv. Funct. Mater. 25 (2015) 481−489. https://doi.org/10.1002/adfm.201402833

[70] B. Wang, B. Luo, X. L. Li, L. J. Zhi, The dimensionality of Sn anode in Li ion batteries, Mater. Today 15 (2012) 544-552. https://doi.org/10.1016/S1369-7021(13)70012-9

[71]   P. G. Bruce, B. Scrosati, J. M. Tarascon, Nanomaterials for rechargeable lithium batteries, Angew. Chem. Int. Ed. 47 (2008) 2930-2946.
https://doi.org/10.1002/anie.200702505

[72]   H. Ying W. Q. Han, Metallic Sn based anode materials:Application in high performance lithium ion and sodium ion batteries, Adv. Science 4 (2017) 1700298.
https://doi.org/10.1002/advs.201700298

[73]   N. E. Motl, A. K. P. Mann, S. E. Skrabalak, Aerosol assisted synthesis and assembly of nanoscale building blocks, J. Mater. Chem. A 1 (2013) 5193-5202.
https://doi.org/10.1039/c3ta01703f

[74]   C. Boissiere, D. Grosso, A. Chaumonnot, L. Nicole, C. Sanchez, Aerosol route to functional nanostructured inorganic and hybrid porous materials, Adv. Mater. 23 (2011) 599-623. https://doi.org/10.1002/adma.201001410

[75]   Y. Liu, N. Zhang, L. Jiao, J. Chen, $MnFe_2O_4@C$ nanofibers as high performance anode for sodium ion batteries, Nano Lett. 16(5) (2016) 3321-3328.
https://doi.org/10.1021/acs.nanolett.6b00942

[76]   X. Y. Tao, R. Wu, Y. Xia, H. Huang, W. C. Chai, T. Feng, Y. P. Gan, W. K. Zhang, Supercritical fluid assisted biotemplating synthesis of Si-O-C microspheres from microalgae for advanced Li ion batteries, ACS Appl. Mater. Interfaces 6 (2014) 3696. batteries, ACS Appl. Mater. Interfaces 6 (2014) 3696.

[77]   Y. Yu, Q. Yang, D. Teng, X. Yang, S. Ryu, Reticular Sn nanoparticles dispersed PAN based carbon nanofibers for anode material in rechargeable lithium ion batteries, Electrochem. Commun. 12 (2010) 1187-1190.
https://doi.org/10.1016/j.elecom.2010.06.015

[78]   S. D. Seo, G. H. Lee, A. H. Lim, K. M. Min, J. C. Kim, H. W. Shim, K. S. Park, D. W. Kim, Direct assembly of tin MWCNT 3D networked anode for rechargeable lithium ion batteries, RSC Adv. 2 (2012) 3315-3320. https://doi.org/10.1039/c2ra00943a

[79]   Y. Liu, N. Zhang, L. Jiao, J. Chen, The nanodots encapsulated in porous nitrogen doped carbon nanofibers as a free standing anode for advanced sodium ion batteries, Adv. Mater. 27 (2015) 6702-6707. https://doi.org/10.1002/adma.201503015

[80]   V. Chabot, D. Higgins, A. P. Yu, X. C. Xiao, Z. W. Chen, J. J. Zhang, A review of graphene and graphene oxide sponge: material synthesis and applications to energy and the environment, Energy Environ. Sci. 7 (2014) 1564-1596.
https://doi.org/10.1039/c3ee43385d

[81]   M. Naguib, V. N. Mochalin, M. W. Barsoum, Y. Gogotsi, Mxene: a new family of two dimensional materials, Adv. Mater. 26 (2014) 992-1005. https://doi.org/10.1002/adma.201304138

[82]   F. Pan, W. Zhang, J. J. Ma, N. Yao, L. Xu, Y. S. He, X. W. Yang, Z. F. Ma, Integrating insitu solvothermal approach synthesized nanostructured tin anchored on graphene sheets into film anodes for sodium ion batteries, Electrochim. Acta 196 (2016) 572-578. https://doi.org/10.1016/j.electacta.2016.02.204

[83]   H. Ying, F. Xin, W. Han, Structure evolution of 3D nano-Sn/reduced graphene oxide composite from a sandwich like structure to a curly Sn@carbon nanocage like structure during lithiation/delithiation, Adv. Mater. Interfaces 3 (2016) 1600498. https://doi.org/10.1002/admi.201600498

[84]   N. Li, H. Song, H. Cui, C. Wang, Sn@graphene grown on vertically aligned graphene for high capacity, high rate and long life lithium storage, Nano Energy 3 (2014) 102-112. https://doi.org/10.1016/j.nanoen.2013.10.014

[85]   H. Ying, S. Zhang, Z. Meng, Z. Sun, W. Q. Han, Ultrasmall Sn nanodots embedded inside N-doped carbon microcages as high performance lithium and sodium ion battery anodes, J. Mater. Chem. A 5 (2017) 8334-8342. https://doi.org/10.1039/C7TA01480E

[86]   D. Deng, J. Y. Lee Reversible storage of lithium in a rambutan like tin-carbon electrode, Angew. Chem., Int. Ed. 48 (2009) 1660-1663. https://doi.org/10.1002/anie.200803420

[87]   S. Chen, Z. Ao, B. Sun, X. Xie, G. Wang, Porous carbon nanocages encapsulated with tin nanoparticles for high performance sodium ion batteries, Energy Storage Mater. 5 (2016) 180-190. https://doi.org/10.1016/j.ensm.2016.07.001

[88]   J. Qin, C. N. He, N. Q. Zhao, Z. Y. Wang, C. S. Shi, E. Z. Liu, J. J. Li, Graphene networks anchored with Sn@graphene as lithium ion battery anode, ACS Nano 8 (2014) 1728-1738. https://doi.org/10.1021/nn406105n

[89]   Y. H. Xu, Y. J. Zhu, F. D. Han, C. Luo, C. S. Wang, Si/C fiber paper electrodes fabricated using a combined electrospray/electrospinning technique for Li ion batteries, Adv. Energy Mater. 5 (2015) 1400753. https://doi.org/10.1002/aenm.201400753

[90]   J. Liu, P. Kopold, P. A. van Aken, J. Maier, Y. Yu, Uniform yolk shell Sn_4P_3@C nanosphere as high capacity and cycle stable anode materials for sodium ion batteries, Energy Environ. Sci. 8 (2015) 3531-3538. https://doi.org/10.1039/C5EE02074C

[91]   W. Chen, D. Deng, Deflated carbon nanospheres encapsulating tin cores decorated on layered 3D carbon structures for low cost sodium ion batteries, ACS Sustainable Chem. Eng. 3 (2015) 63-70. https://doi.org/10.1021/sc500543u

[92]   X. Li, A. Dhanabalan, L. Gu, C. L. Wang, Three dimensional porous core shell Sn@carbon composite anodes for high performance lithium ion battery applications, Adv. Energy Mater. 2 (2012) 238-244. https://doi.org/10.1002/aenm.201100380

[93]   H. Kim, G. Jeong, Y. U. Kim, J. H. Kim, C. M. Park, H. J. Sohn, Metallic anodes for next generation secondary batteries, Chem. Soc. Rev. 42 (2013) 9011-9034. https://doi.org/10.1039/c3cs60177c

[94]   M. Shimizu, H. Usui, H. Sakaguchi, Electrochemical Na-insertion/extraction properties of SnO thick film electrodes prepared by gas-deposition, J. Power Sources 248 (2014) 378-382. https://doi.org/10.1016/j.jpowsour.2013.09.046

[95]   J. Park, J. W. Park, J. H. Han, S. W. Lee, K. Y. Lee, H. S. Ryu, K. W. Kim, G. Wang, J. H. Ahn, H. J. Ahn, Charge discharge properties of tin oxide for sodium ion battery, Mater. Res. Bull. 58 (2014) 186-189. https://doi.org/10.1016/j.materresbull.2014.04.051

[96]   D. Su, C. Wang, H. Ahn, G. Wang, Octahedral tin oxide nanocrystals as high capacity anode materials for Na-ion batteries, Phy. Chem. Chem. Phys. 15 (2013) 12543-12550. https://doi.org/10.1039/c3cp52037d

[97]   H. Zhu, Z. Jia, Y. Chen, N. Weadock, J. Wan, O. Vaaland, X. Han, T. Li, L. Hu, Tin anode for sodium ion batteries using natural wood fiber as a mechanical buffer and electrolyte reservoir, Nano Lett. 13 (2013) 3093-3100. https://doi.org/10.1021/nl400998t

[98]   Z. Li, J. Ding, H. Wang, K. Cui, T. Stephenson, D. Karpuzov, D. Mitlin, High rate SnO_2-graphene dual aerogel anodes and their kinetics of lithiation and sodiation, Nano Energy 15 (2015) 369-378. https://doi.org/10.1016/j.nanoen.2015.04.018

[99]   X. Xia, M. N. Obrovac, J. R. Dahn, Comparision of the reactivity of Na_xC_6 and Li_xC_6 with nonaqueous solvent and electrolyte, Electrochem. Solid-State Lett. 14 (2011) A130-A133. https://doi.org/10.1149/1.3606364

[100]   J. Zhao, L. Zhao, N. Dimov, S. Okada and T. Nishida, Electrochemical and thermal properties of $\propto$-$NaFeO_2$ cathode for Na-ion batteries, J. Electrochem. Soc. 160 (2013) A3077-A3081. https://doi.org/10.1149/2.007305jes

[101]   J. Zhao, J. Xu, D. H. Lee, N. Dimov, Y. S. Meng, S. Okada, Electrochemical and thermal properties of P2-type $Na_{2/3}Fe_{1/3}Mn_{2/3}O_2$ for Na ion batteries, J. Power Sources 264 (2014) 235-239. https://doi.org/10.1016/j.jpowsour.2014.04.048

[102] X. Xia, J. R. Dahn, A study of the reactivity of deintercalated NaNi0.5Mn0.5O2 with nonaqueous solvent and electrolyte by accelerating rate calorimetry, J. Electrochem. Soc. 159 (2012) A1048-A1051. https://doi.org/10.1149/2.060207jes

[103] Y. Lee, H. Lim, S. O. Kim, H. S. Kim, K. J. Kim, K. Y. Lee, W. Choi, Thermal stability of Sn anode material with nonaqueous electrolytes in sodium ion batteries, J. Mater. Chem. A 6 (2018) 20383-20392. https://doi.org/10.1039/C8TA07854H

[104] I. A. Profatilova, S. S. Kim, N. S. Choi, Electrochim. Acta Enhanced thermal properties of the solid electrolyte interphase formed on graphite in an electrolyte with fluoroethylene carbonate, 54 (2009) 4445-4450. https://doi.org/10.1016/j.electacta.2009.03.032

[105] T. Li, U. Gulzar, X. Bai, M. Lenocini, M. Prato, K. E. Aifantis, C. Capigila, R. P. Zaccaria, Insight on failure mechanism of Sn electrodes for sodium ion batteries:Evidence of pore formations during sodiation while crack formation during desodiation, Appl. Energy Mater. 2 (2019) 860-866. https://doi.org/10.1021/acsaem.8b01934

[106] S. P. Ong, V. L. Chevrier, G. Hautier, A. Jain, C. Moore, S. Kim, X. Ma, G. Ceder, Voltage, stability and diffusion barrier differences between sodium-ion and lithium-ion intercalation materials, Energy Environ. Sci. 4 (2011) 3680–3688. https://doi.org/10.1039/c1ee01782a

[107] T. Ramireddy, N. Sharma, T. Xing, Y. Chen, J. Leforestier, A.M. Glushenkov, Size and composition effects in Sb-carbon nanocomposites for sodium ion batteries, ACS Appl. Mater. Interfaces 8 (2016) 30152-30164. https://doi.org/10.1021/acsami.6b09619

[108] J. Zhu, D. Deng, Amorphous bimetallic Co_3Sn_2 nanoalloys are better than crystalline counterparts for sodium storage, J. Phys. Chem. C 119 (2015) 21323-21328. https://doi.org/10.1021/acs.jpcc.5b05232

[109] Y. S. Choi, Y. W. Byeon, J. P. Ahn, J.C. Lee, Formation of zintl ions and their configurational change during sodiation in Na-sn battery, Nano Lett. 17 (2) (2017) 679-686. https://doi.org/10.1021/acs.nanolett.6b03690

[110] K. Li, H. Liu, G. Wang, Sb_2O_3 nanowires as anode material for sodium ion batteries, Arabian J. Sci. Eng. 39 (2014) 6589-6593. https://doi.org/10.1007/s13369-014-1194-4

[111] M. Hu, Y. Jiang, W. Sun, H. Wang, C. Jin, M. Yan, Reversible conversion alloying of Sb_2O_3 as a high capacity, high rate, and durable anode for sodium ion

batteries, ACS Appl. Mater. Interfaces 6 (2014) 19449-19455.
https://doi.org/10.1021/am505505m

[112] Y. Cheng, J. Huang, R. Li, Z. Xu, L. Cao, H. Ouyang, J. Li, H. Qi, C. Wang, Enhanced cycling performances of hollow Sn compared to solid Sn in Na-ion battery, Electrochim. Acta 180 (2015) 227–233. https://doi.org/10.1016/j.electacta.2015.08.125

[113] D. H. Nam, T. H. Kim, K.S. Hong, H.S. Kwon, Template free electrochemical synthesis of Sn nanofibers as high-performance anode materials for Na-ion batteries, ACS Nano 8 (2014) 11824–11835. https://doi.org/10.1021/nn505536t

[114] S. Goriparti, E. Miele, F. De Angelis, E. Di Fabrizio, R.P. Zaccaria, C. Capiglia, Review on recent progress of nanostructured anode materials for Li-ion batteries, J. Power Sources, 257 (2014) 421-443. https://doi.org/10.1016/j.jpowsour.2013.11.103

[115] T. Ishihara, M. Nakasu, M. Yoshio, H. Nishiguchi, Y. Takita, Carbon nanotube coating silicon doped with Cr as a high capacity anode, J. Power Sources 146 (2005) 161-165. https://doi.org/10.1016/j.jpowsour.2005.03.110

[116] H. Usui, S. Yoshioka, K. Wasada, M. Shimizu, H. Sakaguchi, Nb doped rutile TiO2: A potential anode material for Na ion battery, ACS Appl. Mater. Interfaces 7 (2015) 6567-6573. https://doi.org/10.1021/am508670z

[117] Y. W. Byeon, Y. S. Choi, J. P. Ahn, J. C. Lee, Origin of high coulombic loss during sodiation in Na-Sn battery, J. Power Sources 343 (2017) 513-519. https://doi.org/10.1016/j.jpowsour.2017.01.089

[118] A. Ponrouch, D. Monti, A. Boschin, B. Steen, P. Johansson and M. R. Palacin, Non-aqueous electrolytes for sodium ion batteries, J. Mater. Chem. A 3 (2015) 22–42. https://doi.org/10.1039/C4TA04428B

Sodium-Ion Batteries: Materials and Applications
Materials Research Foundations **76** (2020) 159-182

Materials Research Forum LLC
https://doi.org/10.21741/9781644900833-7

Chapter 7

Conducting Polymer Electrodes for Sodium-Ion Batteries

Shubham Singh[1], Sheenam Thatai[1*], Parul Khurana[2], Christine Jeyaseelan[1] and Dinesh Kumar[3]

[1]Amity Institute of Applied Sciences, Amity University, Noida 201313, India

[2]Department of Chemistry, Guru Nanak Khalsa College, Mumbai University, Mumbai

[3]School of Chemical Sciences, Central University of Gujarat, Gandhinagar 382030, Gujarat

*sthatai@amity.edu

Abstract

Sodium ion batteries (SIBs) are the sign of the future success available as another charge storage element due to the availability of plenty of sodium resources. Conducting polymer provides a great opportunity in the development of electrode material for these batteries. They are oligomer, easily processable, flexible, low weight and composed of units such as an aromatic or heteroaromatic ring. They are active redox species and are highly thermally stable up to 600°C. Conductive polymers well as its derivative have long cycle life, high surface area, high energy and power densities. In this chapter the development and properties of these batteries are discussed.

Keywords

SIBs, Conducting Polymer, Electrode Material, Polyacetylene, Polyaniline, Polyphenylene

Contents

1. Introduction

The chemistry of SIBs shows superior performance to existing batteries based on the sodium could be the catalyst to enhance the mass production of the new technologies for energy storage for the applications, including research and industrial sites. This chapter provides a summary of recent research and development of SIBs containing conductive polymers [1]. It is known that 90% of the equipment in the world totally depend on electrical energy, but the source of electrical energy is different either it may be generated from solar energy or from other sources such as nuclear power plants, hydropower plants, wind energy, etc. Due to the increase in higher demand of electrical energy worldwide, the whole world invested in producing electrical energy by renewable and sustainable sources like solar and wind energy [2]. Due to the massive growth and development in the renewable sources of energy, there is increased demand for energy storage devices which store energy for further use due to which they are most important for development related to renewable sources of energy [3]. We all know that lithium-ion batteries dominant the batteries market. But it is also a truth that the resources of lithium-ion batteries are limited. The relative opulence in the earth crust only about 25-65 ppm [4,5]. Due to the increase in demand, the prices are getting higher and higher, and it just like a new "gold"[6]. Thus, it is the most important thing to develop new technology in the area which is related to energy storage devices to overcome higher price crisis for energy deposition [7]. There are different types of energy storage technologies which are discussed further.

2. Types of Energy depository technologies in static application

- Pumped Hydroelectric Depository (PHD)
- Compressed Air Energy Depository (CAED)
- Electrochemical Energy Depository (EED)

2.1 Pump hydroelectric depository (PHD)

The pumped hydroelectric depository is also known as a hydroelectric power plant where electrical energy can be generated by using the potential of water. In this method, the potential energy of the water can be extracted by using a turbine. This method generates about 99% of the net global capacity. The construction of the new PHD requires a large amount of area and requires a higher velocity water flow. The construction of PHD requires a well geographical area where the flow of water is very high [8,9].

Sodium-Ion Batteries: Materials and Applications   Materials Research Forum LLC
Materials Research Foundations **76** (2020) 159-182   https://doi.org/10.21741/9781644900833-7

2.2   Compressed air energy depository (CAED)

In CAED, the electrical energy can be generated with the help of high-pressure air through which the rotor of the turbine can be rotated at a very high speed. The construction of CAED needs plateaus area where mountains contain an air gap between their plates. Due to these high pressures, the kinetic energy of air can be converted into electrical energy [10,11].

2.3   Electrochemical energy storage (EED)

There are different types of energy depository devices which came under EED like lead acid, Na/S, lithium-ion and many other batteries. The LABs are relatively low cost, but the major drawback of LIBs is its lifetime. The LIBs have short cycle life; therefore, these batteries need replacement from time to time [9]. So, another type of batteries came into approach i.e. Li^+ and Na^+ ion batteries which have been discussed in the next part.

Table 1. Comparison of Physical Properties and Resources of Lithium and Sodium Metal [13].

Property	Lithium	Sodium
Atomic radii (pm)	76	102
Atomic mass (amu)	6.941	22.999
Ionic charge	1+	1+
Standard potential V vs. SHE	-3.04	-2.71
Electronegativity	1.0	0.9
Density (g/cm^3)	0.53	0.97
Melting point ($^\circ$C)	180.5	97.5
Boiling point ($^\circ$C)	1342	883
Crustal opulence (ppm)	20	23,600
Oceanic opulence (mg/L)	0.1	10,800
Resources (MT)	40	Unlimited

3.   Lithium-ion batteries (LIBs)

The LIBs provide higher energy density and based on the greater electropositive character of the lithium is lightweight material. The LIBs have been used as a renewable source because of their good efficiency, which approximately equal to 90%. The energy density of modern LIBs is about 160-220 W.h.kg^{-1}. The discharge voltage of the lithium-ion battery is about 3.7 V. The mass production LIBs are about a few billion in a year, which is a huge amount and is ever overgrowing whereas the rapid growth in the production of LIBs causes the lack of lithium resources [12].

For this reason, the question arose how to overcome these problems, is there any other alternative technology which eliminates the LIBs batteries from the market. With the help of these questions, we found that there is another material which can be used instead of lithium-ion is sodium ion. Therefore, new technology has been developing to replace LIBs. Some physical properties of the sodium metal and lithium metal are given in Table 1. The properties discussed in Table 1 indicate sodium ion containing batteries (SIBs) are better than LIBs.

4.   Beginning of new technology in the field of energy storage

Due to the higher cost and lack of material for resources of LIBs, there is a new technology invented and have been developed in the area of energy depository equipment. The most abundant material which is found in the earth crust is sodium; therefore, it has been taken for the development of battery technology. SIBs are the best alternative energy depository device to LIBs due to cheap and metal resource. The SIBs are firstly investigated with lithium-ion batteries in the 1970s to 1980s [14]. The initial development of SIBs was stopped in the 1990s because of the industrial success of the lithium ion containing batteries. The reason behind the stalled in SIBs technology was the material used in the LIBs as an anode was carbonaceous, and cathode material was lithium cobalt oxide. The recent cathode material used in the LIBs is graphite. Graphite has very little tendency to store sodium ions because sodium ions have high ionic radius comparatively lithium ions [6,15]. In the 2000s, Stevens and Dahe discovered that hard carbon shows a reversible Na^+ ion capacity of about 300 m.Ah.g^{-1} comparatively to graphitic carbon of the LIBs [16]. This research and the above challenges for LIBs have led to resurgence in the research and development technologies of SIBs [17,18].

It is known that sodium is a low-cost material, readily available as well as easy to use. The SIBs are abundant, and it is eco-friendly. Due to the shortage of electrode material for the SIBs, there is a lack of development in the technology of SIBs. There are different types of materials used in the SIBs but, the use of conductive polymer in the SIBs is due to specific properties such as strength, stability, durability, flexibility, elasticity, easy to handle, mouldability, etc. In comparison with the saturated polymer, the conductive polymer has better electrical properties and optical properties because the conjugation of the π electrons taking place with carbon chain which act as backbone in the polymer are present in the polymeric chain of the polymer. The electrical conductivity of conductive and semi conductive polymeric material is due to the conjugation of the π electrons.

Sodium-Ion Batteries: Materials and Applications   Materials Research Forum LLC
Materials Research Foundations **76** (2020) 159-182   https://doi.org/10.21741/9781644900833-7

4.1   Electrode material for SIBs

There are different types of materials used in SIBs, such as graphene-based material, MXenes, conducting polymers, NASICON electrodes, polyanions, etc. [19-22]. The recent electrode materials for the SIBs are inorganic compound based on transition metals. Several inorganic materials are toxic in nature and causing various types of environmental pollution and environmental contamination. Due to this, the inorganic compounds undergo insertion type sodium depository mechanism. During interjection and emission of sodium ions (102 pm), there is a huge change in the volume and irreversible phase transition may occur. Due to this, there are low reversible capacities and poor cycling performance [23]. Due to various intrinsic advantages, organic materials which are redox active in nature attract research interest. The materials which do not have metal ions, exhibit increasingly high demand in the development of green energy depository devices [24]. These are low-cost materials because they are produced from renewable natural resources [25].

Table 2. Comparison of different electrode materials for SIBs.

Electrode material		Advantages	Disadvantages
Inorganic electrode materials		Good conductors of electricity	Huge volume change, Environmental contamination, Irreversible, and phase transition
Organic materials	Small organic molecules	Environment-friendly, Low cost, Tenability, Structural diversity, and Flexibility	High solubility in electrolytes and poor conductivity
	Polymers	Environment-friendly, Low cost, Tenability, Structural diversity, Flexibility, and Lower solubility	Poor conductivity except conducting polymer and potential solubility

5.   Polymer electrode material for the SIBs

The organic electrode materials contain tiny organic molecules and polymers. Most of the organic compounds which are smaller suffer from rapid dissolution into the organic electrolytes, which leads to the short life cycle. The process of polymerization is the best method to overcome this problem [26]. The advantages and disadvantages of polymeric material for SIBs are listed in Table 2.

In reduction-oxidation reaction, the polymer electrode material for SIBs is mainly three types. The first reaction involves in the functional group-containing C=O species, polyimides and polyquinones. The second functional group is containing C=N species. The third type is conductive polymer followed by doping reaction, like conjugated conductive polymer and non-conjugated conductive radical containing polymer.

5.1 Polyimides

The polyimide is a polymer of imide monomer, and it is very famous due to their high heat resistance. A simple polyimide is Kapton, which is synthesized by the condensation process of pyromellitic dianhydride and oxide aniline.

There are two main methods involves in the synthesis of polyimides.

1. Treatment of dianhydrides with diamines is most used and a common method.

2. Treatment of dianhydride with the di-isocyanate.

The synthesis of polyimide was done by using diamines and dianhydrides in the polar aprotic solvent to produce respective polyamic acid. In the next process, the polyamic acid is allowed to react with the dehydrating agent to form the polyimides. It can also be formed by one step solvothermal method in which polycondensation and iridizations occur spontaneously [27-29]. Fig. 1 represents the reaction between dianhydride and diamines in a polar aprotic solvent.

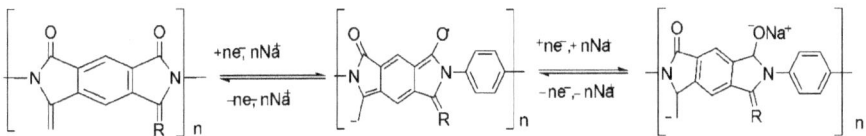

Figure 1. A schematic diagram represents the reaction between dianhydride and diamines in the polar aprotic solvent [30]

The nature of polyimides can be determined by the precursors of diamines and dianhydrides. There are several polyimides which can be synthesized by its corresponding dianhydrides and diamines. Structures of various polyimides which are suitable for SIBs are shown in Fig. 2.

A.

B.

C.

D.

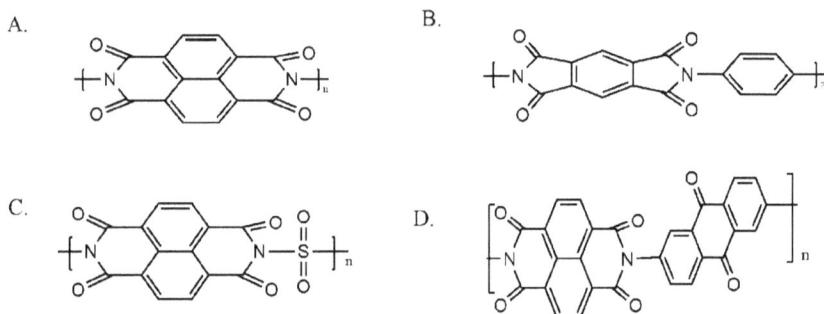

*Figure 2. Structures of various polyimides which are suitable for SIBs; A.
2,7dimethylbenz[3,8]phenanthroline-1,3,6,8(2H,7H)-tetraone, B. 2-methyl-6-p-
tolylpyrrolo[3,4] isoindole-1,3,5,7(2H,6H)-tetraone, C. 2-methyl-7-
(methylsulfonyl)benzo[3,8]phenanthroline-1,3,6,8(2H,7H)-tetraone D. 2-methyl-7-(6-
methyl-9,10-dioxo-8a,9,10,10a-tetrahydroanthracen-2-yl)benzo[3,8]phenanthroline-
1,3,6,8(2H,7H)-tetraone.*

The polyimides have lots of strength and some limitations also. The strength of polyimides is excellent chemical resistance, transparency in many microwave application, mechanical performance is very high, radiation resistance and good temperature adaptability.

The limitations due to which polyimides are replaced with other material are high manufacturing cost, sensitive to alkali and acid attacks and high-temperature requirement in the processing.

To overcome these problems, we use conductive polymers on the place of polyimides due to low-cost manufacturing and having good electrical conductivities.

6. Conducting polymers

The first conducting polymer named polyaniline which was firstly described in mid of 19th century by an English man named Henry Letheby. He identified the electrochemical and oxidized products of aniline in acidic medium. He found that the oxidized form was deep blue, and the reduced form was colourless. Another conducting polymer named polyacetylene (PA) was found by the Shirakawa et al. in 1977s. He observed that the doping in PA allowed conducting polymeric materials similar to metals consisting electrical conductivity as large as 10^5 Scm^{-1}. It was the beginning of a new era of conducting polymers. Conducting polymers are not only for the energy depository devices but it can also be use independently. Their mechanical properties are like plastic, which allows them to be a twist, contract and stiff [31-33]. Conductive polymer electrodes can be introduced ideally into electronic paper, textile industries, structure board, and etc. and even not including the cylindrical cell [33, 34]. The conducting polymer also to be used directly combines with plastics power and polymer to produce cheaper material related to electronic equipment [35].

Conducting polymer is an ideal material for the SIBs due to their conductive backbone. The conducting polymers have high electrical conductivity and columbic efficiency, due to these properties conducting polymer can cycle up to 1000 times with minute decomposition [36]. Conductivity nature of CPs is induced by exchanging both C-C and C=C with the polymer material. The properties like electrical and chemical reversibility are totally depending upon the doping process. Overcoming the problems of conductivity and cyclability, conducting polymers blend with inorganic materials by the synthetic optimization and mechanical flexibility processes led to its property of plasticity property which intern helps in designing of different other instruments [37,38].

Conducting polymers are also used as cathode/anode due to their exchange of ions and their doping nature. The concentration of the electrolytes can achieve the specific capacity; however; some batteries having energies of high specificity related to deposition of potential up to 3.5 V [39].

When CPs present in doped form then they can be used in application such as batteries, used to make diodes, triodes, FETs, photovoltaic solar cells, OLEDs, thin layer transistor, electromagnetic interference (EMI), capacitors, supercapacitors, etc. The various fields where conducting polymers are to be used given below:

6.1 Conducting polymer can provide electromagnetic shielding of electronic devices

The main purpose of the electromagnetic shielding is to protect the electromagnetic interference from the different electronic devices which are very sensitive in nature. A mesh of conductive polymer has been used as the inner material of the electronic devices to separate the two components of the electronic devices [40].

6.2 It absorbs microwaves by using stealth technology

The conductive polymer is capable of absorbing the microwave from the environment. Conductive polymers have intrinsic character due to which it absorbs microwave and converts it into thermal energy. Mainly, the phenomena radar detection having lower frequencies is known as stealth technology. This invention is currently used by the American aircraft F-117A [41].

Sodium-Ion Batteries: Materials and Applications   Materials Research Forum LLC
Materials Research Foundations **76** (2020) 159-182   https://doi.org/10.21741/9781644900833-7

6.3   It can be used as a hole injecting electrode for OLEDs

The first OLEDs were introduced in 1965 by the use of anthracene, but the performance of OLEDs was not good. The ultra-thin film of the conductive polymer is used as the hole injection layer for OLEDs. Conductive polymer plays a main role in the inserting holes in anode to the electronic devices [42].

6.4   Some conducting polymers are promising for field effect transistor (FET)

Conducting polymer has application in electronic elements such as field effect transistor. Gregorian et al. discussed the modification of the P3HT films in the planner stichometry of an organic FET at a lower level when this happens under electric field and the changes occurred seems to be transformable in nature [43,44].

6.5   It can be used in display technology due to their electroluminescent property

The property of electroluminescence can be defined as when electric current flow through the material such as the conducting polymers then the material emits light either they are kept in strong electric and magnetic field or the electricity passed through the material [45].

7.   Types of conductive polymer

The polypyrrole, polyacetylene, polyindole, polyaniline and corresponding copolymers are the main class of conductive polymer as shown in Table 3. The double bonds and aromatic polymers are PA and PPV respectively.

Table 3. Aromatic organic conductive polymers and their compositions.

No heteroatom	Heteroatom	
	Nitrogen	Sulfur
Poly(fluorine)s Polynapthalenes Polyphenylenes Polypyrenes Polyazulenes	Poly(pyrrole)s (PPY), Polyindoles, Polycarbazoles, Polyazepines, Polyanilines (PANI)	Poly(thiophene)s (PT), Poly(3,4-ethylenedioxythio phene) (PEDOT) Poly(p-phenylene sulfide) (PPS)

7.1   Electrically conducting polymer

The properties, such as electrical conductivity of the material, depend on electronic morphology of the material. The conductivity of conjugated polymer happens by conjugating pi electrons including carbon framework. Both planner and linear polymer can be linked different sp^2 C atom linked with an alternating single and double bond. In

Sodium-Ion Batteries: Materials and Applications   Materials Research Forum LLC
Materials Research Foundations **76** (2020) 159-182   https://doi.org/10.21741/9781644900833-7

sigma bonding orbital there are three electrons linked, and the last is in a delocalized p_z-orbital. The p_z-orbital of adjacent carbon overlapped to form π band, enhanced all single lined polymer chain, thus, providing a 1-D delocalized system. The topmost orbital contains filled pi band whereas the bottom orbital contains empty pi^* band. The difference between these bands is said to be the band gap energy, which is denoted by E_g.

7.2 Doping in conductive polymer

To enhance the various properties of the conducting polymers like thermal stability and electrical conductivities doping is essential. The doping may be either p-type or n-type based on the behaviour of dopant.

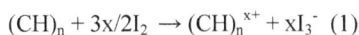

$$(CH)_n + 3x/2I_2 \rightarrow (CH)_n^{x+} + xI_3^- \quad (1)$$

$$(CH)_n + xNa \rightarrow (CH)_n^{x-} + xNa^+ \quad (2)$$

In the above reaction (1) there is a **p**-type of doping taking place in polyacetylene in which dopant is iodide ion. In the second reaction (2) there is **n**-type of doping taking place. However, although chemical doping in the polymers is a useful and significant charge transport method, it is tough to control that the process of doping is homogeneous. The electrochemical doping such that introduction to adequate amount of both charges (+ve and -ve) on polymer chain, this led to alteration of the electronic morphology and electrical properties of the doped polymeric material. The introduction of impurity in trans polyacetylene by doping led to the induces the exposure of the new band at low energy of 0.7 eV, and degree of intensity of the π-$\pi*$ band which is at the energy of 1.7 eV should be brought to a lower energy. These change in band can be explained by the formation of the different localized states which are located inside the band gap because the electrons from the CPs are either added or removed with the help of dopant. Impurity added in the form of doping in conjugated polymer involves varieties of defected responsible for the electrical properties of CPs along with its alteration also. Structures of different types of conductive polymers are given in Fig. 3.

7.3 Polyacetylene and polyphenylene as electrode material for the SIBs

Polyacetylene and poly para-phenylene are the type of conjugated polymers. They are crystalline. The polymers are either partially oxidized or reduced electrochemically. This method results in the formation of an excellent conducting complex between the oxidized or reduced polymer and the suitable counterion from the electrolyte. The electrochemical reaction which has occurred in the batteries containing PA and PPP electrodes involves in the insertion or removal of the electrons from [46].

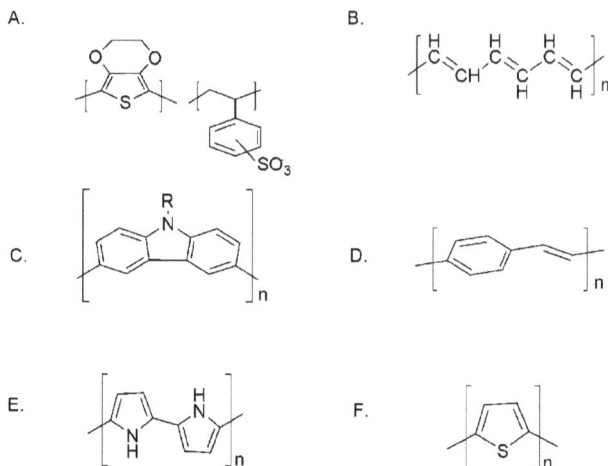

Figure 3. Structure of different types of conducting polymers (A) PEDOT, (B) Polyacetylene, (C) Polycarbazole (D) Polyphenylene-vinylene (E) Polypyrrole (F) Polythiophene.

PA, PPP and other conjugated conductive polymers have been used in the application in nonaqueous-electrolyte secondary batteries, where they may function as either cathode or anode based on the properties of the polymer. The main characteristics of the PA and PPP have been discussed as how these polymeric materials are used as the electrodes for SIBs. The below reaction shows that the reduced polymers for the ion insertion mechanism have been represented by

$$[P]_x + xyC^+ \leftrightarrow [P^{-y}C_y^+]_x$$

Where P denotes the repetitive unit of the polymer composition, C^+ is the inserted cation, y denotes the fractional charge per repeat unit, and x is the degree of the polymerization. In the above reaction, the P denotes the monomeric unit of the polyacetylene and polyphenylene.

There is two most important point of consideration relating to the use of polyacetylene and polyphenyl in the batteries for the charge depository purpose that the charge depository capacity (maximum value of the y) and the value of the voltage at which polymer have been operated. When the whole electrochemical reaction have been performed by a reduction in the more cathodically or anodically stable electrolytes.

Polyacetylene and polyphenylene have high gravimetric charge depository capacity of 0.34 Ah.g^{-1} for PA and 0.15Ah.g^{-1} for PPP, respectively [47,48].

7.4 Conjugated conductive polymer and charge storage mechanism

The conjugated conductive polymer involves in p doping as well as n doping because they are also known to be a biopolymer. In p-doping, the state of polymer changes to positive by adding PF$_6^-$. In n-doping the state of polymer changes to negative due to insertion of sodium ion as a dopant material. The p-doping and n-doping of the conjugated conductive polymer are shown in Fig. 4.

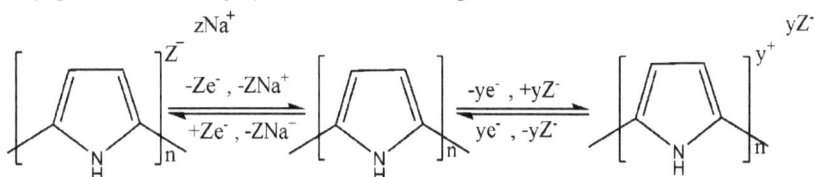

Figure 4. The above reaction shows that the p-doping and n-doping of conjugate conductive polymer.

The conjugated conductive polymer had been firstly studied for the SIBs by the Shacklette et al. in 1985s [47]. It has found that the polymer polyparaphenylene (PPP) and polyacetylene (PA) are considered for the anode material for the SIBs. Both the biopolymers are capable of performing redox reaction, and also, they are active redox polymers. The PPP and PA have the capacity in all energy depository devices because they act like both cathodes as well as anode [49,50].

A conjugated conductive polymer such as PANI and PPy when followed by p-type of doping, then the polymer can be used as the cathode for the SIBs [50]. Similarly, when polymer followed by the n-type doping, then the polymer can be used as an anode for the SIBs [51]. The p-type of doping can be enhanced by the addition of electron withdrawing group. For example, by addition of electron withdrawing group –SO$_3$Na into polyaniline chain, the sulfonated polymer experienced either insertion or extraction of sodium ions rather than the p-type doping reaction [52]. A similar approach was executed in the case of poly(diphenylaminesulfoonic acid sodium) and poly(pyrrole-lyl) propane sulfonate) [53].

The significance of conductive polymer over any other polymer or substance like semiconductors or metals is due to their better electrical conductivities [54]. The capacity of the conductive polymer is low but depends on the degree of doping. The capacity of conducting polymer can be enhanced by adding high capacity redox active species. By inserting the o-nitro aniline into polyaniline polymer, the reversible capacity of the polymer can be increased to 180 m.Ah.g^{-1} and retained at 173 m.Ah.g^{-1} at the 50th cycle [55-57]. Similarly, when we performed doping with Ferro-cyanides and diphenylamine sulfonate anions, there is an increase in capacities of anion doped polypyrrole due to the effect of anion doped redox active species. Thus, to develop a high capacity of conductive polymer electrode material for SIBs, it is more important to focus on the insertion of other redox active material onto the chain of the polymer [52,58]. The polypyrrole has excellent electrochemical property due to their unique size, morphology and structure due to this there is an increase in electrical contact between the polymer molecule and cause penetration of electrolyte into the material [39].

7.5 Non-conjugated conductive radical polymer

The famous non-conjugated radical polymers for the energy depository application are nitroxide radical polymers [59]. The nitroxide radical polymer is not only reversible n-doped aminoxy anions in cathodes reactions at high voltage but also p-doped to oxo ammonium cations in anodic reaction at relatively low voltage [60]. Molecular structure of poly [norborn-2,3-endo, exo-(COO TEMPO-4)$_2$] and PTMA are two non-conjugated radical conductive polymers are shown in Fig. 5.

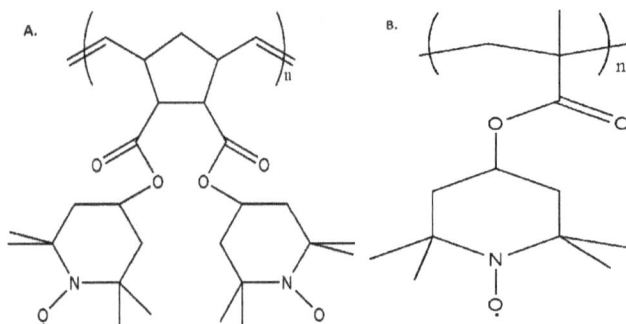

Figure 5. Molecular structure of (A) poly[norborn-2,3-endo-exo-(COO-4-TEMPO)$_2$] and (B) PTMA are two non-conjugated radical conductive polymers.

The first application of SIBs reported in the 2010s by Dai et al. was a poly-norbornene derivative radical polymer, poly[norborn-2,3-endo, exo-(COO TEMPO-4) [61] and PTMA as activated cathodic material [62].

The significance of radical non-conjugated conductive polymer involves in rapid kinetics during the process of oxidation, stable cell voltage due to the unique and stabilized the structure as well as good processability. The disadvantage of radical polymer is spontaneous discharge taking place due to dissolution in the electrolytic solution, which behaves as a redox shuttle. To overcome this problem, some advanced electrolyte has been developed [63].

The charging and discharging of the radical conducting polymer are shown in Fig. 6.

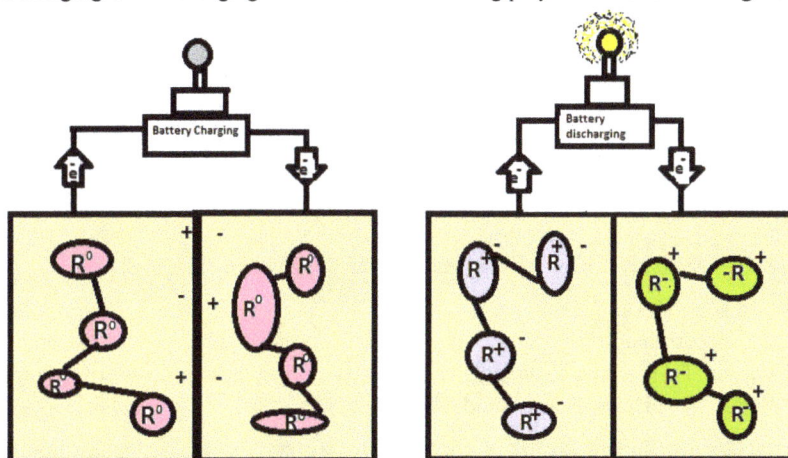

Figure 6. Charging and discharging of the organic radical battery containing two different radical conducting polymer functions as the cathode active and active anode material.

7.6   Inorganic nanoparticles-conducting polymer composite based battery electrodes

Inorganic nanoparticles CPs composite materials are an important element which has been used for batteries due to its unusual mixing of properties deriving from the different elementary blocks. The basic concept of designing the composites starts from the nano-sized combination of polymers, metals and functionality molecules. The classification of the composite materials based on their interactivity (a) functionalized molecules or metallic predecessor undergo very powerful interplay with ionic as well as covalent bonds [64] (b) inorganic nanoparticles embedded polymer have stabilization in charge and discharge characteristics as well as higher electrode capacities [65].

8.   Why conducting polymer?

Conductive polymers have high electrical conductivity, high redox active, low-cost material, higher thermal stability, unique structure, flexibility, mouldability, ease to handle, elasticity, structural anisotropy, etc. Many materials have these properties but the properties which distinguish between the conductive polymer and other materials which is structural anisotropy. The behaviour of conducting polymer is structural anisotropy; therefore, it is considered as electrode material for SIBs. Conducting polymer used in full range of wide area due to its electronic occurrence and can be used for a large number of specific applications. For many of these applications, the valuable function is processability. Recent, CPs are hard to control, but also these polymers are unstable when they are kept in the open environment conditions. The conducting polymer caught by de-doping and undergoes decomposition; then the polymer is not able to be used in several applications. Many routes have been developed to enhance the solubility in some organic solvents, water and one of the best successful routes consists of the grafting on non-aromatic (aliphatic) chains on the conjugated conducting polymer skeletal.

Thin film coatings can be done easily and fast by casting from the electrolytic solution and also electrochemically. There are various areas found where it can be used for coatings and also in nanoelectronics. It can be used to manufacture integrated circuit and technologies like display elements and inkjet printing.

Generally, the main property of a conductive polymer is electrical conductivity and flexibility fallowed by the minimum cost manufacturing lead to the conducting polymer in our everyday life. Due to its identical properties in the electrical conductivities, it is used as an electrode material for the SIBs.

9.   Functions of CPs

Firstly, we have to investigate the general properties of CPs. The energy and conducting ability depend on electronic properties of the CPs it is not hold by the non-aromatic carbon of the polymer, which can be produced by the overlapping of the adjacent p orbitals. Like inorganic semiconductors, the backbone of the conductive polymer and molecular orbital provides a band structure. The more effective associate dimension is an optimum point where no additional monomeric unit added to the polymeric chain which affects the electrical properties of the material. The conduction band HOMO and LUMO

Sodium-Ion Batteries: Materials and Applications
Materials Research Foundations **76** (2020) 159-182

Materials Research Forum LLC
https://doi.org/10.21741/9781644900833-7

values of the CPs can help to determine the properties of stability, optical absorption wavelength and the polymer to be doped. On increasing the conjugation length, the HOMO increases and the lowest occupied molecular orbital decreases. The polymer is more susceptible to electrophiles because of their higher HOMO, and therefore, it is more reactive [66].

9.1 Merits and demerits of the conducting polymer

There is lots of research been done and many research is on conducting polymers such as PANI, PPy and PT due to specific properties like chemically stable, greater theoretical capacity, ease synthesis and electrochemical reversibility [67]. The electrode which is made up of PANI makes equilibrium reaction between fully reduced salt and partially oxidized salt state. Due to the cycling effect emeraldine salt without any change oxidized to per-nigraniline base because of the pathway to per nigraniline salt is not helped as the high potential [68]. However, there are many merits to utilize the PANI stem from insolubility, issues related to the diffusion, cycling and stability. The term intractability and insolubility considered as major obstructing its uses in energy storage [69].

There are a set of ideas to overcome such issue. Polyanions with the PANI nanofiber using template has allowed easy solution suspension for PANI and composed with composites of carbon has been helped to maintain the properties such as conductivity when PANI reduced to a nonconductive state for favorable charge depository [69-71]. It can also agonize by the lower capacity vs active redox materials because of non-Faradic nature of the charge depository [64,72]. PANI nanostructures can be made under cheaper cost and quickly used when new ideas, hard techniques were absent [73]. PANI is valuable for the composites involves inorganic material and carbon-containing materials which have high capacities and allows for facile solution.

CPs which can be doped by itself have ionic side chain, response is rapid and able to improved processability but they are also agonize by low conductivities [74].

Conclusion

The chemical properties of the conducting polymer show an important part in the electrode material for the SIBs. The SIBs are a key to advancement in the development of new technology in the energy depository devices. Research on SIBs materials has increased for the last ten years. Various materials have been considered for the different elements of SIBs. From the above data and discussion, we found that the conducting polymer is suitable for electrode material. It is a low-cost material, having greater thermal stability and strength. Due to the anisotropic structural property, the conductive polymer is different from any other material. Due to the conjugation of π electron, the electrical conductivity of conducting polymer is very high in comparison with semiconductors or even metal. The consideration of conducting polymer is also because many conductive polymers are biopolymer since they can be used as both cathodes as well as anode material for the SIBs. The electrical conductivity of the polyacetylene, when doped with AsF_3, is near about 10^5 $S.cm^{-1}$ which is higher compared to semiconductors as well as some metals such as copper and silver. A conductive polymer such as polyphenylene vinylene, polyaniline, polycarbazole, polythiophene and polypyrrole, represent new advance material for the development of electrode material for SIBs.

Acknowledgement

Dinesh Kumar is thankful DST, New Delhi for financial support to this work (sanctioned vide project Sanction Order F. No. DST/TM/WTI/WIC/2K17/124(C).

References

[1] B. Häupler, A. Wild, U.S. Schubert: Carbonyls: Powerful organic materials for secondary batteries, Adv. Energy. Mater. 5 (2015) 1402034. https://doi.org/10.1002/aenm.201402034

[2] H. Pan, Y.S. Hu, L. Chen: Room-temperature stationary sodium-ion batteries for large-scale electric energy storage, Energy Environ. Sci. 6 (2013) 2338-2360. https://doi.org/10.1039/c3ee40847g

[3] V. Palomares, P. Serras, I. Villaluenga, K.B. Hueso, J. Carretero-González, T. Rojo Na-ion batteries, recent advances and present challenges to become low cost energy storage systems, Energy Environ. Sci. 5 (2012) 5884-5901. https://doi.org/10.1039/c2ee02781j

[4] C. Grosjean, P.H. Miranda, M. Perrin, P. Poggi, Assessment of world lithium resources and consequences of their geographic distribution on the expected development of the electric vehicle industry, Renew. Sust. Energ. Rev. 16 (2012) 1735-1744. https://doi.org/10.1016/j.rser.2011.11.023

[5] A. Yaksic, J.E. Tilton, Using the cumulative availability curve to assess the threat of mineral depletion, The case of lithium, Resour. Pol. 34 (2009) 185-194. https://doi.org/10.1016/j.resourpol.2009.05.002

[6] J. Tarascon, Is lithium the new gold? Nat. Chem. 2 (2010) 510. https://doi.org/10.1038/nchem.680

[7] M Armand, J.M. Tarascon, Building better batteries, Nature 7179 (2008) 451-652. https://doi.org/10.1038/451652a

[8] I. Hadjipaschalis, A Poullikkas, V. Efthimiou, Overview of current and future energy storage technologies for electric power applications, Renew. Sust. Energ. Rev. 13 (2009) 1513-1522. https://doi.org/10.1016/j.rser.2008.09.028

[9] D. EPRI, Electricity energy storage technology options-a white paper primer on applications, costs and benefits, Rept. 1020676 (2010) 1-170.

[10] J. Tzeng, R. Emerson, P. Moy, Composite flywheels for energy storage, Compos. Sci. Technol. 66 (2006) 2520-2527. https://doi.org/10.1016/j.compscitech.2006.01.025

[11] A. Castillo, D.F. Gayme, Grid-scale energy storage applications in renewable energy integration, A Surv. Energy Conv. Mgmt. 87 (2014) 885-894. https://doi.org/10.1016/j.enconman.2014.07.063

[12] M. Smart, B. Ratnakumar, Effects of electrolyte composition on lithium plating in lithium-ion cells, J. Electrochem. Soc. 158 (2011) A379-A389. https://doi.org/10.1149/1.3544439

[13] S. Futamura, A. Zhang, H. Einaga, H. Kabashima, Involvement of catalyst materials in nonthermal plasma chemical processing of hazardous air pollutants, Catal. Today 72 (2002) 259-265. https://doi.org/10.1016/S0920-5861(01)00503-X

[14] A.S. Nagelberg, W.L. Worrell, A thermodynamic study of sodium-intercalated TaS_2 and tis2, J.Solid. State. Chem. 29 (1979) 345-354. https://doi.org/10.1016/0022-4596(79)90191-9

[15] P. Gruber, P.A. Medina, A. Keoleian, S.E. Kesler, M.P. Everson, T.J. Wallington, Global lithium availability, J. Ind. Ecol. 15 (2011) 760-775. https://doi.org/10.1111/j.1530-9290.2011.00359.x

[16] D. Stevens, J. Dahn, High capacity anode materials for rechargeable sodium-ion batteries, J. Electrochem. Soc. 147 (2000) 1271-1273. https://doi.org/10.1149/1.1393348

[17] M.S. Balogun, Y. Luo, W. Qiu, P. Liu, Y. Tong, A review of carbon materials and their composites with alloy metals for sodium ion battery anodes, C. 98 (2016) 162-178. https://doi.org/10.1016/j.carbon.2015.09.091

[18] J. Górka, C .Vix-Guterl, C. M. Ghimbeu, Recent progress in design of biomass-derived hard carbons for sodium ion batteries, C J. Carbon Res. 24 (2016) 1-17. https://doi.org/10.3390/c2040024

[19] L.P. Wang, L. Yu, X. Wang, M. Srinivasan, Z.J. Xu, Recent developments in electrode materials for sodium-ion batteries, J. Mater. Chem. A 3 (2015) 9353-9378. https://doi.org/10.1039/C4TA06467D

[20] M.H. Han, E. Gonzalo, G. Singh, T. Rojo, A comprehensive review of sodium layered oxides: Powerful cathodes for Na-ion batteries, Energy Environ. Sci. 8 (2015) 81-102. https://doi.org/10.1039/C4EE03192J

[21] H. Kim, Z. Ding, M.H. Lee, K. Lim, G. Yoon, K. Kang, Recent progress in electrode materials for sodium-ion batteries, Adv. Energ. Mater. 6 (2016) 1600943. https://doi.org/10.1002/aenm.201600943

[22] S.W. Kim, D.H. Seo, X. Ma, G. Ceder, K. Kang, Electrode materials for rechargeable sodium-ion batteries: Potential alternatives to current lithium-ion batteries, Adv. Energ. Mater. 2 (2012) 710-721. https://doi.org/10.1002/aenm.201200026

[23] P.F. Wang, Y. You, Y.X. Yin, Y.G. Guo, Layered oxide cathodes for sodium-ion batteries, Phase transition, air stability, and performance, Adv. Energ. Mater. 8 (2018) 1701912. https://doi.org/10.1002/aenm.201701912

[24] Z. Song, H. Zhou, Towards sustainable and versatile energy storage devices, An overview of organic electrode materials, Energy Environ. Sci. 6 (2013) 2280-2301. https://doi.org/10.1039/c3ee40709h

[25] R. Emanuelsson, M. Sterby, M. Strømme, M. Sjödin, An all-organic proton battery, J. Am. Chem. Soc. 139 (2017) 4828-4834. https://doi.org/10.1021/jacs.7b00159

[26] Y. Zhang, J. Wang, S.N. Riduan, Strategies toward improving the performance of organic electrodes in rechargeable lithium (sodium) batteries, J. Mater. Chem. 4 A (2016) 14902-14914. https://doi.org/10.1039/C6TA05231B

[27] G. Cheng, M. Aponte, C.A. Ramírez, Degradable polymides, Google Patents, (2008).

[28] B. Baumgartner, M.J. Bojdys, M.M. Unterlass, Geomimetics for green polymer synthesis: Highly ordered polyimides via hydrothermal techniques, Polym. Chem. 5 (2014), 3771-3776. https://doi.org/10.1039/C4PY00263F

[29] Z. Song, H. Zhan, Y. Zhou, Polyimides: Promising energy-storage materials, Angew. Chem. Int. Ed. 49 (2010), 8444-8448. https://doi.org/10.1002/anie.201002439

[30] H. Wang, S. Yuan, D.l. Ma, X.l. Huang, F.l. Meng, X.B. Zhang, Tailored aromatic carbonyl derivative polyimides for high-power and long-cycle sodium-organic batteries, Adv. Energ. Mater. 4 (2014) 1301651 1-18. https://doi.org/10.1002/aenm.201301651

[31] S. Prakash, C.R. Rao, M. Vijayan, Polyaniline–polyelectrolyte–gold (0) ternary nanocomposites: Synthesis and electrochemical properties, Electrochim. Acta 5424 (2009) 5919-5927. https://doi.org/10.1016/j.electacta.2009.05.059

[32] S. Prakash, C.R. Rao, M. Vijayan, New polyaniline (PANI)-polyelectrolyte (pddmac) composites: Synthesis and applications, Electrochimi. Acta 53 (2008) 5704-5710. https://doi.org/10.1016/j.electacta.2008.03.036

[33] L. Wei, G. Yushin, Nanostructured activated carbons from natural precursors for electrical double layer capacitors, Nano Energy 1 (2012) 52-56. https://doi.org/10.1016/j.nanoen.2012.05.002

[34] S. Seyedin, P. Zhang, M. Naebe, S. Qin, J. Chen, X. Wang, J.M. Razal, Textile strain sensors: A review of the fabrication technologies, performance evaluation and applications, Mater. Horiz. 6 (2019) 19-249. https://doi.org/10.1039/C8MH01062E

[35] S.R. Forrest, The path to ubiquitous and low-cost organic electronic appliances on plastic, Nature 428 (2004) 6986. https://doi.org/10.1038/nature02498

[36] L. Wu, X. Hu, J. Qian, F. Pei, F. Wu, R Mao, X. Ai, H. Yang, Y. Cao, Sb-C nanofibers with long cycle life as an anode material for high-performance sodium-ion batteries, Energy Environ. Sci. 7 (2014) 323-328. https://doi.org/10.1039/C3EE42944J

[37] L. Shao, J.W. Jeon, J.L. Lutkenhaus, Polyaniline/vanadium pentoxide layer-by-layer electrodes for energy storage, Chem. Mater. 24 (2011) 181-189. https://doi.org/10.1021/cm202774n

[38] Y.H. Huang, J.B. Goodenough, High-rate $LiFePO_4$ lithium rechargeable battery promoted by electrochemically active polymers, Chem. Mater. 20 (2008) 7237-7241. https://doi.org/10.1021/cm8012304

[39] S Liu, F Wang, R Dong, T Zhang, J Zhang, X Zhuang, Y Mai, X Feng, Dual-template synthesis of 2d mesoporous polypyrrole nanosheets with controlled pore size, Adv. Mater. (2016) 28(38):8365-8370. https://doi.org/10.1002/adma.201603036

[40] J. Dubois, O. Sagnes, F. Henry, Polyheterocyclic conducting polymers and composites derivates, Synth. Met. 28 (1989) 871-878. https://doi.org/10.1016/0379-6779(89)90616-4

[41] L.C. Folgueras, M.C. Rezende, Multilayer radar absorbing material processing by using polymeric nonwoven and conducting polymer, Mater. Res. 11 (2008) 245-249. https://doi.org/10.1590/S1516-14392008000300003

[42] J. Margolis, Conductive polymers and plastics, Springer Sci. Busi. Med., (2012).

[43] F. Garnier, R. Hajlaoui, A. Yassar, P. Srivastava, All-polymer field-effect transistor realized by printing techniques, Science 265 (1994) 1684-1686. https://doi.org/10.1126/science.265.5179.1684

[44] S. Grigorian, D. Tranchida, D. Ksenzov, F. Schäfers, H. Schönherr, U. Pietsch, Structural and morphological changes of p3ht films in the planar geometry of an ofet device under an applied electric field, Eur. Polym. J. 47 (2011) 2189-2196. https://doi.org/10.1016/j.eurpolymj.2011.09.003

[45] R. Friend, R. Gymer, A. Holmes, J. Burroughes, R. Marks, C. Taliani, D. Bradley, D. Dos Santos, J. Bredas, M. Lögdlund, Electroluminescence in conjugated polymers, Nature 397 (1999) 1-21. https://doi.org/10.1038/16393

[46] P.J. Nigrey, D. MacInnes, D.P. Nairns, A.G. MacDiarmid, A.J. Heeger, Lightweight rechargeable storage batteries using polyacetylene,(CH)x as the cathode-active material, J. Electrochem. Soc. 128 (1981) 1651-1654. https://doi.org/10.1149/1.2127704

[47] L. Shacklette, J. Toth, N. Murthy, R. Baughman, Polyacetylene and polyphenylene as anode materials for nonaqueous secondary batteries, J. Electrochem. Soc. 132 (1985) 1529-1535. https://doi.org/10.1149/1.2114159

[48] A.G. MacDiarmid, A.J. Heeger, Organic metals and semiconductors, The chemistry of polyacetylene,(CH) x, and its derivatives. Synth. Met. (1980) 101-118. https://doi.org/10.1016/0379-6779(80)90002-8

[49] X. Zhu, R. Zhao, W. Deng, X. Ai, H. Yang, Y. Cao, An all-solid-state and all-organic sodium-ion battery based on redox-active polymers and plastic crystal electrolyte, Electrochim. Acta 178 (2015) 55-59. https://doi.org/10.1016/j.electacta.2015.07.163

[50] G. Farrington, B. Scrosati, D .Frydrych, J. DeNuzzio, The electrochemical oxidation of polyacetylene and its battery applications, J. Electrochem. Soc. 131 (1984) 7-12. https://doi.org/10.1149/1.2115550

[51] W. Luo, F. Shen, C. Bommier, H. Zhu, X. Ji, L. Hu, Na-ion battery anodes: Materials and electrochemistry, Acc. Chem. Res. 49 (2016) 231-240. https://doi.org/10.1021/acs.accounts.5b00482

[52] Y Shen, D Yuan, X Ai, H Yang, M Zhou, Poly (diphenylaminesulfonic acid sodium) as a cation-exchanging organic cathode for sodium batteries. Electrochem. Commun. 49 (2014) 5-8. https://doi.org/10.1016/j.elecom.2014.09.016

[53] L Zhu, Y Shen, M Sun, J Qian, Y Cao, X Ai, H Yang, Self-doped polypyrrole with ionizable sodium sulfonate as a renewable cathode material for sodium ion batteries, Chem. Commun. (2013) 49(97):11370-11372. https://doi.org/10.1039/c3cc46642f

[54] H. Shirakawa, E.J. Louis, A.G. MacDiarmid, C.K. Chiang, A.J. Heeger, Synthesis of electrically conducting organic polymers: Halogen derivatives of

polyacetylene,$(CH)x$. J. Chem. Soc, Chem. Commun. (1977) 578-580. https://doi.org/10.1039/c39770000578

[55] R. Zhao, L. Zhu, Y. Cao, X. Ai, H.X. Yang, An aniline-nitroaniline copolymer as a high capacity cathode for na-ion batteries, Electrochem. Commun. 21 (2012) 36-38. https://doi.org/10.1016/j.elecom.2012.05.015

[56] M. Zhou, L. Zhu, Y. Cao, R. Zhao, J. Qian, X. Ai, H. Yang, Fe (cn) 6− 4-doped polypyrrole, A high-capacity and high-rate cathode material for sodium-ion batteries, RSC. Adv. 2 (2012) 5495-5498. https://doi.org/10.1039/c2ra20666h

[57] M. Zhou, Y. Xiong, Y. Cao, X. Ai, H. Yang, Electroactive organic anion-doped polypyrrole as a low cost and renewable cathode for sodium-ion batteries. J. Polym Sci Part B, PolyM. Phy. 51 (2013) 114-118. https://doi.org/10.1002/polb.23184

[59] D. Su, J. Zhang, S. Dou, G. Wang, Polypyrrole hollow nanospheres, Stable cathode materials for sodium-ion batteries, Chem. Commun. 51 (2015) 16092-16095. https://doi.org/10.1039/C5CC04229A

[59] K. Nakahara, S Iwasa, M Satoh, Y Morioka, J Iriyama, M Suguro, E Hasegawa, Rechargeable batteries with organic radical cathodes. Cheml Phy Lett. 359 (2002) 351-354. https://doi.org/10.1016/S0009-2614(02)00705-4

[60] Y. Dai, Y. Zhang, L. Gao, G. Xu, J. Xie, A sodium ion based organic radical battery, Electrochem. Solid-State Lett. 13 (2010) A22-A24. https://doi.org/10.1149/1.3276736

[61] J.K. Kim, Y. Kim, S. Park, H. Ko, Y. Kim, Encapsulation of organic active materials in carbon nanotubes for application to high-electrochemical-performance sodium batteries, Energ. Environ. Sci. 9 (2016) 1264-1269. https://doi.org/10.1039/C5EE02806J

[62] T. Janoschka, M.D. Hager, U.S. Schubert, Powering up the future: Radical polymers for battery applications, Adv. Mater. 24 (2012) 397-6409. https://doi.org/10.1002/adma.201203119

[63] L. Bugnon, C.J. Morton, P. Novak, J. Vetter, P. Nesvadba, Synthesis of poly (4-methacryloyloxy-tempo) via group-transfer polymerization and its evaluation in organic radical battery, Chem. Mater. 19 (2007) 2910-2914. https://doi.org/10.1021/cm063052h

[64] Z-F Li, H Zhang, Q Liu, Y Liu, L Stanciu, J Xie, Novel pyrolyzed polyaniline-grafted silicon nanoparticles encapsulated in graphene sheets as li-ion battery anodes, ACS. Appl. Mater. Interfaces. 6 (2014) 5996-6002. https://doi.org/10.1021/am501239r

[65] K.S. Park, S.B. Schougaard, J. Goodenough:Conducting-polymer/iron-redox-couple composite cathodes for lithium secondary batteries, Adv. Mater. 19 (2007) 848-851. https://doi.org/10.1002/adma.200600369

[66] J. Roncali, Electrogenerated functional conjugated polymers as advanced electrode materials, J. Mater. Chem. 9 (1999) 1875-1893. https://doi.org/10.1039/a902747e

[67] D.A. Pasquier, I. Plitz, S. Menocal, G. Amatucci, A comparative study of li-ion battery, supercapacitor and nonaqueous asymmetric hybrid devices for automotive applications, J. Power Sources 115 (2003) 171-178. https://doi.org/10.1016/S0378-7753(02)00718-8

[68] D.A. Pasquier, A. Laforgue, P. Simon, G.G. Amatucci, J.F. Fauvarque, A nonaqueous asymmetric hybrid li4ti5 o 12/poly (fluorophenylthiophene) energy storage device, J. Electrochem. Soc.149 (2002) A302-A306. https://doi.org/10.1149/1.1446081

[69] S. Bhadra, D. Khastgir, N.K. Singha, J.H. Lee, Progress in preparation, processing and applications of polyaniline, Prog. Polym. Sci. 34 (2009) 783-810. https://doi.org/10.1016/j.progpolymsci.2009.04.003

[70] W. Stockton, M. Rubner, Molecular-level processing of conjugated polymers Layer-by-layer manipulation of polyaniline via hydrogen-bonding interactions, Macrom. 30 (1997) 2717-2725. https://doi.org/10.1021/ma9700486

[71] D. Li, J. Huang, R.B. Kaner, Polyaniline nanofibers: A unique polymer nanostructure for ersatile applications, Acc. Chem. Res. 42 (2009) 135-145. https://doi.org/10.1021/ar800080n

[72] I. Dumitrescu, P.R. Unwin, J.V. Macpherson, Electrochemistry at carbon nanotubes: Perspective and issues, Chem. Commun. 45 (2009) 6886-6901. https://doi.org/10.1039/b909734a

[73] H.D. Tran, D. Li, R.B. Kaner, One-dimensional conducting polymer nanostructures: Bulk synthesis and applications, Adv. Mater. 21 (2009) 1487-1499. https://doi.org/10.1002/adma.200802289

[74] H. Ghenaatian, M. Mousavi, S. Kazemi, M. Shamsipur, Electrochemical investigations of self-doped polyaniline nanofibers as a new electroactive material for high performance redox supercapacitor, Synth. Metals, 159 (2009) 1717-1722. https://doi.org/10.1016/j.synthmet.2009.05.014

Sodium-Ion Batteries: Materials and Applications
Materials Research Foundations 76 (2020) 183-204

Materials Research Forum LLC
https://doi.org/10.21741/9781644900833-8

Chapter 8

Recent Progress in Electrode Materials for Sodium Ion Batteries

Mesut Yıldız[1], Haydar Göksu[2]*, Husnu Gerengi[1], Kubilay Arıkan[3], Mohd Imran Ahamed[4],
Fatih Şen[3]*

[1] Corrosion Research Laboratory, Department of Mechanical Engineering, Faculty of
Engineering, Duzce University, 81620, Duzce, Turkey

[2] Kaynasli Vocational College, Duzce University, 81900 Duzce, Turkey

[3] Sen Research Group, Department of Biochemistry, Dumlupınar University, 43100 Kütahya,
Turkey

[4] Department of Chemistry, Faculty of Science, Aligarh Muslim University, Aligarh-202 002,
India

haydargoksu@duzce.edu.tr, fatih.sen@dpu.edu.tr

Abstract

Fossil fuels, which meet most of the energy needs and have a limited reserve, are known
to be exhausted. Due to the increase in fossil fuel prices and adverse environmental
impacts, renewable energy sources such as water, sea wave, solar, wind, and geothermal
energy are gaining importance. Renewable energy sources are of great importance since
they are accessible and inexhaustible. Various devices are used for the storage of these
renewable energy sources. Lithium batteries, called Li-ion batteries (LIBs), are the most
widely used devices for storing this energy. The demand for portable electronic devices
has increased with the developing technology. In addition, the importance of electric cars
in the transport sector and the research and development activities on lithium-ion
batteries are becoming increasingly important. LIBs are preferred by designers and
consumers due to their low CO_2 emissions. However, in addition to these advantages,
different studies have been carried out due to the high cost of production of LIBs and
limited lithium reserves. There are also sodium-ion batteries (SIBs) that can be
alternative to LIBs and have large-scale energy storage capacity due to their widespread
use and low costs. SIBs have essential advantages such as good cycle life, high energy
density, and self-discharge. In this chapter, SIBs were compared with the LIB, and the
history of SIB, the various anode materials used in SIB was examined.

Keywords

Magnetic Nanomaterial, Nanoparticle, Ion Batteries, Sodium

Contents

1.   Introduction

Ozone layer depletion, acid rain, and global warming are observed due to fossil fuels such as oil, coal, and natural gas was used extensively in the 20th century. It is also known that fossil fuels, which supplies most of the energy needs and have a finite reserve, will be exhausted day by day. Because of the increase in the prices of fossil fuels and the negative environmental impacts, renewable energy sources such as water, sea wave, solar, wind, and geothermal energy are increasing in importance. Renewable energy sources are of great importance, as they are accessible and inexhaustible. In recent years, the storage of energy has an important place in daily life and industrial area owing to increasing energy needs. The energy produced from renewable energy sources, which vary to time, place, and season, should be stored. Batteries, fuel cells, and supercapacitors are known as the most efficient electrochemical energy storage and conversion systems. These devices are known to be environmentally friendly as they do not cause pollution when converting chemical energy into electrical energy[1]. The terms specific energy (Wh / kg), energy density (Wh / L), specific power (W / kg) and power density (W / L) are used to compare the energy content and speed capacity of a system. Ragone graphics have been developed to compare the energy and power properties of different systems. The Ragone graph in Figure 1 shows that the batteries have moderate energy and power characteristics. Moreover, it has been seen that none of the electrochemical energy storage or conversion systems in its own right can reach the level of combustion engines and gas turbines.

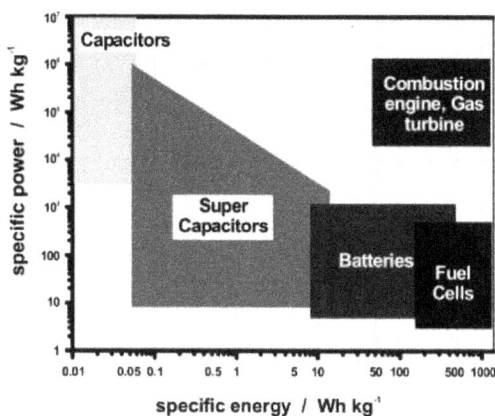

Figure 1. A typical Ragone diagram to compare the energy and power characteristics of various systems [2]

The batteries which produce high energy due to their large energy capacity and low power form the basis of the electrical storage systems. Lithium batteries called Li-ion batteries (LIB) are currently the most widely used devices. Research and development activities on lithium-ion batteries are gaining increasing importance with increasing demand for portable electronic devices, as well as the importance of electric cars in the transport sector. LIBs are used in mobile phones, laptops, and small household appliances due to their high energy density and non-toxicity. It is also preferred by designers and consumers because of low CO_2 emissions. However, researches on sodium-ion batteries (NIB) have increased due to the high cost of production of LIBs and limited lithium reserves [3–5]. NIBs can be used as an alternative large-scale energy storage device to LIBs depending on the common usage and low cost [6]. Figure 2 shows the amounts of the elements in the earth's crust.

Figure 2 shows that the sources of lithium are less than sodium source [8,9]. Sodium, the fourth most abundant element in the Earth's crust, is the second lightest and the smallest alkali metal after lithium. The redox potential of sodium is only 0.3 V greater than lithium ($ENa^+/Na = -2,71$ V vs. standard hydrogen electrode). Furthermore, NIBs may be preferred as an alternative to LIBs since the electrochemical properties of sodium are similar to lithium [10]. SIBs have important advantages such as good cycle life, high energy density, and self-discharge [11–16] . The excellent performance of an energy storage device depends on the properties of the electrode materials [17–22]. Table 1 shows the comparison of the properties of sodium and lithium. NIBs have lower gravimetric and volumetric densities because sodium has a larger ionic radius and has less reduction potential.

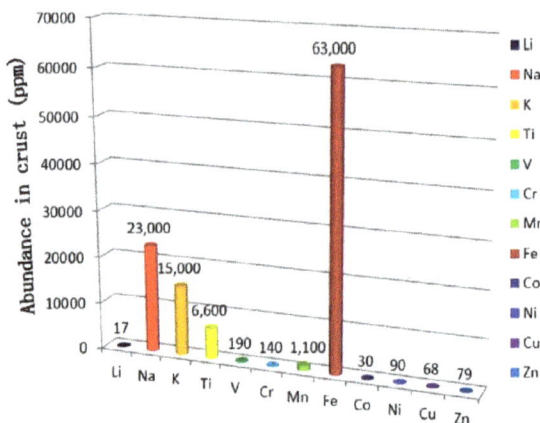

Figure 2. The amounts of the elements in the earth's crust [7]

Table 1. The comparison of properties of sodium and lithium

	Sodium	Lithium
Cation radius	97 pm	68 pm
Relative atomic mass	23 g mol^{-1}	6.94g mol^{-1}
E_0 vs. SHE	-2.7 V	-3.04 V
Price of carbonates	0.07-0.37 €/kg	4.11-4.49 €
Capacity density	1165 mAhg^{-1}	3829 mAhg^{-1}
Melting point	97.7 ^{0}C	180.5 ^{0}C
Coordination preference	Octahedral and prismatic	Octahedral and tetrahedral

2.   History and working principal of SIB

SIBs were first proposed before LIBs in the late 1970s [23]. The SIBs, which were examined with LIBs [24,25] remained behind LIBs due to their low energy density, short battery life, and safety concerns [26–33]. It is known that Na ion compared to lithium has a larger size, higher redox potential, lower capacity density, and two times cheaper [34].

However, SIB reviews were interrupted because the commercial success in LIB was higher [35]. Therefore, researchers have focused on SIB due to a large amount of sodium and low-cost production. In 2014, Aquion Energy, which produces SIBs and electrical storage systems, introduced the SIB with a similar cost and capacity to the lead acid batter. SIBs contain an anode-cathode electrode pair, porous separator with liquid or solid polymer electrolyte [36]. Both electrodes of SIBs consist of Na-containing composition and electrolyte consists of anhydrous sodium salt solution [37].

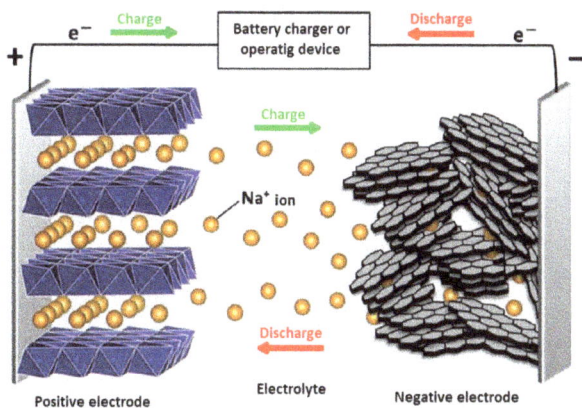

Figure 3. Schematic representation of SIB [38]

During the cycle energy charge, the formed energy in the battery supply the transfer of electrons from the cathode electrode to the anode electrode. The conversion of sodium atoms into positive and negative ions occurs in the cathode electrode (Eq. 1).

$$Na \rightarrow Na^+ + e^- \tag{1}$$

The formed positive sodium ions flow to the anode electrode thanks to an electrical potential through the ionic electrolyte, and the electrons flow to an external conducting system. This combination system consisted of electrons and positive sodium ions compose a stable sodium form (Eq. 2).

$$Na^+ + e^- \rightarrow Na \tag{2}$$

These reactions are reversed upon an empty state of the system occurs; hence, the energy stored is released (Figure 3) [39]. After the SIB is discharged, the sodium ions leave the anode and enter the region of the cathode. The application of SIBs is limited due to large volume expansion during low electrical conductivity, low cycle stability, the sodium ion intercalation / deintercalation process, and slower kinetic diffusion. Therefore, the selection of anode, cathode, and electrolyte is of great importance for obtaining high-performance SIBs [40–42].

3.   Anode Materials for SIB

It is necessary to connect the rechargeable sodium-ion batteries which can be used as an energy storage system in large-scale applications with a suitable anode because of their wide distribution, natural abundance and not being expensive [43,44]. Initially, hard carbon materials were used as anodic materials of SIBs. The cyclic stability and initial current yields of these anodes are low compared to LIBs [45]. Compared with metallic sodium-based batteries and Lithium metal anodes, the first has a lower specific capacity (1166 mAhg^{-1}) and higher standard reduction potential. However, metallic sodium has disadvantages such as dendrite formation on the surface and low cycle performance [46]. The use of traditional graphite in SIBs as an anode material is not suitable because of poor bonding between sodium and graphite and the small inter-layer distance [47,48]. Therefore, in recent years, anode materials with high electrical conductivity, low cost, and high sodium storage capacity attract attention in SIB [49–55].
Carbon-based Anode Materials

There are many studies on the use of carbon-based materials as anode material of SIBs [56–60]. Metal-organic frames (MOFs) with high surface area, adjustable structure, very regular pore, and uniform metal zones attract attention for the diversification of nanostructured materials such as porous carbon-based materials, metallic kalcogenic, metallic carbide, metallic phosphite, metallic oxide and their composites [61].

Carbon nanofibers (CNFs) are used as a direct anode or conductive substrates to improve the electrochemical performance of the active substances [62–64]. In a study, the porous MoSe$_2$/C composite was synthesized using the nano-casting technique together with the heat treatment. It was found that the MoSe$_2$/C composite, which was examined as an anode material for SIBs, have good storage capacity for sodium ions [65]. Additives are used to improve the cycle performance of hard carbon, the popular anode material for SIBs. As a result of the study, it has been observed that the fluoroethylene carbonate and succinic anhydride have improved the cycle performance of hard carbon symmetric cells [66]. In the SIBs, porous and spherical nano Sb@ hollow carbon microspheres (Sb@HCMs) were used as anode material. In the study, the electrochemical performance was investigated, and it was determined that Sb@HCMs electrode showed superior cycle and speed performance, and it could be used as an anode in developed SIBs [67]. When carbons, called tetraethoxysilane (TEOS) and triblock copolymer surfactant PEO140-PPO39-PEO140 (F108) used as an anode for SIB, exhibit a behavior without conductive additives [68].

3.1 Metal Oxide Anode Materials

Metal oxides with rich material resources and high theoretical capacity are used as anodes for SIBs. Composite materials obtained by combining graphene and metal oxides with composite nanostructure and composition significantly affect the performance of SIBs [69]. The electrode materials used in SIBs are limited in their applications because of their humidity and air sensitivity and high storage costs. In a study, $Na_{0.76}Mn_{0.48}Ti_{0.44}O_2$ was used as the anode of SIBs. According to the results of the study, it is stated that the charging capacity and performance show good results. Furthermore, when the storage properties were examined, it was observed that the materials had a significant performance even exposed to the environment for more than 100 days. The composite material used is promising in electrode material design for the storage and electrochemical performance of SIBs [70]. The $NiCo_2O_4$ nanotubes used as an anode for SIBs exhibit good speed capacity, high initial discharge capacity, excellent electrochemical performance, and cycle stability [71].

3.2 Alloy Anode Materials

The anode determines the operating voltage of the battery together with the cathode and the electrolyte. Therefore, it is important for the development of SIB's with a high rate of ability and superior durability [72]. Nickel tin alloy nanoparticles were synthesized and used as anode material for SIBs. As a result of the study, nickel tin-based electrodes showed significant performance over 200mAhg^{-1} at 0.1Ag^{-1} but faded over 120 cycles [73]. In a study in which the bimetallic Sb_2Te_3-TiC compound was used as an anodic material in SIBs, performance-enhancing adhesion ability, and passivation capabilities of a functional binding agent were emphasized [74]. The bimetallic single-phase nanopore SnSb alloy was synthesized and used as an anode for SIBs. The results show that the NP-SnSb alloy has exceptional conversion stability, ultra-high rate capability, and high specific capacity. In addition, it has been determined that $Na_3V_2(PO_4)_3$ cathode and NP-SnSb anode exhibit excellent electrochemical performance in cells [75]. Sulfur-doped carbon (SC) were synthesized for use as anode material for SIBs because of pyrolysis of freeze-dried agarose and sulfur powder gel. The synthesized sulfur-doped carbon showed excellent capacity and long-term cyclic stability and the storage mechanism of it was examined with the help of various characterization methods [76]. The usability of the $Bi_4Ge_3O_{12}NSs@NF$ synthesized by the solvothermal method was investigated as an anode in sodium ion batteries. The result of the study shows that $Bi_4Ge_3O_{12}NSs@NF$ is a potential candidate for next-generation energy storage devices [77].

4. Cathode Materials for SIBs

The cathode materials used determine the cyclic stability, capacity, and potential range of SIBs. Since the choice of electrolytes and additives has a significant effect on the electrochemical behavior of cathode materials, it is necessary to provide a long cycle life or to separate the electrolyte solution at the negative electrode [78]. The potential of materials used as cathodes in SIBs varies between 0-1 V vs. Na^+/Na. Cathode materials serve as the main material for sodium in SIB. Polyanionic compounds and layered oxide materials are the two basic materials used as cathode materials. High-performance cathode materials such as polyanionic compounds [79,80], metal hexacyano metallates [81], organic compounds [82], and layered oxide materials [83,84] have been developed for SIBs. Recently, researchers have obtained efficient results with the use of these materials as cathode material [46].

4.1 Layered Oxide Cathode Materials

The most popular cathodes for sodium ion batteries are layered oxides containing sodium. In a study, it is stated that the stabilized P3 type stacked electrode has excellent cycle performance, superior speed, and high energy efficiency. In addition, the main differences in the material design of lithium and sodium ion batteries were pointed out in the study [85]. It was determined that the initial reversible capacities of P2 type $Na_{0.67}[Ni_{0.167}Co_{0.167}Mn_{0.67}]_{1-x}Ti_xO_2$ ($0{\leq}x{\leq}0.4$), which was synthesized as a cathode material, decreased with increasing Ti substitution. It was reported that the material exhibited an initial discharge capacity of 138 mAhg^{-1} at 20 mAhg^{-1} between 2 and 4.5 V when x = 0.2 and provided a capacity of 89.4% after 300 cycles [86]. Metal oxide materials are promising cathode materials due to their compositions and high specific capacities. In particular, metal oxide materials having a lower material cost, lower toxicity, and simple synthesis method have great potential for completing or replacing SIBs [87]. In a study in which lithium doped $Na_{0.67}Li_{0.1}Fe_{0.4}Mn_{0.5}O_2$ microspheres were synthesized, improved cycle stability and rate performance were obtained with dual modification morphology control and lithium additive. The results showed that the applied modification enhanced the electrochemical performance of the P2-type Fe / Mn-based oxide [88]. Oxides containing sodium are the most popular cathodes for SIB. In a study, it is stated that the stabilized P3 type stacked electrode has excellent cycle performance superior rate and high energy efficiency. In addition, the main differences in the material design of lithium and sodium ion batteries were noted [85]. $Na_{0.9}Ni_{0.45}Ti_{0.55}O_2$ was synthesized as a novel bipolar material for SIB. In the study, the material used as anode and cathode provides an idea about the design and synthesis of new electrode materials for SIBs and some other applications [89]. β-V_2O_5 produced at high temperature, and the pressure was investigated as cathode material for SIBs. It is stated that the phase transformation increases the electrochemical performance by decreasing the charge-transfer resistance [90]. P2 type $Na_{0.67}Fe_{0.5}Mn_{0.5}O_2$ with economical and environmentally friendly metal elements is used as a cathode for SIBs. In the study, the P2 phase was transformed into O3 phase to improve the cyclic characteristics. It is stated that the new composition exhibits an exceptional rate and cycle performance [91].

4.2 Polyanionic Cathode Materials

Polyanionic compounds, as compared to layered oxides, are one of the major cathode active agents that are structurally stable and support their cycle life. Moreover, they are thermally stable and against oxidation at high potentials because of the strong covalent bonding of the oxygen atom. The weakness of electronic conductivity limits the electrochemical properties of these compounds. Reduction of particle size or carbon coating of particle surfaces increases the kinetic performance of sodium storage [87]. Phosphates, pyrophosphates, fluorophosphates, and sulfates are a variety of polyanionic compounds. The layered $NaVOPO_4$ synthesized as a cathode material for SIB exhibited high discharge capacity, high voltage, and cyclability with 67% over 1000 cycles. It was reported that the results were due to high Na^+ ion diffusion rate and the behavior of sodiation/desodiation [92]. Qi et al. have synthesized nanosized $Na_3(VO_{1-x}PO_4)_2F_{1+2x}$ ($0 \leq x \leq 1$) material as a cathode material for SIBler by solvothermal method at low temperatures. It was determined that this compound exhibited the best sodium storage performance with long cycle stability and high-speed capacity over 1200 cycles [93]. Rangasamy et al. investigated the potential of the carbon-coated Na_2CoSiO_4 compound as a cathode material for SIB. The performance of Na2CoSiO4, which had lower polarization compared to Na_2FeSiO_4 and Na_2MnSiO_4, increased about 50% when multi-walled carbon nanotubes were included [94]. NaFeSO$_4$F-CNT, which is prepared by solid-state technique, has a capacity of about $110mAg^{-1}$ at 0.1 C as a cathode material and it is reported that this material is a high sulfate-based cathode for SIBs [95]. $Na_{2+2x}Fe_{2-x}(SO_4)_3$ with low electronic conductivity was synthesized by spray drying method with graphene oxide composites. It has been reported that the synthesized material exhibits excellent ionic and electronic conductivity and has powerful electrochemical properties [96]. Co was added to Na_2FePO_4F, a cathode material in SIBs to improve the electrochemical properties of it and it was found that Co-doped-Na_2FePO_4F can be used in applications. Furthermore, it was compared to $Na_2Fe_{0.94}Co_{0.06}PO_4F/C$ synthesized by the sol-gel method to determine the validity of the metal doping [97]. Graphene layers are coated to improve the electrochemical performance of the NaVPO4F cathode. It has determined that the electrochemical properties of NaVPO4F composite, detected to be covered with graphene by surface imaging techniques have improved. In addition, EIS measurements showed that sodium-ion diffusion was faster, and its resistance was lower [98]. As a half-cell cathode, the NVP / C porous microspheres have excellent speed capability and superior cyclic stability. In the study, an approach is proposed to improve the electrochemical performance of battery materials [99]. Xu et al. investigated the morphology and physicochemical properties of nano and micro sized Vanadium fluorophosphate $Na_3V_2(PO_4)_2O_2F$, which is a high voltage cathode material. In addition, it is stated that battery and speed capacity depends on cathode morphology. The causes of deterioration of cycle performance in the form of interconnection reactions and wide operating voltage windows are discussed by taking into consideration the battery charge discharge behavior, electrochemical impedance analysis, in situ morphology measurements and material models [100]. As a summary, nanomaterials were used in variety of applications such as hydrogen storage, fuel cells, batteries, sensors etc [101–118].

Conclusion

SIBs are promising candidates for both portable and stationary energy storage because of concerns about the cost of lithium and the sustainability of resources. SIBs are attracting great attention because of their low costs for potential applications in rich sodium resources and large-scale energy stores. The use of advanced anode and cathode materials with exceptional electrochemical performance is very important for next-generation rechargeable batteries. In addition, the selection and structural design of anode and cathode materials are important for the development of high-performance batteries.

References

[1] A. Manthiram, A. Vadivel Murugan, A. Sarkar, T. Muraliganth, Nanostructured electrode materials for electrochemical energy storage and conversion, Energy Environ. Sci. 1 (2008) 621. doi:10.1039/b811802g.

[2] M. Winter, R.J. Brodd*, What Are Batteries, Fuel Cells, and Supercapacitors?,

(2004). doi:10.1021/CR020730K.

[3] D. Kundu, E. Talaie, V. Duffort, L.F. Nazar, The Emerging Chemistry of Sodium Ion Batteries for Electrochemical Energy Storage, Angew. Chemie Int. Ed. 54 (2015) 3431–3448. doi:10.1002/anie.201410376.

[4] M.S. Islam, C.A.J. Fisher, Lithium and sodium battery cathode materials: computational insights into voltage, diffusion and nanostructural properties, Chem. Soc. Rev. 43 (2014) 185–204. doi:10.1039/C3CS60199D.

[5] V. Palomares, M. Casas-Cabanas, E. Castillo-Martínez, M.H. Han, T. Rojo, Update on Na-based battery materials. A growing research path, Energy Environ. Sci. 6 (2013) 2312. doi:10.1039/c3ee41031e.

[6] W. Zhou, Y. Li, S. Xin, J.B. Goodenough, Rechargeable Sodium All-Solid-State Battery, ACS Cent. Sci. 3 (2017) 52–57. doi:10.1021/acscentsci.6b00321.

[7] Y. Li, Y. Lu, C. Zhao, Y.-S. Hu, M.-M. Titirici, H. Li, X. Huang, L. Chen, Recent advances of electrode materials for low-cost sodium-ion batteries towards practical application for grid energy storage, Energy Storage Mater. 7 (2017) 130–151. doi:10.1016/J.ENSM.2017.01.002.

[8] S. Guo, H. Yu, P. Liu, Y. Ren, T. Zhang, M. Chen, M. Ishida, H. Zhou, High-performance symmetric sodium-ion batteries using a new, bipolar O3-type material, $Na_{0.8}Ni_{0.4}Ti_{0.6}O_2$, Energy Environ. Sci. 8 (2015) 1237–1244. doi:10.1039/C4EE03361B.

[9] Y.-L. Ding, P. Kopold, K. Hahn, P.A. van Aken, J. Maier, Y. Yu, A Lamellar Hybrid Assembled from Metal Disulfide Nanowall Arrays Anchored on a Carbon Layer: In Situ Hybridization and Improved Sodium Storage, Adv. Mater. 28 (2016) 7774–7782. doi:10.1002/adma.201602009.

[10]   X. Xiong, G. Wang, Y. Lin, Y. Wang, X. Ou, F. Zheng, C. Yang, J.-H. Wang, M. Liu, Enhancing Sodium Ion Battery Performance by Strongly Binding Nanostructured Sb_2S_3 on Sulfur-Doped Graphene Sheets, ACS Nano. 10 (2016) 10953–10959. doi:10.1021/acsnano.6b05653.

[11]   J. Xu, J. Ma, Q. Fan, S. Guo, S. Dou, Recent Progress in the Design of Advanced Cathode Materials and Battery Models for High-Performance Lithium-X ($X = O_2$, S, Se, Te, I_2, Br_2) Batteries, Adv. Mater. 29 (2017) 1606454. doi:10.1002/adma.201606454.

[12]   W. Luo, F. Shen, C. Bommier, H. Zhu, X. Ji, L. Hu, Na-Ion Battery Anodes:

Materials and Electrochemistry, Acc. Chem. Res. 49 (2016) 231–240. doi:10.1021/acs.accounts.5b00482.

[13] J. Mao, T. Zhou, Y. Zheng, H. Gao, H. kun Liu, Z. Guo, Two-dimensional nanostructures for sodium-ion battery anodes, J. Mater. Chem. A. 6 (2018) 3284–3303. doi:10.1039/C7TA10500B.

[14] J. Xu, Y. Dou, Z. Wei, J. Ma, Y. Deng, Y. Li, H. Liu, S. Dou, Recent Progress in Graphite Intercalation Compounds for Rechargeable Metal (Li, Na, K, Al)-Ion Batteries, Adv. Sci. 4 (2017) 1700146. doi:10.1002/advs.201700146.

[15] Q. Wang, J. Xu, W. Zhang, M. Mao, Z. Wei, L. Wang, C. Cui, Y. Zhu, J. Ma, Research progress on vanadium-based cathode materials for sodium ion batteries, J. Mater. Chem. A. 6 (2018) 8815–8838. doi:10.1039/C8TA01627E.

[16] P. Simon, Y. Gogotsi, Capacitive Energy Storage in Nanostructured Carbon–Electrolyte Systems, Acc. Chem. Res. 46 (2013) 1094–1103. doi:10.1021/ar200306b.

[17] W. Li, J. Liu, D. Zhao, Mesoporous materials for energy conversion and storage devices, Nat. Rev. Mater. 1 (2016) 16023. doi:10.1038/natrevmats.2016.23.

[18] R. Raccichini, A. Varzi, S. Passerini, B. Scrosati, The role of graphene for electrochemical energy storage, Nat. Mater. 14 (2015) 271–279. doi:10.1038/nmat4170.

[19] F. Bonaccorso, L. Colombo, G. Yu, M. Stoller, V. Tozzini, A.C. Ferrari, R.S. Ruoff, V. Pellegrini, Graphene, related two-dimensional crystals, and hybrid systems for energy conversion and storage, Science (80-.). 347 (2015) 1246501. doi:10.1126/science.1246501.

[20] E. Pomerantseva, Y. Gogotsi, Two-dimensional heterostructures for energy storage, Nat. Energy. 2 (2017) 17089. doi:10.1038/nenergy.2017.89.

[21] P.G. Bruce, B. Scrosati, J.-M. Tarascon, Nanomaterials for Rechargeable Lithium Batteries, Angew. Chemie Int. Ed. 47 (2008) 2930–2946. doi:10.1002/anie.200702505.

[22] C. Liu, F. Li, L.-P. Ma, H.-M. Cheng, Advanced Materials for Energy Storage, Adv. Mater. 22 (2010) E28–E62. doi:10.1002/adma.200903328.

[23] D. Chao, C. Zhu, P. Yang, X. Xia, J. Liu, J. Wang, X. Fan, S. V. Savilov, J. Lin, H.J. Fan, Z.X. Shen, Array of nanosheets render ultrafast and high-capacity Na-ion storage by tunable pseudocapacitance, Nat. Commun. 7 (2016) 12122. doi:10.1038/ncomms12122.

[24] C. Fouassier, C. Delmas, P. Hagenmuller, Evolution structurale et proprietes physiques des phases AXMO2 (A = Na, K; M = Cr, Mn, Co) (x ⩽ 1), Mater. Res. Bull. 10 (1975) 443–449. doi:10.1016/0025-5408(75)90166-X.

[25] J.J. Braconnier, C. Delmas, P. Hagenmuller, Etude par desintercalation electrochimique des systemes NaxCrO2 et NaxNiO2, Mater. Res. Bull. 17 (1982) 993–1000. doi:10.1016/0025-5408(82)90124-6.

[26] M.S. Whittingham, Chemistry of intercalation compounds: Metal guests in chalcogenide hosts, Prog. Solid State Chem. 12 (1978) 41–99. doi:10.1016/0079-6786(78)90003-1.

[27] A.S. Nagelberg, W.L. Worrell, A thermodynamic study of sodium-intercalated TaS2 and TiS2, J. Solid State Chem. 29 (1979) 345–354. doi:10.1016/0022-4596(79)90191-9.

[28] C. Delmas, J. Braconnier, C. Fouassier, P. Hagenmuller, Electrochemical intercalation of sodium in NaxCoO2 bronzes, Solid State Ionics. 3–4 (1981) 165–169. doi:10.1016/0167-2738(81)90076-X.

[29] J. Molenda, C. Delmas, P. Hagenmuller, Electronic and electrochemical properties of NaxCoO2−y cathode, Solid State Ionics. 9–10 (1983) 431–435. doi:10.1016/0167-2738(83)90271-0.

[30] L.W. Shacklette, T.R. Jow, L. Townsend, Rechargeable Electrodes from Sodium Cobalt Bronzes, J. Electrochem. Soc. 135 (1988) 2669. doi:10.1149/1.2095407.

[31] J.M. Tarascon, G.W. Hull, Sodium intercalation into the layer oxides NaxMo2O4, Solid State Ionics. 22 (1986) 85–96. doi:10.1016/0167-2738(86)90062-7.

[32] T.R. Jow, L.W. Shacklette, M. Maxfield, D. Vernick, The Role of Conductive Polymers in Alkali-Metal Secondary Electrodes, J. Electrochem. Soc. 134 (1987) 1730. doi:10.1149/1.2100746.

[33] K. West, B. Zachau-Christiansen, T. Jacobsen, S. Skaarup, Sodium insertion in vanadium oxides, Solid State Ionics. 28–30 (1988) 1128–1131. doi:10.1016/0167-2738(88)90343-8.

[34] L. Wu, H. Lu, L. Xiao, X. Ai, H. Yang, Y. Cao, Improved sodium-storage performance of stannous sulfide@reduced graphene oxide composite as high capacity anodes for sodium-ion batteries, J. Power Sources. 293 (2015) 784–789. doi:10.1016/J.JPOWSOUR.2015.06.015.

[35] M. Armand, J.-M. Tarascon, Building better batteries, Nature. 451 (2008) 652–

657. doi:10.1038/451652a.

[36] J. Tang, A.D. Dysart, V.G. Pol, Advancement in sodium-ion rechargeable batteries, Curr. Opin. Chem. Eng. 9 (2015) 34–41. doi:10.1016/j.coche.2015.08.007.

[37] Shoshen Umeed Maaroof Alı, Ulusal Tez Merkezi, Yüksek Öğretim Kurumu. (2018) 63. https://tez.yok.gov.tr/UlusalTezMerkezi/tezSorguSonucYeni.jsp (accessed July 9, 2019).

[38] K. Kubota, S. Komaba, Review—Practical Issues and Future Perspective for Na-Ion Batteries, J. Electrochem. Soc. 162 (2015) A2538–A2550. doi:10.1149/2.0151514jes.

[39] R. Mogensen, D. Brandell, R. Younesi, Solubility of the Solid Electrolyte Interphase (SEI) in Sodium Ion Batteries, ACS Energy Lett. 1 (2016) 1173–1178. doi:10.1021/acsenergylett.6b00491.

[40] J.-Y. Hwang, S.-T. Myung, Y.-K. Sun, Sodium-ion batteries: present and future, Chem. Soc. Rev. 46 (2017) 3529–3614. doi:10.1039/C6CS00776G.

[41] X. Chen, K. Du, Y. Lai, G. Shang, H. Li, Z. Xiao, Y. Chen, J. Li, Z. Zhang, In-situ carbon-coated Na 2 FeP 2 O 7 anchored in three-dimensional reduced graphene oxide framework as a durable and high-rate sodium-ion battery cathode, J. Power Sources. 357 (2017) 164–172. doi:10.1016/j.jpowsour.2017.04.075.

[42] C. Zhu, F. Xu, H. Min, Y. Huang, W. Xia, Y. Wang, Q. Xu, P. Gao, L. Sun, Identifying the Conversion Mechanism of NiCo 2 O 4 during Sodiation-Desodiation Cycling by In Situ TEM, Adv. Funct. Mater. 27 (2017) 1606163. doi:10.1002/adfm.201606163.

[43] K. Zhang, Z. Hu, X. Liu, Z. Tao, J. Chen, FeSe 2 Microspheres as a High-Performance Anode Material for Na-Ion Batteries, Adv. Mater. 27 (2015) 3305–3309. doi:10.1002/adma.201500196.

[44] H. Yu, Y. Ren, D. Xiao, S. Guo, Y. Zhu, Y. Qian, L. Gu, H. Zhou, An ultrastable anode for long-life room-temperature sodium-ion batteries., Angew. Chem. Int. Ed. Engl. 53 (2014) 8963–9. doi:10.1002/anie.201404549.

[45] J. Qian, X. Wu, Y. Cao, X. Ai, H. Yang, High Capacity and Rate Capability of Amorphous Phosphorus for Sodium Ion Batteries, Angew. Chemie Int. Ed. 52 (2013) 4633–4636. doi:10.1002/anie.201209689.

[46] M.D. Slater, D. Kim, E. Lee, C.S. Johnson, Sodium-Ion Batteries, Adv. Funct. Mater. 23 (2013) 947–958. doi:10.1002/adfm.201200691.

[47]   U. Halim, C.R. Zheng, Y. Chen, Z. Lin, S. Jiang, R. Cheng, Y. Huang, X. Duan, A rational design of cosolvent exfoliation of layered materials by directly probing liquid–solid interaction, Nat. Commun. 4 (2013) 2213. doi:10.1038/ncomms3213.

[48]   L. Shi, T. Zhao, Recent advances in inorganic 2D materials and their applications in lithium and sodium batteries, J. Mater. Chem. A. 5 (2017) 3735–3758. doi:10.1039/C6TA09831B.

[49]   J. Shi, Q. Ji, Z. Liu, Y. Zhang, Recent Advances in Controlling Syntheses and Energy Related Applications of MX_2 and MX_2/Graphene Heterostructures, Adv. Energy Mater. 6 (2016) 1600459. doi:10.1002/aenm.201600459.

[50]   C. Guerrero-Bermea, L.P. Rajukumar, A. Dasgupta, Y. Lei, Y. Hashimoto, S. Sepulveda-Guzman, R. Cruz-Silva, M. Endo, M. Terrones, Two-dimensional and three-dimensional hybrid assemblies based on graphene oxide and other layered structures: A carbon science perspective, Carbon N. Y. 125 (2017) 437–453. doi:10.1016/j.carbon.2017.09.082.

[51]   S. Wu, Y. Du, S. Sun, Transition metal dichalcogenide based nanomaterials for rechargeable batteries, Chem. Eng. J. 307 (2017) 189–207. doi:10.1016/J.CEJ.2016.08.044.

[52]   M. Pumera, Z. Sofer, A. Ambrosi, Layered transition metal dichalcogenides for electrochemical energy generation and storage, J. Mater. Chem. A. 2 (2014) 8981–8987. doi:10.1039/C4TA00652F.

[53]   C.N.R. Rao, H.S.S. Ramakrishna Matte, U. Maitra, Graphene Analogues of Inorganic Layered Materials, Angew. Chemie Int. Ed. 52 (2013) 13162–13185. doi:10.1002/anie.201301548.

[54]   M. Samadi, N. Sarikhani, M. Zirak, H. Zhang, H.-L. Zhang, A.Z. Moshfegh, Group 6 transition metal dichalcogenide nanomaterials: synthesis, applications and future perspectives, Nanoscale Horizons. 3 (2018) 90–204. doi:10.1039/C7NH00137A.

[55]   W. Kang, Y. Wang, J. Xu, Recent progress in layered metal dichalcogenide nanostructures as electrodes for high-performance sodium-ion batteries, J. Mater. Chem. A. 5 (2017) 7667–7690. doi:10.1039/C7TA00003K.x"

[56]   Y. Wen, K. He, Y. Zhu, F. Han, Y. Xu, I. Matsuda, Y. Ishii, J. Cumings, C. Wang, Expanded graphite as superior anode for sodium-ion batteries, Nat. Commun. 5 (2014) 4033. doi:10.1038/ncomms5033.

[57]   Y. Liu, F. Fan, J. Wang, Y. Liu, H. Chen, K.L. Jungjohann, Y. Xu, Y. Zhu, D. Bigio, T. Zhu, C. Wang, In Situ Transmission Electron Microscopy Study of Electrochemical Sodiation and Potassiation of Carbon Nanofibers, Nano Lett. 14 (2014) 3445–3452. doi:10.1021/nl500970a.

[58]   H. Hou, C.E. Banks, M. Jing, Y. Zhang, X. Ji, Carbon Quantum Dots and Their Derivative 3D Porous Carbon Frameworks for Sodium-Ion Batteries with Ultralong Cycle Life, Adv. Mater. 27 (2015) 7861–7866. doi:10.1002/adma.201503816.

[59]   W. Li, M. Zhou, H. Li, K. Wang, S. Cheng, K. Jiang, A high performance sulfur-doped disordered carbon anode for sodium ion batteries, Energy Environ. Sci. 8 (2015) 2916–2921. doi:10.1039/C5EE01985K.

[60]   H. Hou, L. Shao, Y. Zhang, G. Zou, J. Chen, X. Ji, Large-Area Carbon Nanosheets Doped with Phosphorus: A High-Performance Anode Material for Sodium-Ion Batteries., Adv. Sci. (Weinheim, Baden-Wurttemberg, Ger. 4 (2017) 1600243. doi:10.1002/advs.201600243.

[61]   M. Zhong, L. Kong, N. Li, Y.-Y. Liu, J. Zhu, X.-H. Bu, Synthesis of MOF-derived nanostructures and their applications as anodes in lithium and sodium ion batteries, Coord. Chem. Rev. 388 (2019) 172–201. doi:10.1016/j.ccr.2019.02.029.

[62]   M. Wang, Z. Yang, W. Li, L. Gu, Y. Yu, Superior Sodium Storage in 3D Interconnected Nitrogen and Oxygen Dual-Doped Carbon Network, Small. 12 (2016) 2559–2566. doi:10.1002/smll.201600101.

[63]   W. Li, L. Zeng, Z. Yang, L. Gu, J. Wang, X. Liu, J. Cheng, Y. Yu, Free-standing and binder-free sodium-ion electrodes with ultralong cycle life and high rate performance based on porous carbon nanofibers, Nanoscale. 6 (2014) 693–698. doi:10.1039/C3NR05022J.

[64]   Y. Zhu, X. Han, Y. Xu, Y. Liu, S. Zheng, K. Xu, L. Hu, C. Wang, Electrospun Sb/C Fibers for a Stable and Fast Sodium-Ion Battery Anode, ACS Nano. 7 (2013) 6378–6386. doi:10.1021/nn4025674.

[65]   F. Qin, H. Hu, Y. Jiang, K. Zhang, Z. Fang, Y. Lai, J. Li, Mesoporous MoSe2/C composite as anode material for sodium/lithium ion batteries, J. Electroanal. Chem. 823 (2018) 67–72. doi:10.1016/J.JELECHEM.2018.05.023.

[66]   D.-H. Kim, B. Kang, H. Lee, Comparative study of fluoroethylene carbonate and succinic anhydride as electrolyte additive for hard carbon anodes of Na-ion batteries, J. Power Sources. 423 (2019) 137–143. doi:10.1016/j.jpowsour.2019.03.047.

[67] W. Meng, M. Guo, X. Liu, J. Chen, Z. Bai, Z. Wang, Spherical nano Sb@HCMs as high-rate and superior cycle performance anode material for sodium-ion batteries, J. Alloys Compd. 795 (2019) 141–150. doi:10.1016/j.jallcom.2019.04.285.

[68] V. Velez, G. Ramos-Sánchez, B. Lopez, L. Lartundo-Rojas, I. González, L. Sierra, Synthesis of novel hard mesoporous carbons and their applications as anodes for Li and Na ion batteries, Carbon N. Y. 147 (2019) 214–226. doi:10.1016/J.CARBON.2019.02.083.

[69] L. Wang, Z. Wei, M. Mao, H. Wang, Y. Li, J. Ma, Metal oxide/graphene composite anode materials for sodium-ion batteries, Energy Storage Mater. 16 (2019) 434–454. doi:10.1016/j.ensm.2018.06.027.

[70] R. Zhang, Y. Cui, W. Fan, G. He, X. Liu, Ambient stable Na0.76Mn0.48Ti0.44O2 as anode for Na-ion battery, Electrochim. Acta. 295 (2019) 181–186. doi:10.1016/j.electacta.2018.10.126.

[71] L. Li, Y. Ding, D. Yu, L. Li, S. Ramakrishna, S. Peng, Electrospun NiCo2O4 nanotubes as anodes for Li- and Na-ion batteries, J. Alloys Compd. 777 (2019) 1286–1293. doi:10.1016/J.JALLCOM.2018.11.115.

[72] L. Wang, J. Światowska, S. Dai, M. Cao, Z. Zhong, Y. Shen, M. Wang, Promises and challenges of alloy-type and conversion-type anode materials for sodium–ion batteries, Mater. Today Energy. 11 (2019) 46–60. doi:10.1016/J.MTENER.2018.10.017.

[73] J. Li, X. Xu, Z. Luo, C. Zhang, X. Yu, Y. Zuo, T. Zhang, P. Tang, J. Arbiol, J. Llorca, J. Liu, A. Cabot, Compositionally tuned NixSn alloys as anode materials for lithium-ion and sodium-ion batteries with a high pseudocapacitive contribution, Electrochim. Acta. 304 (2019) 246–254. doi:10.1016/J.ELECTACTA.2019.02.098.

[74] V.M. Nagulapati, Y.H. Yoon, D.S. Kim, H. Kim, W.S. Lee, J.H. Lee, K.H. Kim, J. Hur, I.T. Kim, S.G. Lee, Effect of binders and additives to tailor the electrochemical performance of Sb2Te3-TiC alloy anodes for high-performance sodium-ion batteries, J. Ind. Eng. Chem. 76 (2019) 419–428. doi:10.1016/J.JIEC.2019.04.008.

[75] W. Ma, K. Yin, H. Gao, J. Niu, Z. Peng, Z. Zhang, Alloying boosting superior sodium storage performance in nanoporous tin-antimony alloy anode for sodium ion batteries, Nano Energy. 54 (2018) 349–359. doi:10.1016/J.NANOEN.2018.10.027.

[76] P. Feng, W. Wang, K. Wang, S. Cheng, K. Jiang, A high-performance carbon with sulfur doped between interlayers and its sodium storage mechanism as anode material for sodium ion batteries, J. Alloys Compd. 795 (2019) 223–232.

doi:10.1016/J.JALLCOM.2019.04.338.

[77]   J. Xu, W. Wei, X. Zhang, L. Liang, M. Xu, Lotus-stalk Bi4Ge3O12 as binder-free anode for lithium and sodium ion batteries, Chinese Chem. Lett. 30 (2019) 1341–1345. doi:10.1016/J.CCLET.2019.03.005.

[78]   N. Yabuuchi, K. Kubota, M. Dahbi, S. Komaba, Research Development on Sodium-Ion Batteries, Chem. Rev. 114 (2014) 11636–11682. doi:10.1021/cr500192f.

[79]   J.B. Goodenough, H..-P. Hong, J.A. Kafalas, Fast Na+-ion transport in skeleton structures, Mater. Res. Bull. 11 (1976) 203–220. doi:10.1016/0025-5408(76)90077-5.

[80]   W. Song, X. Ji, Z. Wu, Y. Yang, Z. Zhou, F. Li, Q. Chen, C.E. Banks, Exploration of ion migration mechanism and diffusion capability for Na3V2(PO4)2F3 cathode utilized in rechargeable sodium-ion batteries, J. Power Sources. 256 (2014) 258–263. doi:10.1016/J.JPOWSOUR.2014.01.025.

[81]   Q. Yang, W. Wang, H. Li, J. Zhang, F. Kang, B. Li, Investigation of iron hexacyanoferrate as a high rate cathode for aqueous batteries: Sodium-ion batteries and lithium-ion batteries, Electrochim. Acta. 270 (2018) 96–103. doi:10.1016/J.ELECTACTA.2018.02.171.

[82]   T. Huang, D. Lu, L. Ma, X. Xi, R. Liu, D. Wu, A hit-and-run strategy towards perylene diimide/reduced graphene oxide as high performance sodium ion battery cathode, Chem. Eng. J. 349 (2018) 66–71. doi:10.1016/J.CEJ.2018.05.078.

[83]   G.K. Veerasubramani, Y. Subramanian, M.-S. Park, B. Senthilkumar, A. Eftekhari, S.J. Kim, D.-W. Kim, Enhanced sodium-ion storage capability of P2/O3 biphase by Li-ion substitution into P2-type Na0.5Fe0.5Mn0.5O2 layered cathode, Electrochim. Acta. 296 (2019) 1027–1034. doi:10.1016/J.ELECTACTA.2018.11.160.

[84]   X. Song, T. Meng, Y. Deng, A. Gao, J. Nan, D. Shu, F. Yi, The effects of the functional electrolyte additive on the cathode material Na0.76Ni0.3Fe0.4Mn0.3O2 for sodium-ion batteries, Electrochim. Acta. 281 (2018) 370–377. doi:10.1016/J.ELECTACTA.2018.05.185.

[85]   S. Guo, Y. Sun, P. Liu, J. Yi, P. He, X. Zhang, Y. Zhu, R. Senga, K. Suenaga, M. Chen, H. Zhou, Cation-mixing stabilized layered oxide cathodes for sodium-ion batteries, Sci. Bull. 63 (2018) 376–384. doi:10.1016/J.SCIB.2018.02.012.

[86]   S. Bao, S. Luo, Z. Wang, S. Yan, Q. Wang, Improving the electrochemical performance of layered cathode oxide for sodium-ion batteries by optimizing the titanium content, J. Colloid Interface Sci. 544 (2019) 164–171.

doi:10.1016/J.JCIS.2019.02.094.

[87] Y. Lyu, Y. Liu, Z.-E. Yu, N. Su, Y. Liu, W. Li, Q. Li, B. Guo, B. Liu, Recent advances in high energy-density cathode materials for sodium-ion batteries, Sustain. Mater. Technol. 21 (2019) e00098. doi:10.1016/J.SUSMAT.2019.E00098.

[88] K. Tang, Y. Wang, X. Zhang, S. Jamil, Y. Huang, S. Cao, X. Xie, Y. Bai, X. Wang, Z. Luo, G. Chen, High-performance P2-Type Fe/Mn-based oxide cathode materials for sodium-ion batteries, Electrochim. Acta. 312 (2019) 45–53. doi:10.1016/J.ELECTACTA.2019.04.183.

[89] X. Li, S. Zhou, Q. Wu, X. Wang, T. Yao, Y. Zhang, B. Song, Na0.9Ni0.45Ti0.55O2 as novel bipolar material for sodium ion batteries, Solid State Ionics. 334 (2019) 14–20. doi:10.1016/J.SSI.2019.01.033.

[90] R. Córdoba, A. Kuhn, J.C. Pérez-Flores, E. Morán, J.M. Gallardo-Amores, F. García-Alvarado, Sodium insertion in high pressure β-V2O5: A new high capacity cathode material for sodium ion batteries, J. Power Sources. 422 (2019) 42–48. doi:10.1016/J.JPOWSOUR.2019.03.018.

[91] D. Zhou, W. Huang, X. Lv, F. Zhao, A novel P2/O3 biphase Na0.67Fe0.425Mn0.425Mg0.15O2 as cathode for high-performance sodium-ion batteries, J. Power Sources. 421 (2019) 147–155. doi:10.1016/J.JPOWSOUR.2019.02.061.

[92] Y. Fang, Q. Liu, L. Xiao, Y. Rong, Y. Liu, Z. Chen, X. Ai, Y. Cao, H. Yang, J. Xie, C. Sun, X. Zhang, B. Aoun, X. Xing, X. Xiao, Y. Ren, A Fully Sodiated NaVOPO4 with Layered Structure for High-Voltage and Long-Lifespan Sodium-Ion Batteries, Chem. 4 (2018) 1167–1180. doi:10.1016/J.CHEMPR.2018.03.006.

[93] Y. Qi, L. Mu, J. Zhao, Y.-S. Hu, H. Liu, S. Dai, Superior Na-Storage Performance of Low-Temperature-Synthesized Na$_3$(VO$_{1-x}$PO$_4$)$_2$F$_{1+2x}$ $(0 \leq x \leq 1)$ Nanoparticles for Na-Ion Batteries, Angew. Chemie Int. Ed. 54 (2015) 9911–9916. doi:10.1002/anie.201503188.

[94] V.S. Rangasamy, S. Thayumanasundaram, J.-P. Locquet, Solvothermal synthesis and electrochemical properties of Na2CoSiO4 and Na2CoSiO4/carbon nanotube cathode materials for sodium-ion batteries, Electrochim. Acta. 276 (2018) 102–110. doi:10.1016/J.ELECTACTA.2018.04.166.

[95] R. Rajagopalan, Z. Wu, Y. Liu, S. Al-Rubaye, E. Wang, C. Wu, W. Xiang, B. Zhong, X. Guo, S.X. Dou, H.K. Liu, A novel high voltage battery cathodes of Fe2+/Fe3+ sodium fluoro sulfate lined with carbon nanotubes for stable sodium

batteries, J. Power Sources. 398 (2018) 175–182.
doi:10.1016/J.JPOWSOUR.2018.07.066.

[96]   X. Liu, L. Tang, Q. Xu, H. Liu, Y. Wang, Ultrafast and ultrastable high voltage cathode of Na2+2xFe2-x(SO4)3 microsphere scaffolded by graphene for sodium ion batteries, Electrochim. Acta. 296 (2019) 345–354.
doi:10.1016/J.ELECTACTA.2018.11.064.

[97]   D. Jin, H. Qiu, F. Du, Y. Wei, X. Meng, Co-doped Na2FePO4F fluorophosphates as a promising cathode material for rechargeable sodium-ion batteries, Solid State Sci. 93 (2019) 62–69. doi:10.1016/J.SOLIDSTATESCIENCES.2019.04.014.

[98]   Y.-L. Ruan, K. Wang, S.-D. Song, X. Han, B.-W. Cheng, Graphene modified sodium vanadium fluorophosphate as a high voltage cathode material for sodium ion batteries, Electrochim. Acta. 160 (2015) 330–336.
doi:10.1016/J.ELECTACTA.2015.01.186.

[99]   X. Cao, A. Pan, B. Yin, G. Fang, Y. Wang, X. Kong, T. Zhu, J. Zhou, G. Cao, S. Liang, Nanoflake-constructed porous Na3V2(PO4)3/C hierarchical microspheres as a bicontinuous cathode for sodium-ion batteries applications, Nano Energy. 60 (2019) 312–323. doi:10.1016/J.NANOEN.2019.03.066.

[100] J. Xu, J. Chen, L. Tao, Z. Tian, S. Zhou, N. Zhao, C.-P. Wong, Investigation of Na3V2(PO4)2O2F as a sodium ion battery cathode material: Influences of morphology and voltage window, Nano Energy. 60 (2019) 510–519.
doi:10.1016/J.NANOEN.2019.03.063.

[101] N. Lolak, E. Kuyuldar, H. Burhan, H. Goksu, S. Akocak, F. Sen, Composites of Palladium–Nickel Alloy Nanoparticles and Graphene Oxide for the Knoevenagel Condensation of Aldehydes with Malononitrile, ACS Omega. 4 (2019) 6848–6853. doi:10.1021/acsomega.9b00485.

[102] H. Göksu, B. Çelik, Y. Yıldız, F. Şen, B. Kılbaş, Superior Monodisperse CNT-Supported CoPd (CoPd@CNT) Nanoparticles for Selective Reduction of Nitro Compounds to Primary Amines with NaBH4 in Aqueous Medium, ChemistrySelect. 1 (2016) 2366–2372. doi:10.1002/slct.201600509.

[103] ‡ and Fatih Şen†, † Gülsün Gökağaç*, Different Sized Platinum Nanoparticles Supported on Carbon: An XPS Study on These Methanol Oxidation Catalysts, (2007). doi:10.1021/JP068381B.

[104] B. Sen, S. Kuzu, E. Demir, S. Akocak, F. Sen, Highly monodisperse RuCo nanoparticles decorated on functionalized multiwalled carbon nanotube with the

Sodium-Ion Batteries: Materials and Applications   Materials Research Forum LLC
Materials Research Foundations **76** (2020) 183-204   https://doi.org/10.21741/9781644900833-8

highest observed catalytic activity in the dehydrogenation of dimethylamine–borane, Int. J. Hydrogen Energy. 42 (2017) 23292–23298. doi:10.1016/J.IJHYDENE.2017.06.032.

[105] E. Demir, A. Savk, B. Sen, F. Sen, A Novel Monodisperse Metal Nanoparticles Anchored Graphene Oxide as Counter Electrode for Dye-Sensitized Solar Cells, Nano-Structures and Nano-Objects. 12 (2017) 41–45. doi:10.1016/j.nanoso.2017.08.018.

[106] R. Ayranci, B. Demirkan, B. Sen, A. Şavk, M. Ak, F. Şen, Use of the monodisperse Pt/Ni@rGO nanocomposite synthesized by ultrasonic hydroxide assisted reduction method in electrochemical nonenzymatic glucose detection., Mater. Sci. Eng. C. Mater. Biol. Appl. 99 (2019) 951–956. doi:10.1016/j.msec.2019.02.040.

[107] B. Sen, A. Şavk, F. Sen, Highly Efficient Monodisperse Pt Nanoparticles Confined in The Carbon Black Hybrid Material for Hydrogen Liberation, J. Colloid Interface Sci. 520 (2018) 112–118. doi:10.1016/j.jcis.2018.03.004.

[108] S. Ertan, F. Şen, S. Şen, G. Gökağaç, Platinum nanocatalysts prepared with different surfactants for C1–C3 alcohol oxidations and their surface morphologies by AFM, J. Nanoparticle Res. 14 (2012) 922–934. doi:10.1007/s11051-012-0922-5.

[109] B. Şen, B. Demirkan, A. Savk, R. Kartop, M.S. Nas, M.H. Alma, S. Sürdem, F. Şen, High-performance graphite-supported ruthenium nanocatalyst for hydrogen evolution reaction, J. Mol. Liq. 268 (2018) 807–812. doi:10.1016/j.molliq.2018.07.117.

[110] R. Ayranci, G. Baskaya, M. Guzel, S. Bozkurt, M. Ak, A. Savk, F. Sen, Activated Carbon Furnished Monodisperse Pt Nanocomposites as a Superior Adsorbent for Methylene Blue Removal from Aqueous Solutions, Nano-Structures and Nano-Objects. 11 (2017) 13–19. doi:10.1016/j.nanoso.2017.05.008.

[111] R. Ayranci, G. Baskaya, M. Guzel, S. Bozkurt, M. Ak, A. Savk, F. Sen, Enhanced optical and electrical properties of PEDOT via nanostructured carbon materials: A comparative investigation, Nano-Structures & Nano-Objects. 11 (2017) 13–19. doi:10.1016/j.nanoso.2017.05.008.

[112] B. Şen, A. Aygün, T.O. Okyay, A. Şavk, R. Kartop, F. Şen, Monodisperse Palladium Nanoparticles Assembled on Graphene Oxide with The High Catalytic Activity and Reusability in The Dehydrogenation of Dimethylamine-borane. International Journal of Hydrogen Energy, 43 (2018) 20176–20182. https://doi.org/10.1016/j.ij, Int. J. Hydrogen Energy. (2018).

doi:10.1016/j.ijhydene.2018.03.175.

[113] F. Sen, Y. Karatas, M. Gulcan, M. Zahmakiran, Amylamine stabilized platinum(0) nanoparticles: Active and reusable nanocatalyst in the room temperature dehydrogenation of dimethylamine-borane, RSC Adv. (2014). doi:10.1039/c3ra43701a.

[114] Y. Koskun, A. Şavk, B. Şen, F. Şen, Highly Sensitive Glucose Sensor Based on Monodisperse Palladium Nickel/Activated Carbon Nanocomposites, Anal. Chim. Acta. 1010 (2018) 37–43. doi:10.1016/j.aca.2018.01.035.

[115] S. Günbatar, A. Aygun, Y. Karataş, M. Gülcan, F. Şen, Carbon-nanotube-based Rhodium Nanoparticles as Highly-Active Catalyst for Hydrolytic Dehydrogenation of Dimethylamineborane at Room Temperature, J. Colloid Interface Sci. 530 (2018) 321–327. doi:10.1016/j.jcis.2018.06.100.

[116] B. Şen, A. Aygün, A. Şavk, S. Akocak, F. Şen, Bimetallic Palladium–iridium Alloy Nanoparticles as Highly Efficient and Stable Catalyst for The Hydrogen Evolution Reaction, Int. J. Hydrogen Energy. 43 (2018) 20183–20191. doi:10.1016/j.ijhydene.2018.07.081.

[117] B. Şen, A. Aygün, A. Şavk, S. Akocak, F. Şen, Bimetallic palladium–iridium alloy nanoparticles as highly efficient and stable catalyst for the hydrogen evolution reaction, Int. J. Hydrogen Energy. 43 (2018) 20183–20191. doi:10.1016/J.IJHYDENE.2018.07.081.

[118] Y. Yıldız, S. Kuzu, B. Sen, A. Savk, S. Akocak, F. Şen, Different ligand based monodispersed Pt nanoparticles decorated with rGO as highly active and reusable catalysts for the methanol oxidation, Int. J. Hydrogen Energy. 42 (2017) 13061–13069. doi:10.1016/j.ijhydene.2017.03.230.

Sodium-Ion Batteries: Materials and Applications   Materials Research Forum LLC
Materials Research Foundations 76 (2020) 205-228   https://doi.org/10.21741/9781644900833-9

Chapter 9

Electrolytes for Na-O₂ Batteries: Towards a Rational Design

Iñigo Lozano[1,2], Idoia Ruiz de Larramendi[2*], Nagore Ortiz-Vitoriano[1,3*]

[1]Center for Cooperative Research on Alternative Energies (CIC EnergiGUNE), Basque Research and Technology Alliance (BRTA), Parque Tecnológico de Alava, Albert Einstein 48, 01510, Vitoria-Gasteiz, Spain

[2]Departamento de Química Inorgánica, Facultad de Ciencia y Tecnología, Universidad del País Vasco (UPV/EHU) Barrio Sarriena s/n, 48940 Leioa - Bizkaia, Spain

[3]Ikerbasque, Basque Foundation for Science, María Díaz de Haro 3, 48013 Bilbao, Spain

idoia.ruizdelarramendi@ehu.eus, nortiz@cicenergigune.com

Abstract

Energy storage is a critical challenge for modern society, with batteries being the predominant technology of choice. Within this area, sodium-oxygen batteries present advantages such as low cost and high energy density. In order to facilitate their use, the development of targeted approaches to dealing with the technology's unique chemistry is required. Electrolytes, consisting of a salt and a non-aqueous solvent, are a key component of any optimized system. The parameters affecting electrolyte physicochemistry are, therefore, critical to battery performance, lifetime and safety, yet the field of non-aqueous solvation chemistry remains relatively unexplored even though it plays a critical role in applications as wide ranging as supercapacitors, batteries, catalysis and chemical synthesis.

Keywords

Na-O₂ Batteries, Electrolytes, Redox Mediators, Superoxide Stability, Singlet Oxygen

Contents

Sodium-Ion Batteries: Materials and Applications   Materials Research Forum LLC
Materials Research Foundations **76** (2020) 205-228   https://doi.org/10.21741/9781644900833-9

1.   Introduction

Among all the electrochemical systems, batteries are considered the most promising alternative for energy storage since the 19th century with the development of the first battery prototypes. Scientific interest in these devices is due to the direct energy conversion (absence of intermediate steps), high efficiency, absence of mobile parts, low pollution and ease in transportation. At the end of the 20th century, lithium ion batteries (LiBs) revolutionize the portable electronic device industry with their commercialization and implementation in mobile phones and laptops. However, new energy challenges (e.g., automotive field) have arisen which require a higher energy density than that provided by LiBs. In this sense, R&D community has focused its efforts on developing "beyond lithium" battery technologies to augment, or in certain situations replace, LiBs. In this scenario, metal-O_2 (M-O_2) batteries have the capability to play an important role in the development of electric vehicles, due to their high theoretical energy density compared to current systems [1,2].

At the beginning of the 21st century, a new technology emerges in the energy storage scenario: the Li-O_2 battery. This Li-based system presents the highest theoretical energy density among the M-O_2 batteries. Li-O_2 batteries consist of three fundamental components: an electrolyte and two electrodes (cathode and anode), where an oxygen cathode coupled to a lithium anode results in the following discharge reactions:

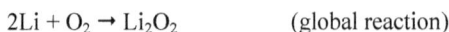

$$2Li^+ + O_2 + 2e^- \rightarrow Li_2O_2 \quad \text{(cathodic reaction)}$$

$$2Li \rightarrow 2Li^+ + 2e^- \quad \text{(anodic reaction)}$$

$$2Li + O_2 \rightarrow Li_2O_2 \quad \text{(global reaction)}$$

Throughout the discharge reaction oxygen is consumed at the cathode and during charge this O_2 is released. Therefore, they are also known as *breathing batteries*. During discharge and charge this battery exhibits poor kinetics of the oxygen reduction reaction (ORR) and the oxygen evolution reaction (OER), being the key limitations of these systems. For this reason, the development of new catalytically effective materials for both the ORR and OER has become a highly relevant research area [3,4]. Furthermore, during

Sodium-Ion Batteries: Materials and Applications   Materials Research Forum LLC
Materials Research Foundations **76** (2020) 205-228   https://doi.org/10.21741/9781644900833-9

the battery operation, several fundamental mechanisms were identified that stopped the system from achieving its potential [5]. These limitations are partially due to stability effects such as: (i) short circuits caused by the metallic Li anode during cycling (Li dendrite growth); (ii) reaction of the metallic Li with air pollutants (e.g., CO_2 and H_2O) and with certain electrolytes; (iii) side reactions due to the reactivity of oxygen with Li_2O_2; and (iv) stability of the electrolyte with reaction intermediates. Moreover, the Li-O_2 battery has shown to provide low coulombic efficiencies and poor cyclability.

In addition, the extensive use of lithium-based batteries has exposed some problems related to its abundance and geographical distribution. The problems that the specific location and scarce abundance of strategic products cause, such as oil or rare earths, are well known. In recent years, therefore, sodium-based batteries have received increasing attention from the scientific community due to their good performance and lower cost [6,7]. Na-O_2 batteries present high potential due to the replacement of Li by Na which leads to important advantages. In this way, sodium, exhibits an energy density of 1108 or 1600 Wh kg^{-1} (based on NaO_2 or Na_2O_2 discharge product, respectively, favors the stable formation of the desired discharge product (usually NaO_2), which does not occur in Li-based systems and results in a lower charge potential when compared to Li [8]. However, Na-O_2 batteries are a novel technology which requires further development before widespread commercialization is possible.

2.   Na-O_2 Batteries

The configuration of non-aqueous Na-O_2 batteries is similar to its Li-O_2 counterpart, comprising a metallic Na anode, a Na-ion conducting electrolyte and an electronic conductor cathode. During ORR the anode is electrochemically oxidized to Na$^+$ which migrates through the electrolyte to the cathode. The reaction mechanism is thought to involve the dissolution of the oxygen gas (O_2) in the electrolyte through the porous cathode (usually a carbonaceous material) in order to form the superoxide species (O_2^-).

Thus, the reaction between the Na$^+$ cations and the superoxide anion yields the formation of sodium superoxide, NaO_2, (Na$^+$ + O_2 + e$^-$ → NaO_2) as the final discharge product (shown in Fig. 1). This product is then decomposed during OER. Sodium peroxide (Na_2O_2) and sodium peroxide dihydrate ($Na_2O_2 \cdot 2H_2O$) have also been reported as other discharge products. A recent study has shown that a combination of NaO_2 and Na_2O_2 can be achieved when different basal planes of platinum single crystal are analyzed as cathode surfaces [9].

Although the theoretical energy density is lower than that reported for Li-O_2 batteries (~ 1108 Wh kg^{-1} based on NaO_2 discharge product; see Table 1), reduced overpotential (<

100 mV) than that observed in Li systems (~ 1000 mV) is attained [10]. In addition, another important benefit of Na-O_2 batteries lies in their high coulombic efficiency (> 95%) [11]. The price of replacing lithium with sodium provides a number of extra benefits: (i) the sodium salts used in the formulation of the electrolyte are more abundant comparing with the equivalent lithium salts [12], and (ii) unlike lithium, sodium is not reactive to aluminum, allowing thin aluminum foil to be used as a collector anode current, much lighter and cheaper than copper or nickel required in Li systems. Table 1 shows the main discharge products observed in Li and Na-O_2 batteries.

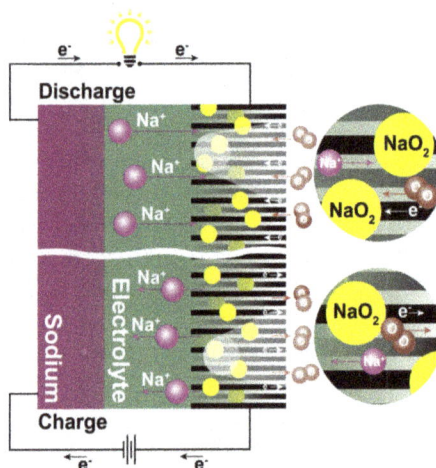

Figure 1.   Operation principle of the Na-O_2 battery.

The interest in Na-O_2 devices has unquestionably increased since 2011 when Peled et al. reported the use of Na melted anode in the first Na-O_2 battery [17]. The limitations associated with the high impedance of the cell and the formation of dendrites at the anode were overcome by analyzing the system at 100 °C, obtaining efficiencies of 85%. In 2012, the group led by Prof. Janek developed the first Na-O_2 battery at room temperature, showing cubic NaO_2 crystals as the only discharge product [18]. However, NaO_2 is not the only product mentioned in the literature. Other discharge products have been described such as Na_2O_2 or its hydrate phase, $Na_2O_2 \cdot 2H_2O$ [19,20]. This lack of agreement in the discharge product formation has motivated several works focused on clarifying such controversy. Kang et al. argued using theoretical calculations (density functional theory, DFT) that due to the similarity between the equilibrium potentials of both NaO_2 (2.27 V)

Sodium-Ion Batteries: Materials and Applications   Materials Research Forum LLC
Materials Research Foundations **76** (2020) 205-228   https://doi.org/10.21741/9781644900833-9

and Na_2O_2 (2.33 V), NaO_2 preferentially grows at the nanoscale and Na_2O_2 is the most stable in bulk phase [21]. Lee et al., however, differed from these results and predicted greater stability for the NaO_2 at 300 K and 1 atm [22]. Experimentally, Yadegari et al. obtained both discharge products (superoxide and peroxide) during ORR [23]. In their study, the charge profile revealed the presence of three plateaus, where superoxide is the first degraded compound, while oxidation of peroxide occurs at higher potentials. In contrast, the studies carried out by the Lutz, Janek and Shao-Horn groups concluded that NaO_2 was the sole product [11,24,25]. Moreover, different carbon electrodes were studied by Janek et al. finding no evidence of the Na_2O_2 formation [24]. McCloskey et al. confirmed without a doubt that both the ORR and OER are exchange processes of 1 e^-/O_2 molecule, demonstrating the formation and decomposition of NaO_2 as the only discharge product [11]. The Shao-Horn group also detected NaO_2 using carbon nanotube electrodes [25]. However, they also identified the presence of $Na_2O_2 \cdot 2H_2O$, by exposing the electrodes to air. Despite the evidence provided by this work, it cannot be ensured that the formation of the $Na_2O_2 \cdot 2H_2O$ product is always caused by water vapor. What is evident is that the system's overpotential as well as instability are smaller when the discharge product formed is NaO_2 versus $Na_2O_2 \cdot 2H_2O$. Likewise, it has been verified that discharge products are unstable in the battery environment [26]. The existing controversy regarding the discharge product formation makes clear the need to develop new methods for the *in-situ* analysis of the involved mechanisms (chemical and/or electrochemical).

Table 1.   *Main characteristics of Na-O_2 and Li-O_2 batteries* [13–16].

	Discharge product	Discharge reaction	Cell voltage (V)	Theoretical capacity (mAh g^{-1})	Energy density (Wh kg^{-1})
Na-based	NaO_2	$Na + O_2 \rightarrow NaO_2$	2.27	488	1108
	Na_2O_2	$2\,Na + O_2 \rightarrow Na_2O_2$	2.33	689	1605
	Na_2O	$2\,Na + \frac{1}{2}O_2 \rightarrow Na_2O$	1.95	867	1687
	$NaOH$	$4\,Na + O_2 + 2H_2O \rightarrow 4NaOH \cdot H_2O$	2.77	462	1281
Li-based	Li_2O_2	$2Li^+ + O_2 + 2e^- \rightarrow Li_2O_2$	2.96	1168	3458

Concerning the reaction mechanisms, Janek et al. proposed two possible explanations for the growth of NaO_2 [27]. The first route, known as solution-mediated, involves the formation of superoxide ions on the surface of the air electrode. These ions dissolve in the electrolyte where they interact with the sodium cations, giving rise to the formation of small nuclei that, as they grow, end up precipitating at the electrolyte/electrode interface forming

Sodium-Ion Batteries: Materials and Applications   Materials Research Forum LLC
Materials Research Foundations **76** (2020) 205-228   https://doi.org/10.21741/9781644900833-9

the discharge products. The second hypothesis is the surface-mediated route which implies that the ORR occurs directly at the electrolyte/electrode interface, which leads to the direct formation of the discharge products on that surface. The last mechanism requires the transport of the electrons to the NaO_2/electrolyte interface from the electrode surface. Theoretical studies, however, have revealed the insulating nature of NaO_2 [28,29]. Based on these results, the mechanism through the solution prevails as the most realistic explanation for the growth of NaO_2. The grow process has a direct relationship with the premature death of the battery. It has been found that by applying low current densities, pore blockage of the air electrode occurs due to the growth of discharge products in the form of large, cubic NaO_2 crystals [30]. This blockage is, therefore, responsible for the death of the battery. The situation changes with increasing current density, where the formation of a thin NaO_2 film occur which becomes the limiting factor for the capacity due to the passivation of the surface.

Finally, the cathode design is another important element that affects battery efficiency. During ORR the oxygen transport has been identified as the limiting factor (i.e., the discharge capacity) [18]. From this study it is possible to conclude that the areas of the electrode in direct contact with the oxygen gas during ORR will present higher discharge product concentrations, whereas less exposed regions will show negligible concentrations [31–33]. This fact emphasizes the need to optimize the design of the cathode microstructure in order to achieve an extensive use of the material and thus maximize the discharge capacity.

A clear proof of the great potential of metal-O_2 batteries is the recent (January 2017) grant of a patent to the company Tesla Motors, Inc. related to the charging technology of metal-O_2 batteries which paves the way towards a new generation of hybridization with lithium-ion batteries [34]. This new hybrid device uses the LiB in the short trips while for those requiring longer distances the metal-O_2 battery recharges the LiB.

3.   Instability of electrolyte

As already mentioned, in Na-O_2 batteries it is possible to identify different discharge products, which in turn can be generated through different mechanisms: solution or surface-mediated. The determining factor that leads to each discharge product or mechanism has not yet been fully understood, although some parameters that influence these growth mechanisms have been established [35]. In this sense, the scientific community recognizes the importance of the electrolyte on the Na-O_2 battery mechanism. For example, different studies have been published regarding the effect of different solvents on the electrochemical response [36–40]. Other groups have focused their analysis on the effect of the presence of water [41,42] or on the influence of the oxygen electrode design [20,43,44]. In fact, it has been demonstrated, by computational and experimental studies, that the nature of the electrolyte is able to control the distribution and morphology of the discharge products. Through thermodynamic calculations and the prediction of crystalline structures, it has been possible to establish the important role of the electrolyte [45]. More specifically, in the design of the correct electrolyte it is of vital importance to analyze the donor number (DN) associated with the electrolyte. The Gutmann's DN for solvents is a parameter that gives an idea of the Lewis basicity of a solvent. For a solvent with high DN will be easier to solvate the Na^+ cations, which means that will dissolve the discharge products. However, other factors contribute during the discharge product formation mechanism such as the solvation barrier that is established once the solvent captures the cation. If the solvent acts as a good chelating agent (tetraglyme (TEGDME), for example), the solvation barrier might be too high and problems to release the trapped cations in order to form the discharge products could appear [46]. In parallel, the acceptor number (AN) that is related to the ability to solvate anions by the solvent is usually analyzed along with the DN. These numbers give an idea of how the interactions between the different species present in the electrolyte take place, although it is a merely indicative value. In fact, DN values differ when looking at different studies. These values are calculated against relatively strong electron pair acceptor (Lewis acid), antimony pentachloride ($SbCl_5$) [47], and the physicochemical differences between the Sb^{5+} and Na^+ cations must be taken into account. In addition, the DN does not reflect the steric hindrance nor the chelating effect of the solvent; for example, a low DN does not imply a weaker solvation (e.g., TEGDME) [46]. Recently, a new strategy has been proposed to measure the solvating power of the electrolyte solvents in which the organic structure of the solvent is also considered [48]. Even so, most of the published studies analyze the stability of the species in the electrolyte as a function of the DN of the solvent. Wang et al. determined that a high DN (> 12.5, which is the situation of the glymes commonly used as solvents) guides the formation of NaO_2, while a DN below that limit can favor the NaO_2-Na_2O_2 mixture [45].

The effect of the presence of water on the reaction mechanism has been analyzed by several authors [16,49–53]. In the same cell, Pinedo et al. were able to form NaO_2 or $Na_2O_2 \cdot 2H_2O$ deliberately [54], by varying the oxygen gas provided to the cell. The main factor controlling the formation of the discharge product was the amount of H_2O available in the gas supply. Water impurities could accumulate over time if a flowing O_2 supply is used, and these impurities cause the formation of $Na_2O_2·2 H_2O$. Ortiz-Vitoriano et al. also detected the formation of hydrated peroxide by exposing NaO_2 to ambient atmosphere, but $Na_2O_2 \cdot 2H_2O$ was not obtained in the Na-O_2 cells, even when 6000 ppm of H_2O was added to the electrolyte [25]. Although up to date all the factors that govern the type of discharge product formed are not completely clear, there is a consensus regarding the impact of the presence of water in the electrolyte among the scientific community. Nazar et al. proposed a novel concept where the proton phase transfer catalysis (PPTC) acts as solubilizer and transporter of the superoxide [55]. They determined that the controlled presence of water in the electrolyte (up to 100 ppm) resulted in an increase in battery capacity, producing larger NaO_2 cubes. The role of remaining water in the electrolyte was evaluated and a mechanism in which oxygen is reduced at the electrode surface was further suggested. Once the oxygen is reduced, HO_2 species are formed by the reaction with the protons, leading to the growth of NaO_2 crystals through the solution. Water, therefore, is considered as a phase transfer catalyst during ORR and OER. A NaO_2 film is produced in absence of water which blocks the electrode surface, lowering the discharge capacity. This effect was attributed to the presence of H^+ in the H_2O that mediated the growth of the crystals, not as solvation but as PPTC. The formation of the HO_2 species as an intermediate (Eq. 1) can reduce the appearance of parasitic reactions, leading to greater capacity [56].

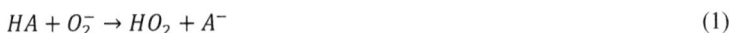

$$HA + O_2^- \rightarrow HO_2 + A^- \tag{1}$$

This PPTC mechanism implies that the proton would act as a superoxide transporter from the surface of the electrode to the electrolyte, where the superoxide would interact with the sodium cations, releasing the proton again through the metathesis reaction (Eq. 2):

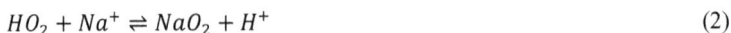

$$HO_2 + Na^+ \rightleftharpoons NaO_2 + H^+ \tag{2}$$

It is also known that H_2O can promote a solution-mediated pathway [57], although it can also degrade the reversibility of aprotic $Li-O_2$ batteries [58].

The electrolyte is also responsible for the instability of the discharge products. Several researchers have studied the composition of the discharge products, their evolution after the electrochemical reduction and their stability by different techniques [26,59–61]. The migration of the electrochemically generated superoxide radical has been identified as the factor limiting the coulombic efficiency of the cell during the first galvanostatic cycle. From the second cycle, the presence of O_2^- in the electrolyte prevents such migration; thus, there will be a higher concentration of superoxide available, increasing the efficiency of the system [26]. In this work, the great instability of NaO_2 in the cell environment was detected, observing an increase of the pressure in the cell during the resting periods after the discharge, which was identified as O_2 release by decomposition of the NaO_2 cubes [26]. Subsequently, by means of transmission X-ray microscopy (TXM) it was possible to analyze both the microstructure and the composition of the cubic-shaped discharge products, observing the formation of the discharge products together with a subsequent passivation layer that was not eliminated in the charge process. A wide range of techniques have been used to study the discharge product growth and dissolution such as electron microscopy (SEM, EDX), spectroscopic techniques (Raman), synchrotron-based oxygen K-edge and nitrogen K-edge X-ray absorption spectroscopy (XAS) and in situ transmission electron microscopy (TEM) [62–65]. The obtained results confirmed the formation of a core-shell type structure where the outer layer was mainly sodium carbonate. The formation of the Na_2CO_3 shell results in an accumulation of insulating side products on the surface of the oxygen electrode throughout cycling [60]. This accumulation leads to a premature death of the device. Different mechanisms have been proposed for the formation of these secondary products. The first hypothesis associates their formation to the parasitic reactions which involves the carbon electrode and a Na_2O_2/NaO_2 mixture [66]. This hypothesis is widely accepted in order to explain the formation of carbonate-based side products on the carbonaceous air electrode during discharge. In parallel, the formation of electrochemically inert species, such as sodium formate or sodium acetate were also found [14,67]. These products exhibit an insulating nature and their deposition leads to passivation of the electrode surface resulting in the capacity fading. Oxalate-based side products have also been detected and appear to result from the breakdown of diglyme on the surface of the NaO_2 cubes [64]. Liu et al. also analyzed the nonaqueous cyclability of $Na-O_2$ batteries and stated that the chemical instability of superoxide species against the electrolyte is probably the most fundamental challenge for this device [68]. In a first attempt to avoid the decomposition of the electrolyte in metal-O_2 batteries, the protons were removed from the solvent and replace them with methyl groups [69]. Following this strategy, Adams et al. were able to minimize the amount of side reactions in Li-O_2 batteries, but the change in the structure of the solvent might also affect the solvation mechanism and the stability of the electrolyte itself.

Although initially the formation of these side products seemed to be linked to the presence of the superoxide species that fostered the decomposition of the electrolyte, it has recently been shown that this decomposition is accelerated by the presence of singlet oxygen ($^1O_2$) as the reaction intermediate [70]. In fact, $^1O_2$ is also related to the main cause of the deactivation of organic redox mediators, leading to the decrease of their catalytic effect during cycling in Li-O_2 batteries [71]. The presence of this strong oxidizing agent ($^1O_2$) was detected for the first time in Li-O_2 batteries through in operando electron paramagnetic resonance (EPR) using 4-Oxo-TEMPO (4- Oxo-2,2,6,6-tetramethyl-1-piperidinyloxy) as a trapping agent [70]. Later, their formation was monitored through fluorescence measurements using DMA (9,10-dimethylanthracene) quencher [72]. In both studies, the degradation of both the electrolyte and the cathode in Li-O_2 was due to the presence of parasitic side reactions related to the existence of $^1O_2$.

This highly reactive specie is formed mainly in the charging processes (> 3.55 V vs Li^+/Li). Regarding Na-O_2 batteries, superoxide cannot be solely responsible for the side reactions; it was observed, using DMA quencher, the formation of $^1O_2$ reactive species - present in higher quantities at high potentials [73]. In aprotic solvents, the 1-electron oxygen electroreduction (Eq. 3) yields the radical anion superoxide which is stabilized by large hydrophobic cations.

$$O_2 + e \rightarrow O_2^- \tag{3}$$

The presence of $^1O_2$ has also been detected in the discharge process, through the reaction of superoxide with traces of water. In the presence of proton donors, HA:

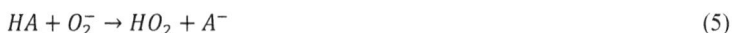

$$2\,O_2^- + 2\,H^+ \rightarrow H_2O_2 + {}^1O_2 \tag{4}$$

$$HA + O_2^- \rightarrow HO_2 + A^- \tag{5}$$

Batteries, both metal ion and metal-O_2, have traditionally used an organic liquid electrolyte due to their high ionic conductivities and exceptional ability to wet the electrodes, which allows a greater efficiency at the electrode/electrolyte interfaces. The use of organic-based electrolyte entails, in turn, a series of disadvantages: low thermal stability, the fact that they are flammable and the risk of leakage, which implies serious safety problems. In this sense, in recent years, the design of solid-state electrolytes has gained a great deal of attention in order to overcome the above-mentioned limitations [74,75]. These electrolytes also help to suppress the growth of sodium dendrites at the anode side [76]. The main challenge, however, is their poor ionic conductivity at room temperature which is an order of magnitude lower than that of liquid-based electrolytes [77]. Solid electrolytes are classified according to their chemical-structural nature in the following categories: polymeric solid electrolytes (PSE), inorganic solid electrolytes (ISE) and inorganic/polymeric hybrid composites (IPHC). PSE are constructed from a polymeric matrix and a sodium salt giving rise to a macromolecular network, which presents high flexibility; thus, avoiding mechanical problems during battery operation. This category includes electrolytes based on the use of polyethylene oxide (PEO) where the effect of the different sodium salts has been analyzed, obtaining promising results in the case of $NaN(SO_2CF_3)_2$ (sodium trifluoromethanesulfonimide, NaTFSI) with ionic conductivities of $\approx 10^{-4}$ S/cm above 80 °C [78]. Inorganic solid electrolytes (ISE) are also an interesting alternative to those based on polymeric matrix due to their higher ionic conductivity at room temperature ($\approx 10^{-3}$ S/cm) and ability to prevent the growth of sodium dendrites. However, they present low (electro)chemical stability that can lead to secondary reactions at the interfaces [79]. Among the most studied ISEs, Na-β/β'' alumina and NASICON ($Na_{1+x}Zr_2Si_xP_{3-x}O_{12}$) can be found [80]. Finally, the last strategy explored towards the concept of *all solid-state battery* focuses on the development of hybrid composites that combines the main features of polymeric and inorganic systems. By using inorganic fillers in polymer matrices, it is possible to increase the ionic conductivity of the polymer and improve the mechanical properties of the composite. In this way, using inorganic nanoparticles of materials such as TiO_2 or SiO_2 it is possible to obtain Na^+ conductivity in the order of $\approx 10^{-3}$ S/cm at room temperature [81]. The development of the *all solid-state battery* concept for the $Na-O_2$ system has potential to revolutionize the energy storage market, and it would be a major step towards improving efficiency and, above all the advantages, electric vehicle safety.

4. The use of additives

There are several strategies that have been explored to improve the performance of the $Na-O_2$ batteries. Some works focus on improving the efficiency of the air electrode through new designs [19,24,43]; others choose to protect the anode in order to prevent the growth of dendrites [82–85]. But possibly the most efficient are those related to improve the electrolyte formulation; through the correct composition it is possible to control the discharge product growth mechanisms. Thus, the appearance of side reactions can be avoided to a large extent and further improvement of the cyclability is possible [37,38,45,86].

Among the variety of additives used in metal-O_2 batteries, redox mediators (RMs) arise as a great alternative. These compounds, dissolved in the electrolyte, act as charge transfer carriers that diffuse from the carbon electrode surface, where they are oxidized, to the discharge product particle surface to oxidize it [87,88]. These RMs have traditionally been used in Li-O_2 batteries to reduce the large overpotential in the charge/discharge processes [89]. In this sense, the RM can act during discharge, charge or both processes. Throughout the discharge, the RM can be reduced by accepting an electron from the air electrode (Eq. 6). The reduced species migrate to the electrolyte where they can react chemically with the dissolved O_2 (Eq. 7). In this way there is no accumulation of discharge products on the surface of the electrode, favoring a more solution-mediated mechanism.

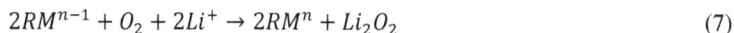

$$RM^n + e^- \rightarrow RM^{n-1} \tag{6}$$

$$2RM^{n-1} + O_2 + 2Li^+ \rightarrow 2RM^n + Li_2O_2 \tag{7}$$

Regarding the charging process, the task of the RM is to facilitate the decomposition of the discharge products. The RM, once electro-oxidized at the electrode surface (Eq. 8), chemically oxidizes the Li_2O_2 particles (Eq. 8).

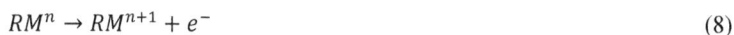

$$RM^n \rightarrow RM^{n+1} + e^- \tag{8}$$

$$2RM^{n+1} + Li_2O_2 \rightarrow 2RM^n + O_2 + 2Li^+ \tag{9}$$

Based on this, a good RM should meet the following requirements [88]:

- Being totally soluble in the electrolyte and stable to the attack of species such as oxygen radicals or anions such as superoxide.

- A redox potential close to that of the O_2/O_2^{2-} pair.

- Must be regenerated after the redox reaction, not being consumed or degraded during the discharge/charge processes.

- Moderate diffusivity in the electrolyte to prevent shuttling.

Although the use of RM has been a strategy widely exploited by different groups in Li-O_2 batteries, in the case of Na systems its use has been limited to a few studies. This is related to the intrinsic property of Na-O_2 batteries which, unlike Li-O_2 systems, have a very low charge overpotential (< 100 mV). Even so, it is possible to find some studies on the effect of RM in Na-O_2 batteries. Fu et al. analyzed the addition of NaI and ferrocene as potential RMs in these devices [90,91]. In these studies, they observed that the addition of NaI helped to efficiently remove the discharge products. Note that in this work the discharge product obtained was Na_2O_2 and not the usually reported discharge product, NaO_2. Regarding the use of ferrocene, they proved that its presence favored the charging process by decomposing the Na_2O_2 generated during discharge; it is surprising, however, that their system was not able to charge without the presence of the RM. The most interesting study analyzes the activity of the ethyl viologen, (RM extensively studied for the Li systems) observing an increase in capacity by 50% and an increase in the discharge voltage by 100 mV [92]. Because of the formation of NaO_2 further away from the electrode surface, the use of this electron shuttle changes the response pathway, mitigating pore clogging and passivation.

Recently, the effect of the addition of lithium and potassium salts into the electrolyte has been evaluated to study the effect of the change of acidity in the electrolyte on the reaction mechanisms [93]. Through this study it was possible to modify the reaction mechanisms by generating different interactions with other cations. Initially the O_2 is reduced to O_2^- and these species are stabilized in different ways depending on the acidity of the electrolyte. In the case of Li^+ in contact with other Na^+ salts, the increase in acidity of the electrolyte led to a rapid deposition of NaO_2 on the surface of the electrode. NaO_2 was capable of reacting chemically with the Li^+ cations dissolved in the electrolyte, forming the LiO_2 species which, through disproportionation reactions, results in the formation of Li_2O_2 deposits. The fact that superoxide had a greater initial tendency to interact with Na^+ than with Li^+ can be predicted by applying Pearson's theory where soft bases have a greater tendency to bind to softer acids. In this case, Na^+, which is larger than Li^+ and presents the same charge, is softer. By adding K^+, they observed that superoxide had the tendency to bind to the K^+ rather than Li^+, because K^+ is softer than Na^+. In the presence of K^+, the high stability of the intermediate $[K^+(diglyme)_m \cdots O_2^-]$ led to low overpotentials and high intensities in the cyclic voltammetry. However, the electrolyte saturates rapidly in O_2^- (due to the difficulty in forming the KO_2 species) and deposits on the surface of the electrode as NaO_2.

The concentration of the sodium salt in the electrolyte is also a fundamental parameter for the performance of these batteries [86,94]. Recently, some researchers have begun to analyze in depth the use of electrolytes with high salt concentrations (> 3M), mainly in Li-ion batteries [95]. By using high salt concentrations, the physico-chemical properties of the electrolyte can change drastically. In fact, a higher concentration of the salt leads to a reinforcement of the interactions between cations, anions and the solvent. If the concentration is sufficiently high, between 3 and 5 M, a peculiar 3D solution structure is stabilized, which can provide greater interfacial stability [96]. Another advantage of the use of this type of electrolyte, is the less corrosion of the electrodes at higher potential due to the less amount of free solvent molecules. Regarding, $Na-O_2$ batteries, He et al. investigated the use of dimethyl sulfoxide solutions in concentrated sodium trifluoromethanesulfonimide salts (> 3 mol kg^{-1}) [97]. Using these concentrated electrolytes, enhanced cyclability and the possibility of using metallic sodium (which normally reacts with Na, being unable to perform any discharge/charge process) was addressed. They concluded that the enhanced Na stability originates from the reduced amount of free DMSO molecules, and the enhanced electron affinity of the anion TFSI which leads to its decomposition and subsequent formation of a protected surface layer.

5. Outlook

Electrolytes are key of any optimized system as they are responsible for the ion transport from the anode to the cathode. At present, $Na-O_2$ systems have mostly been designed according to strategies developed for $Li-O_2$ research which, though a useful initial approach, may have led to the use of suboptimal materials due to the current lack of exploration in this area. The vast majority of the work in this field has been devoted to electrode materials, with little or negligible work on electrolytes which are key components of any optimized $Na-O_2$ battery. The critical parameters affecting electrolyte physiochemistry are the solvent-salt structure and dynamics. Thus, future research on this topic should focus on understanding the thermodynamics and coordination chemistry (salt-solvent interactions), and subsequently linked these to sodium battery electrolyte properties, through the combination of both experimental and computational methodologies. These findings need to be coupled with a thorough understanding of the fundamental mechanisms of the electrode processes, with the ultimate goal to develop a rechargeable Na-O_2 system.

Acknowledgements

The authors acknowledge support from the "Ministerio de Economía y Competitividad" of Spain (under project MAT2016-78266-P), the "Fondo Europeo de Desarrollo Regional" (FEDER) and the Eusko Jaurlaritza/Gobierno Vasco (under projects IT-570-13 and ELKARTEK project CICE17).

References

[1]   M. Armand, J. Tarascon, Building better batteries, Nature 451 (2008) 652–657. https://doi.org/10.1038/451652a

[2]   G. Girishkumar, B. McCloskey, A.C. Luntz, S. Swanson, W. Wilcke, Lithium−Air battery: Promise and challenges, J. Phys. Chem. Lett. 1 (2010) 2193–2203. https://doi.org/10.1021/jz1005384

[3]   Y. Chen, S.A. Freunberger, Z. Peng, O. Fontaine, P.G. Bruce, Charging a Li-O_2 battery using a redox mediator, Nat. Chem. 5 (2013) 489–94. https://doi.org/10.1038/nchem.1646. https://doi.org/10.1038/nchem.1646

[4]   H.G. Jung, Y.S. Jeong, J.B. Park, Y.K. Sun, B. Scrosati, Y.J. Lee, Ruthenium-based electrocatalysts supported on reduced graphene oxide for lithium-air batteries, ACS Nano. 7 (2013) 3532–3539. https://doi.org/10.1021/nn400477d

[5]   M.S. Whittingham, Lithium batteries and cathode materials, Chem. Rev. 104 (2004) 4271–4302. https://doi.org/10.1021/cr020731c

[6]   V. Palomares, P. Serras, I. Villaluenga, K.B. Hueso, J. Carretero-González, T. Rojo, Na-ion batteries, recent advances and present challenges to become low cost energy storage systems, Energy Environ. Sci. 5 (2012) 5884–5901. https://doi.org/10.1039/c2ee02781j

[7]   V. Palomares, M. Casas-Cabanas, E. Castillo-Martínez, M.H. Han, T. Rojo, Update on Na-based battery materials. A growing research path, Energy Environ. Sci. 6 (2013) 2312–2337. https://doi.org/10.1039/c3ee41031e

[8]   P. Hartmann, C.L. Bender, M. Vračar, A.K. Dürr, A. Garsuch, J. Janek, P. Adelhelm, A rechargeable room-temperature sodium superoxide (NaO$_2$) battery, Nat. Mater. 12 (2013) 228–232. https://doi.org/10.1038/nmat3486

[9]   G. Attard, L.J. Hardwick, J.-C. Dong, J.-F. Li, T.A. Galloway, Oxygen reactions on Pt{ hkl } in a non-aqueous Na^+ electrolyte: Site selective stabilisation of a sodium peroxy species , Chem. Sci. (2019). https://doi.org/10.1039/c8sc05489d

[10]   P.G. Bruce, S. a Freunberger, L.J. Hardwick, J.-M. Tarascon, $Li–O_2$ and Li–S batteries with high energy storage, Nat. Mater. 11 (2012) 19–30. https://doi.org/10.1038/NMAT3191

[11]   B.D. McCloskey, J.M. Garcia, A.C. Luntz, Chemical and electrochemical differences in nonaqueous $Li–O_2$ and $Na–O_2$ Batteries, J. Phys. Chem. Lett. 5 (2014) 1230–1235. https://doi.org/10.1021/jz500494s

[12]   K.B. Hueso, M. Armand, T. Rojo, High temperature sodium batteries: Status, challenges and future trends, Energy Environ. Sci. 6 (2013) 734–749. https://doi.org/10.1039/c3ee24086j

[13]   I. Landa-Medrano, C. Li, N. Ortiz-Vitoriano, I. Ruiz De Larramendi, J. Carrasco, T. Rojo, Sodium-oxygen battery: steps toward reality, J. Phys. Chem. Lett. 7 (2016). https://doi.org/10.1021/acs.jpclett.5b02845.

[14]   H. Yadegari, X. Sun, Recent advances on sodium–oxygen batteries: A chemical perspective, Acc. Chem. Res. 51 (2018) 1532–1540. https://doi.org/10.1021/acs.accounts.8b00139

[15]   B. Sun, C. Pompe, S. Dongmo, J. Zhang, K. Kretschmer, D. Schröder, J. Janek, G. Wang, Challenges for developing rechargeable room-temperature sodium oxygen batteries, Adv. Mater. Technol. 1800110 (2018) 1800110. https://doi.org/10.1002/admt.201800110

[16]   Q. Sun, H. Yadegari, M.N. Banis, J. Liu, B. Xiao, X. Li, C. Langford, R. Li, X. Sun, Toward a sodium-"Air" battery: Revealing the critical role of humidity, J. Phys. Chem. C. 119 (2015) 13433–13441. https://doi.org/10.1021/acs.jpcc.5b02673

[17]   E. Peled, D. Golodnitsky, H. Mazor, M. Goor, S. Avshalomov, Parameter analysis of a practical lithium- and sodium-air electric vehicle battery, J. Power Sources. 196 (2011) 6835–6840. https://doi.org/10.1016/j.jpowsour.2010.09.104

[18]   P. Hartmann, C.L. Bender, M. Vračar, A.K. Dürr, J. Janek, P. Adelhelm, Supporting Information on " A rechargeable room-temperature sodium superoxide battery ", Nat. Mater. 12 (2013) 1–11. https://doi.org/10.1038/NMAT3486

[19]   Y. Li, H. Yadegari, X. Li, M.N. Banis, R. Li, X. Sun, Superior catalytic activity of

Materials Research Forum LLC
https://doi.org/10.21741/9781644900833-9

nitrogen-doped graphene cathodes for high energy capacity sodium-air batteries, Chem. Commun. 49 (2013) 11731–11733. https://doi.org/10.1039/c3cc46606j

[20] W. Liu, Q. Sun, Y. Yang, J.Y. Xie, Z.W. Fu, An enhanced electrochemical performance of a sodium-air battery with graphene nanosheets as air electrode catalysts, Chem. Commun. 49 (2013) 1951–1953. https://doi.org/10.1039/c3cc00085k

[21] S. Kang, Y. Mo, S.P. Ong, G. Ceder, Nano-scale stabilization of sodium oxides: Implications for Na-O_2 batteries., Nano Lett. 14 (2014) 1016–1020. https://doi.org/10.1021/nl404557w

[22] B. Lee, D.H. Seo, H.D. Lim, I. Park, K.Y. Park, J. Kim, K. Kang, First-principles study of the reaction mechanism in sodium-oxygen batteries, Chem. Mater. 26 (2014) 1048–1055. https://doi.org/10.1021/cm403163c

[23] H. Yadegari, Y. Li, M.N. Banis, X. Li, B. Wang, Q. Sun, R. Li, T.K. Sham, X. Cui, X. Sun, On rechargeability and reaction kinetics of sodium-air batteries, Energy Environ. Sci. 7 (2014) 3747–3757. https://doi.org/10.1039/c4ee01654h

[24] C.L. Bender, P. Hartmann, M. Vračar, P. Adelhelm, J. Janek, On the thermodynamics, the role of the carbon cathode, and the cycle life of the sodium superoxide (NaO2) battery, Adv. Energy Mater. 4 (2014) 2–11. https://doi.org/10.1002/aenm.201301863

[25] N. Ortiz-Vitoriano, T.P. Batcho, D.G. Kwabi, B. Han, N. Pour, K.P.C. Yao, C. V. Thompson, Y. Shao-Horn, Rate-Dependent Nucleation and Growth of NaO_2 in Na-O_2 Batteries, J. Phys. Chem. Lett. 6 (2015) 2636–2643. https://doi.org/10.1021/acs.jpclett.5b00919

[26] I. Landa-Medrano, R. Pinedo, X. Bi, I. Ruiz de Larramendi, L. Lezama, J. Janek, K. Amine, J. Lu, T. Rojo, New Insights into the Instability of Discharge Products in Na−O2 Batteries, ACS Appl. Mater. Interfaces. 8 (2016) 20120–20127. https://doi.org/10.1021/acsami.6b06577

[27] P. Hartmann, M. Heinemann, C.L. Bender, K. Graf, R.P. Baumann, P. Adelhelm, C. Heiliger, J. Janek, Discharge and charge reaction paths in sodium-oxygen batteries: Does NaO_2 form by direct electrochemical growth or by precipitation from solution?, J. Phys. Chem. C. 119 (2015) 22778–22786. https://doi.org/10.1021/acs.jpcc.5b06007

[28] S. Yang, D.J. Siegel, Intrinsic conductivity in sodium-air battery discharge phases: Sodium superoxide vs sodium peroxide, Chem. Mater. 27 (2015) 3852–3860.

https://doi.org/10.1021/acs.chemmater.5b00285

[29] O. Arcelus, C. Li, T. Rojo, J. Carrasco, Electronic structure of sodium superoxide bulk, (100) surface, and clusters using hybrid density functional: Relevance for Na-O_2 batteries, J. Phys. Chem. Lett. 6 (2015) 2027–2031. https://doi.org/10.1021/acs.jpclett.5b00814

[30] J.E. Nichols, B.D. McCloskey, The sudden death phenomena in nonaqueous Na-O_2 batteries, J. Phys. Chem. C. 181 (2017) 85–96. https://doi.org/10.1021/acs.jpcc.6b09663.

[31] C.L. Bender, W. Bartuli, M.G. Schwab, P. Adelhelm, J. Janek, Toward better sodium-oxygen batteries: A study on the performance of engineered oxygen electrodes based on carbon nanotubes, Energy Technol. 3 (2015) 242–248. https://doi.org/10.1002/ente.201402208

[32] D. Schröder, C.L. Bender, R. Pinedo, W. Bartuli, M.G. Schwab, Ž. Tomović, J. Janek, How to control the discharge product in sodium–oxygen batteries: Proposing new pathways for sodium peroxide formation, Energy Technol. 5 (2017) 1242–1249. https://doi.org/10.1002/ente.201600539

[33] I. Landa-Medrano, R. Pinedo, I.R. De Larramendi, N. Ortiz-Vitoriano, T. Rojo, Monitoring the location of cathode-reactions in Li-O_2 batteries, J. Electrochem. Soc. 162 (2015). https://doi.org/10.1149/2.0191502jes

[34] W.A. Hermann, J.B. Straubel, D.G. Beck, (12) US9559532B2 (2017).

[35] J.E. Nichols, B.D. McCloskey, The Sudden Death Phenomena in Nonaqueous Na-O 2 Batteries, J. Phys. Chem. C. 181 (2017) 85–96. https://doi.org/10.1021/acs.jpcc.6b09663

[36] J. Kim, H.D. Lim, H. Gwon, K. Kang, Sodium-oxygen batteries with alkyl-carbonate and ether based electrolytes, Phys. Chem. Chem. Phys. 15 (2013) 3623–3629. https://doi.org/10.1039/c3cp43225d

[37] K. Li, S.R. Galle Kankanamge, T.K. Weldeghiorghis, R. Jorn, D.G. Kuroda, R. Kumar, Predicting ion association in sodium electrolytes: A transferrable model for investigating glymes, J. Phys. Chem. C. 122 (2018) 4747–4756. https://doi.org/10.1021/acs.jpcc.7b09995

[38] L. Lutz, D. Alves Dalla Corte, M. Tang, E. Salager, M. Deschamps, A. Grimaud, L. Johnson, P.G. Bruce, J.M. Tarascon, Role of electrolyte anions in the na-o2battery: implications for NaO_2 solvation and the stability of the sodium solid

Materials Research Forum LLC
https://doi.org/10.21741/9781644900833-9

electrolyte interphase in glyme ethers, Chem. Mater. 29 (2017) 6066–6075. https://doi.org/10.1021/acs.chemmater.7b01953

[39] I.M. Aldous, L.J. Hardwick, Growth and dissolution of NaO_2 in an ether-based electrolyte as the discharge product in the Na-O_2 cell, Chem. Commun. 54 (2018) 3444–3447. https://doi.org/10.1039/c7cc08201k

[40] R. Tatara, G.M. Leverick, S. Feng, S. Wan, S. Terada, K. Dokko, M. Watanabe, Y. Shao-Horn, Tuning NaO_2 Cube Sizes by Controlling Na^+ and Solvent Activity in Na-O2Batteries, J. Phys. Chem. C. 122 (2018) 18316–18328. https://doi.org/10.1021/acs.jpcc.8b05418

[41] N. Dubouis, A. Serva, E. Salager, M. Deschamps, M. Salanne, A. Grimaud, The Fate of Water at the Electrochemical Interfaces: Electrochemical Behavior of Free Water vs . Coordinating Water, (2018). https://doi.org/10.26434/chemrxiv.7140782.v1

[42] T. Liu, J.T. Frith, G. Kim, R.N. Kerber, N. Dubouis, Y. Shao, Z. Liu, P.C.M.M. Magusin, M.T.L. Casford, N. Garcia-Araez, C.P. Grey, The effect of water on quinone redox mediators in nonaqueous Li-O_2 Batteries, J. Am. Chem. Soc. 140 (2018) 1428–1437. https://doi.org/10.1021/jacs.7b11007

[43] M. Enterría, C. Botas, J.L. Gómez-Urbano, B. Acebedo, J.M. López Del Amo, D. Carriazo, T. Rojo, N. Ortiz-Vitoriano, Pathways towards high performance Na-O_2 batteries: tailoring graphene aerogel cathode porosity & nanostructure, J. Mater. Chem. A. 6 (2018) 20778–20787. https://doi.org/10.1039/c8ta07273f

[44] N. Li, Y. Yin, F. Meng, Q. Zhang, J. Yan, Q. Jiang, Enabling Pyrochlore-Type Oxides as Highly efficient electrocatalysts for high-capacity and stable Na-O_2 Batteries: The synergy of electronic structure and morphology, ACS Catal. 7 (2017) 7688–7694. https://doi.org/10.1021/acscatal.7b02074

[45] B. Wang, N. Zhao, Y. Wang, W. Zhang, W. Lu, X. Guo, J. Liu, Electrolyte-controlled discharge product distribution of Na-O_2 batteries: A combined computational and experimental study, Phys. Chem. Chem. Phys. 19 (2017) 2940–2949. https://doi.org/10.1039/c6cp07537a

[46] L. Lutz, W. Yin, A. Grimaud, D. Alves, D. Corte, M. Tang, L. Johnson, E. Azaceta, V. Salou-Kanin, A.J. Naylor, S. Hamad, J.A. Anta, E. Salager, R. Tena-Zaera, P.G. Bruce, J. Tarascon, High capacity nao batteries – key parameters for solution - mediated discharge, J. Phys. Chem. C. 120 (2016) 20068–20076. https://doi.org/10.1021/acs.jpcc.6b07659

[47] Franco Cataldo, A revision of the Gutmann donor numbers of a series of phosphoramides including TEPA, Eur. Chem. Bull. 4 (2015) 92–97. https://doi.org/10.17628/ECB.2015.4.92

[48] C.C. Su, M. He, R. Amine, T. Rojas, L. Cheng, A.T. Ngo, K. Amine, Solvating power series of electrolyte solvents for lithium batteries, Energy Environ. Sci. 12 (2019) 1249–1254. https://doi.org/10.1039/c9ee00141g

[49] N. Zhao, X. Guo, Cell Chemistry of Sodium-Oxygen Batteries with Various Nonaqueous Electrolytes, J. Phys. Chem. C. 119 (2015) 25319–25326. https://doi.org/10.1021/acs.jpcc.5b09187

[50] Q. Sun, X. Lin, H. Yadegari, W. Xiao, Y. Zhao, K.R. Adair, R. Li, X. Sun, Aligning the binder effect on sodium-air batteries, J. Mater. Chem. A. 6 (2018) 1473–1484. https://doi.org/10.1039/c7ta09028e

[51] D. Sharon, D. Hirshberg, M. Afri, A.A. Frimer, M. Noked, D. Aurbach, Aprotic metal-oxygen batteries: Recent findings and insights, J. Solid State Electrochem. 21 (2017) 1861–1878. https://doi.org/10.1007/s10008-017-3590-7

[52] S. Wu, J. Tang, F. Li, X. Liu, Y. Yamauchi, M. Ishida, H. Zhou, A synergistic system for lithium-oxygen batteries in humid atmosphere integrating a composite cathode and a hydrophobic ionic liquid-based electrolyte, Adv. Funct. Mater. 26 (2016) 3291–3298. https://doi.org/10.1002/adfm.201505420

[53] C.L. Bender, D. Schröder, R. Pinedo, P. Adelhelm, J. Janek, One- or two-electron transfer? the ambiguous nature of the discharge products in sodium-oxygen batteries, Angew. Chem. Int. Ed. 55 (2016) 4640–4649. https://doi.org/10.1002/anie.201510856

[54] R. Pinedo, D.A. Weber, B.J. Bergner, D. Schröder, P. Adelhelm, J. Janek, Insights into the chemical nature and formation mechanisms of discharge Products in Na-O_2 batteries by means of operando X-ray diffraction, J. Phys. Chem. C. 120 (2016) 8472–8481. https://doi.org/10.1021/acs.jpcc.6b00903

[55] C. Xia, R. Black, R. Fernandes, B. Adams, L.F. Nazar, The critical role of phase-transfer catalysis in aprotic sodium oxygen batteries, Nat. Chem. 7 (2015) 496–501. https://doi.org/10.1038/NCHEM.2260

[56] Y. Qiao, S. Wu, J. Yi, Y. Sun, S. Guo, S. Yang, P. He, H. Zhou, From O^{2-} to HO^{2-}: reducing by-products and overpotential in Li-O_2 batteries by water addition, Angew. Chem. Int. Ed. 56 (2017) 4960–4964. https://doi.org/10.1002/anie.201611122

[57] D.G. Kwabi, T.P. Batcho, S. Feng, L. Giordano, C. V. Thompson, Y. Shao-Horn, The effect of water on discharge product growth and chemistry in Li-O$_2$ batteries, Phys. Chem. Chem. Phys. 18 (2016) 24944–24953. https://doi.org/10.1039/c6cp03695c

[58] S. Ma, J. Wang, J. Huang, Z. Zhou, Z. Peng, Unveiling the complex effects of H$_2$O on discharge-recharge behaviors of aprotic Lithium-O$_2$ batteries, J. Phys. Chem. Lett. 9 (2018) 3333–3339. https://doi.org/10.1021/acs.jpclett.8b01333

[59] I. Landa-Medrano, A. Sorrentino, L. Stievano, I. Ruiz de Larramendi, E. Pereiro, L. Lezama, T. Rojo, D. Tonti, Architecture of Na-O$_2$ battery deposits revealed by transmission X-ray microscopy, Nano Energy 37 (2017). https://doi.org/10.1016/j.nanoen.2017.05.021

[60] H. Yadegari, M.N. Banis, B. Xiao, Q. Sun, X. Li, A. Lushington, B. Wang, R. Li, T.K. Sham, X. Cui, X. Sun, Three-dimensional nanostructured air electrode for sodium-oxygen batteries: A mechanism study toward the cyclability of the cell, Chem. Mater. 27 (2015) 3040–3047. https://doi.org/10.1021/acs.chemmater.5b00435

[61] C. Liu, M. Carboni, W.R. Brant, R. Pan, J. Hedman, J. Zhu, T. Gustafsson, R. Younesi, On the stability of NaO$_2$ in Na-O$_2$ batteries, ACS Appl. Mater. Interfaces 10 (2018) 13534–13541. https://doi.org/10.1021/acsami.8b01516

[62] Q. Sun, J. Liu, B. Xiao, B. Wang, M. Banis, H. Yadegari, K.R. Adair, R. Li, X. Sun, Visualizing the Oxidation Mechanism and Morphological Evolution of the Cubic-Shaped Superoxide Discharge Product in Na–Air Batteries, Adv. Funct. Mater. 29 (2019) 1–9. https://doi.org/10.1002/adfm.201808332

[63] W.J. Kwak, L. Luo, H.G. Jung, C. Wang, Y.K. Sun, Revealing the reaction mechanism of Na-O$_2$ batteries using environmental transmission electron microscopy, ACS Energy Lett. 3 (2018) 393–399. https://doi.org/10.1021/acsenergylett.7b01273

[64] H. Yadegari, M. Norouzi Banis, X. Lin, A. Koo, R. Li, X. Sun, Revealing the chemical mechanism of NaO$_2$ decomposition by in situ Raman imaging, Chem. Mater. 30 (2018) 5156–5160. https://doi.org/10.1021/acs.chemmater.8b01704

[65] M.N. Banis, H. Yadegari, Q. Sun, T. Regier, T. Boyko, J. Zhou, Y.M. Yiu, R. Li, Y. Hu, T.K. Sham, X. Sun, Revealing the charge/discharge mechanism of Na-O 2 cells by: In situ soft X-ray absorption spectroscopy, Energy Environ. Sci. 11 (2018) 2073–2077. https://doi.org/10.1039/c8ee00721g

Materials Research Forum LLC
https://doi.org/10.21741/9781644900833-9

[66] Z.E.M. Reeve, C.J. Franko, K.J. Harris, H. Yadegari, X. Sun, G.R. Goward, Detection of electrochemical reaction products from the sodium-oxygen cell with solid-state ^{23}Na NMR spectroscopy, J. Am. Chem. Soc. 139 (2017) 595–598. https://doi.org/10.1021/jacs.6b11333

[67] R. Black, A. Shyamsunder, P. Adeli, D. Kundu, G.K. Murphy, L.F. Nazar, The nature and impact of side reactions in glyme-based sodium – oxygen batteries, ChemSusChem. 9 (2016) 1795–1803. https://doi.org/10.1002/cssc.201600034

[68] T. Liu, G. Kim, M.T.L. Casford, C.P. Grey, Mechanistic insights into the challenges of cycling a nonaqueous Na-O$_2$ battery, J. Phys. Chem. Lett. 7 (2016) 4841–4846. https://doi.org/10.1021/acs.jpclett.6b02267

[69] B.D. Adams, R. Black, Z. Williams, R. Fernandes, M. Cuisinier, E.J. Berg, P. Novak, G.K. Murphy, L.F. Nazar, Towards a stable organic electrolyte for the lithium oxygen battery, Adv. Energy Mater. 5 (2015) 1–11. https://doi.org/10.1002/aenm.201400867

[70] J. Wandt, P. Jakes, J. Granwehr, H.A. Gasteiger, R.A. Eichel, Singlet oxygen formation during the charging process of an aprotic lithium-oxygen battery, Angew. Chem. Int. Ed. 55 (2016) 6892–6895. https://doi.org/10.1002/anie.201602142

[71] W.-J. Kwak, H. Kim, Y.K. Petit, C. Leypold, T.T. Nguyen, N. Mahne, P. Redfern, L.A. Curtiss, H.-G. Jung, S.M. Borisov, S.A. Freunberger, Y.K. Sun, Deactivation of redox mediators in lithium-oxygen batteries by singlet oxygen, Nat. Commun. 10 (2019) 1380. https://doi.org/10.1038/s41467-019-09399-0

[72] N. Mahne, B. Schafzahl, C. Leypold, M. Leypold, S. Grumm, A. Leitgeb, G.A. Strohmeier, M. Wilkening, O. Fontaine, D. Kramer, C. Slugovc, S.M. Borisov, S.A. Freunberger, Singlet oxygen generation as a major cause for parasitic reactions during cycling of aprotic lithium-oxygen batteries, Nat. Energy. 2 (2017) 1–9. https://doi.org/10.1038/nenergy.2017.36

[73] L. Schafzahl, N. Mahne, B. Schafzahl, M. Wilkening, C. Slugovc, S.M. Borisov, S.A. Freunberger, Singlet oxygen during cycling of the aprotic sodium–O$_2$ battery, Angew. Chem. Int. Ed. 56 (2017) 15728–15732. https://doi.org/10.1002/anie.201709351

[74] J.B. Goodenough, P. Singh, Review-Solid electrolytes in rechargeable electrochemical cells, J. Electrochem. Soc. 162 (2015) 5–10. https://doi.org/10.1149/2.0021514jes

[75]   T. Bartsch, F. Strauss, T. Hatsukade, A. Schiele, A.Y. Kim, P. Hartmann, J. Janek, T. Brezesinski, Gas evolution in all-solid-state battery cells, ACS Energy Lett. 3 (2018) 2539–2543. https://doi.org/10.1021/acsenergylett.8b01457

[76]   W. Zhou, Y. Li, S. Xin, J.B. Goodenough, Rechargeable Sodium All-Solid-State Battery, (2017) 0–5. https://doi.org/10.1021/acscentsci.6b00321

[77]   C. Zhao, L. Liu, X. Qi, Y. Lu, F. Wu, J. Zhao, Y. Yu, Y.S. Hu, L. Chen, Solid-State Sodium Batteries, Adv. Energy Mater. 1703012 (2018) 14–16. https://doi.org/10.1002/aenm.201703012

[78]   J.S. Moreno, M. Armand, M.B. Berman, S.G. Greenbaum, B. Scrosati, S. Panero, Composite PEOn:NaTFSI polymer electrolyte: Preparation, thermal and electrochemical characterization, J. Power Sources 248 (2014) 695–702. https://doi.org/10.1016/j.jpowsour.2013.09.137

[79]   R. Khurana, J.L. Schaefer, L.A. Archer, G.W. Coates, Suppression of lithium dendrite growth using cross-linked polyethylene/poly(ethylene oxide) electrolytes: A new approach for practical lithium-metal polymer batteries, J. Am. Chem. Soc. 136 (2014) 7395–7402. https://doi.org/10.1021/ja502133j

[80]   L. Fan, S. Wei, S. Li, Q. Li, Y. Lu, Recent progress of the solid-state electrolytes for high-energy metal-based batteries, Adv. Energy Mater. 1702657 (2018) 1–31. https://doi.org/10.1002/aenm.201702657

[81]   S. Song, M. Kotobuki, F. Zheng, C. Xu, S. V. Savilov, N. Hu, L. Lu, Y. Wang, W.D.Z. Li, A hybrid polymer/oxide/ionic-liquid solid electrolyte for Na-metal batteries, J. Mater. Chem. A. 5 (2017) 6424–6431. https://doi.org/10.1039/C6TA11165C

[82]   H. Wang, C. Wang, E. Matios, W. Li, Facile Stabilization of the Sodium Metal Anode with Additives: Unexpected Key Role of Sodium Polysulfide and Adverse Effect of Sodium Nitrate, Angew. Chemie - Int. Ed. 57 (2018) 7734–7737. https://doi.org/10.1002/anie.201801818

[83]   X. Lin, Q. Sun, H. Yadegari, X. Yang, Y. Zhao, C. Wang, J. Liang, A. Koo, R. Li, X. Sun, On the cycling performance of Na-O$_2$ Cells: revealing the impact of the superoxide crossover toward the metallic Na electrode, Adv. Funct. Mater. 28 (2018) 1–12. https://doi.org/10.1002/adfm.201801904

[84]   B. Sun, P. Li, J. Zhang, D. Wang, P. Munroe, C. Wang, P.H.L. Notten, G. Wang, Dendrite-free sodium-metal anodes for high-energy sodium-metal batteries, Adv. Mater. 30 (2018) 1–8. https://doi.org/10.1002/adma.201801334

[85] L. Fan, X. Li, Recent advances in effective protection of sodium metal anode, Nano Energy. 53 (2018) 630–642. https://doi.org/10.1016/j.nanoen.2018.09.017.

[86] S.R. Galle Kankanamge, K. Li, K.D. Fulfer, P. Du, R. Jorn, R. Kumar, D.G. Kuroda, Mechanism behind the unusually high conductivities of high concentrated sodium ion glyme-based electrolytes, J. Phys. Chem. C. 122 (2018) 25237-25246. https://doi.org/10.1021/acs.jpcc.8b06991

[87] H. Lim, H. Song, J. Kim, H. Gwon, Y. Bae, K. Park, J. Hong, H. Kim, T. Kim, Y.H. Kim, X. Lepró, R. Ovalle-Robles, R.H. Baughman, K. Kang, Superior rechargeability and efficiency of lithium – oxygen batteries : Hierarchical air electrode architecture combined with a soluble catalyst, Angew. Chem. Int. Ed.126 (2014) 4007–4012. https://doi.org/10.1002/anie.201400711

[88] I. Landa-Medrano, I. Lozano, N. Ortiz-Vitoriano, I. Ruiz De Larramendi, T. Rojo, Redox mediators: A shuttle to efficacy in metal-O_2 batteries, J. Mater. Chem. A. 7 (2019) 8746–8764. https://doi.org/10.1039/c8ta12487f

[89] J.B. Park, S.H. Lee, H.G. Jung, D. Aurbach, Y.K. Sun, Redox mediators for Li–O_2 batteries: Status and perspectives, Adv. Mater. 30 (2018) 1–13. https://doi.org/10.1002/adma.201704162

[90] W.-W. Yin, Z. Shadike, Y. Yang, F. Ding, L. Sang, H. Li, Z.-W. Fu, A long-life Na – air battery based on a soluble NaI catalyst, Chem. Commun. 51 (2015) 2324–2327. https://doi.org/10.1039/C4CC08439J

[91] W.-W. Yin, J.-L. Yue, M.-H. Cao, W. Liu, J.-J. Ding, F. Ding, L. Sang, Z.-W. Fu, Dual catalytic behavior of a soluble ferrocene as an electrocatalyst and in the electrochemistry for Na – air batteries, J. Mater. Chem. A. 3 (2015) 19027–19032. https://doi.org/10.1039/C5TA04647E

[92] J.T. Frith, I. Landa-Medrano, I. Ruiz De Larramendi, T. Rojo, J.R. Owen, N. Garcia-Araez, Improving Na-O_2 batteries with redox mediators, Chem. Commun. 53 (2017). https://doi.org/10.1039/c7cc06679a

[93] I. Landa-Medrano, I. Ruiz de Larramendi, T. Rojo, Modifying the ORR route by the addition of lithium and potassium salts in Na-O_2 batteries, Electrochim. Acta. 263 (2018) 102–109. https://doi.org/10.1016/j.electacta.2017.12.141

[94] Y. Zhang, N. Ortiz-Vitoriano, B. Acebedo, L. O'Dell, D.R. MacFarlane, T. Rojo, M. Forsyth, P.C. Howlett, C. Pozo-Gonzalo, Elucidating the impact of sodium salt concentration on the cathode-electrolyte interface of Na-Air Batteries, J. Phys. Chem. C. 122 (2018) 15276–15286. https://doi.org/10.1021/acs.jpcc.8b02004

[95] Y. Yamada, J. Wang, S. Ko, E. Watanabe, A. Yamada, Advances and issues in developing salt-concentrated battery electrolytes, Nat. Energy. (2019). https://doi.org/10.1038/s41560-019-0336-z

[96] J. Zheng, J.A. Lochala, A. Kwok, Z.D. Deng, J. Xiao, Research progress towards understanding the unique interfaces between concentrated electrolytes and electrodes for energy storage applications, Adv. Sci. 4 (2017) 1–19. https://doi.org/10.1002/advs.201700032

[97] M. He, K.C. Lau, X. Ren, N. Xiao, W.D. McCulloch, L.A. Curtiss, Y. Wu, Concentrated electrolyte for the sodium-oxygen battery: Solvation structure and improved cycle life, Angew. Chemie Int. Ed. 128 (2016) 15536–15540. https://doi.org/10.1002/anie.201608607

Sodium-Ion Batteries: Materials and Applications
Materials Research Foundations **76** (2020) 229-250

Materials Research Forum LLC
https://doi.org/10.21741/9781644900833-10

Chapter 10

State-of-the-Art, Future Prospects and Challenges in Sodium-Ion Battery Technology

[1]Kritika S. Sharma, [2]Vaishali Tomar, [3]Rekha Sharma and [1]Dinesh Kumar*

[1]School of Chemical Sciences, Central University of Gujarat, Gandhinagar 382030, India

[2]Formulation, ABH Natures Products, New York, USA

[3]Department of Chemistry, Banasthali Vidyapith, Rajasthan 304022, India

*dsbchoudhary2002@gmail.com

Abstract

Energy is a matter of interest for scientists, business and policymakers. This concern will continue to increase due to the gradual reduction of fossil fuels day by day. Though, lithium-ion batteries (LIBs) have widely been used due to high energy density, little memory effect, and low self-discharge for portable electronics across the world since 1991. Research in the field of sodium-ion batteries (SIBs) is comparatively new and thus has broad scope due to the abundance and low price of sodium precursors. Presently available electrodes and electrolytes are at the beginning stage of development, and more research is needed to produce SIBs at a massive scale. Selenium and selenium-sulfur (Se_xS_y)-based cathode materials for room temperature lithium, and sodium batteries have been developed. This chapter summarizes current research on materials, discuss prospects and challenges in SIBs technology. This will provide valuable understanding to scientific, commercial, and practical opportunity in the development of sodium-ion batteries.

Keywords

Battery, Sodium-Ion, Research, Lithium-Ion, Challenges

Contents

1.   Introduction

A battery stores electrical energy in the form of chemical energy. This stored energy will be useful in case of more demand and electrical energy crisis [1]. Thus, battery technology is essential [2, 3]. Batteries are gaining more research interest and importance. This is due to the increase in use electric vehicles, consumer electronics, and electric network grid steadiness by massive population, hence increase in energy need [1].

Various types of batteries and systems have been available in the market like lead acetate batteries, nickel batteries, high-temperature Na-beta batteries, Li-ion systems, Li-metal solid-state batteries, Li-S systems, Li-air systems, another metal-air system. However, each of the above-mentioned batteries and systems has certain disadvantages and limitations. Thus this fact paves the way for next-generation systems such as Na-ion, Mg-ion, and Zn-ion battries in aqueous and non-aqueous [2]. This chapter explores the various aspects of Na-ion battries (SIBs). In some literature instead of SIBs, abbreviation NIBs is also used for Na-ion system. However, we are going to use SIBs abbreviation throughout in this chapter.

There is a serious need for future outlook in terms of battery because the LIBs are defamed in many previous decennaries. This is due to its disadvantages and limitations, such as intercalation in charge-discharge procedure, high expenditure, 30% charge state shipment (which is a dangerous condition), and storage limit. In batteries, there is a need for development w.r.t. cost, shelf life, safety, and efficiency. Thus, development in batteries will play a crucial part, as a bridge for such technical fissure or limitations.

Thus, the substitution of LIBs with other monovalent cation batteries such as Na is an exciting topic. This is due to their visionary resemblance with LIBs. Growth stationary energy storage paves the way for future battery systems. Substitution of LIBs for improvement w.r.t. power, durability, standard thermal stability and energy storage, expenditure, 0.0 V discharge feasibility, monthly storage, zero-charge state shipment, and hundreds of cycles on restarting etc.

Thus if these unique and distinctive aspects will be achieved in the future, they will further provide a broad scope of implementation. Herein, we are going to discuss SIB latest progress, future scope, and hurdles.

SIB is less expensive than well established LIBs. This is due to the use of low-cost materials as components in SIBs. Additionally, by not using cobalt in the cathode, it is less-toxic and makes health-related distress less severe. Initial results of substitution of the current copper (Cu) collector (as in LIBs) with aluminum (Al) in SIBs look promising. This will eliminate one of the failure mechanisms, although not much is known about safety measures of SIBs as compared to LIBs [2, 4]. More research is thus needed about safety measures of SIB and to improve less energy density of SIB with respect to (w.r.t.) LIBs [4].

With the once more increasing in prominence of SIBs, we discuss the cause of interests in SIBs, the alliance of LIBs and SIBs system and more usage in the future. Manufacture of SIBs will lead to decrease in LIBs production. Hence SIBs are termed as drop-in-technology for LIBs. However, market and industrial (materialistic) feasibility, security, passive voltage, and cell equilibrium characteristics need to be checked for SIBs [4]. Figure 1 shows the charging and discharging mechanism of SIBs.

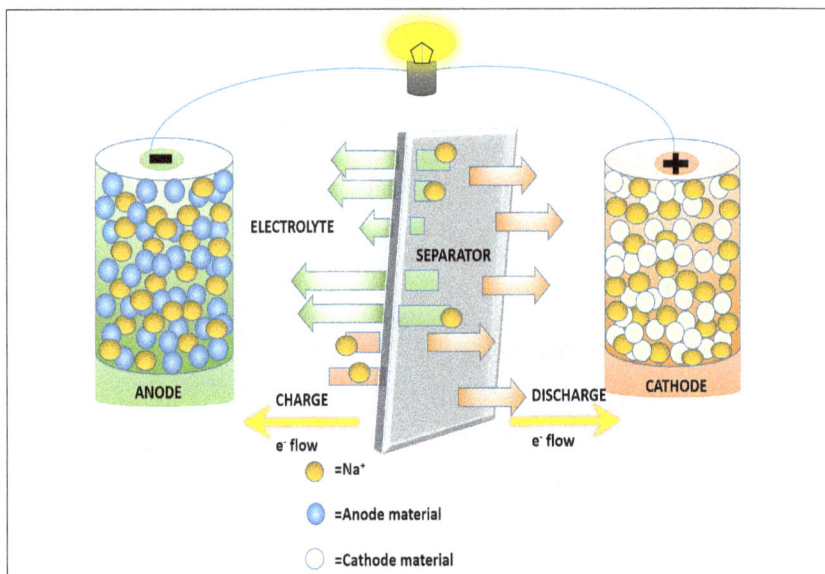

Figure 1. SIBs intercalation working principle, (similar to LIBs).

2.   Background

In the 1970s and 1980s, specifically, SIB and its Na intercalating materials have been studied. However, interest in LIBs sounded because of the commercialization of LIB in the 1990s. And thus, attention shifted to Li intercalation. LIBs are most prominent in various industrial processes. This is because LIBs has broad, diverse use and automobile industry being its most significant consumer. However, LIBs were initially optimized for consumer electronics. With increasing demand, the LIBs usage is speculated to grow with an increment from USD 61 billion (2015) to >USD 95 billion (2025). Whereas, they have various limitations like the unavailability of raw materials, expensive, and safety issues. The resemblance in LIBs and SIBs are precursor components, and cell production or manufacturing procedure as shown in Figure 2.

However, SIBs has first difference as Na^+ charge carrier ion and not Li^+ as in LIBs. SIBs are benefited by different ionic size and atomic weight of Na^+. Na^+ has atomic weight 23 g/mol and ionic radii 102 pm, which is more than Li^+ with 6.9 g/mol, 76 pm. This affects the maximum specific capacities of the materials. And indeed, when subjected to the substitution of these ions with those of another material, the sturdiness of host crystal is different in SIBs.

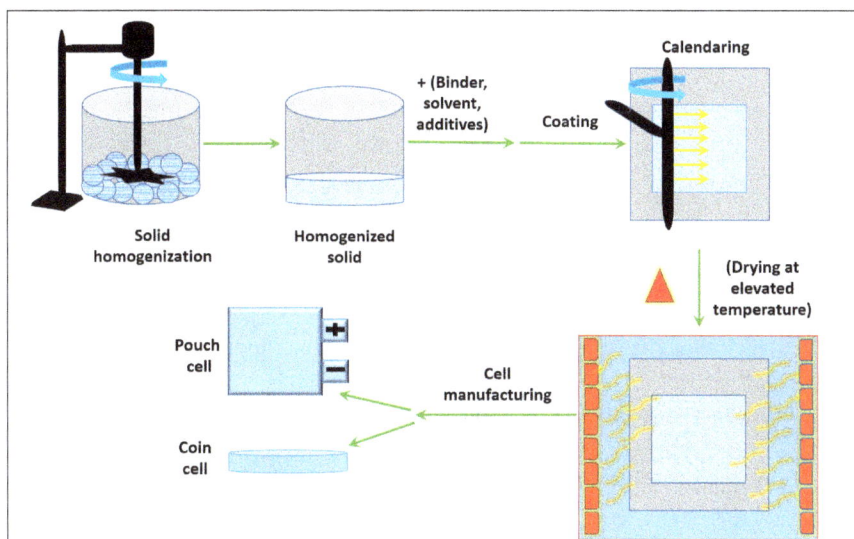

Figure 2. SIBs standard manufacturing procedure, (similar to LIBs).

Low cell voltage for Na material because of more standard electrode potential (SEP) w.r.t. Li (-2.71 V w.r.t SHE cf. -3.02 V w.r.t. SHE) (SHE: Standard Hydrogen Electrode). Irreversible alteration in the crystal structure occurs by withdrawing $>$ ca. 0.5 mol Li from the $Li_{1-x}CoO_2$ crystal structure, and this leads to security distress [5, 6]. In contrast, oxides of the Na layer can show a better sturdiness for $Na_{1-x}MO_2$ [7, 8, 9].

The significant interest of researchers has drawn in the past few years in SIBs. The cost-effective and sustainable precursors for anode and cathode materials have been explored [10, 11]. There is an increase in SIBs research to overcome the higher cost issues of LIBs precursors. These precursors cost about 80% of the overall or final cost of LIBs system. In SIBs results which work in favor of Na^+ diffusion are 3-folds high Na^+ ions diffusion coefficient w.r.t. Li^+, less rigid coordination strength and less energy of solvation. This was seen when Na^+ transport and solvation properties were considered in organic electrolyte solvent like Ethylene Carbonate (E.C.) [12].

Notably, the substitution of components of LIB with Na containing cathode, electrolyte salt, and anode leads to the low cost of SIB. Because of these components compromises of 21%, 8%, and 7%, respectively of the total cell. One such example carbonate salts of Li and Na compounds, Li_2CO_3 (USD 6600/Mt), Na_2CO_3 (USD 60/Mt) shows a high price difference. Al alloys do not amalgamate with Na. However, Al amalgamates with Li below 0.1 V w.r.t. Li/Li^+. Thus, Al may be utilized as an anode and cathode current collector for SIBs. 0 V shipment of SIB is beneficial for security and rate of cell [13]. In contrast to Cu (extensively used anodic current collector for LIB), which raises a safety concern. Thus LIB is shipped at 5-30% charge state to block Cu diffusion and precipitation [14, 15].

3. State-of-the-art or current status of SIBs

As LIBs and SIBs work on the same basic principle of the intercalation reaction mechanism, with a difference of Li^+ with Na^+ into host electrodes for storage. With the current stage of development, SIBs comes in non-aqueous (approximate 250 $W.h.kg^{-1}$, pouch cell) and aqueous electrolyte media (less than 30 $W.h.kg^{-1}$).

Aqueous SIBs are inexpensive due to low-cost material used such as manganese, carbon, water, and sodium sulfate salt. Aqueous SIBs, utilize hard carbon or sodium titanium phosphate as anode material combined with sodium manganese oxide as a cathode, and sodium sulfate water-based as an electrolyte. Aqueous SIBs are generally used in grid storage implementation. In Korea, grid storage sodium-seawater aqueous batteries are in a fully developed state. In Asia, this is a competitive case of SIBs in a high-energy-density grid storage battery, may be used at a larger scale in the future [2].

Non-aqueous SIBs have hard carbon anode and $NaVPO_4F$ or sodium transition metal oxides cathode. Non-aqueous SIBs has symbolic 3.3 V and modest power capability in prismatic and 18650 cylindrical cells.

In the UK, Faradion Ltd., and in France, RS2E and CEA firms were developed the preliminary version of cells. These cells cycled to 3000 cycles at rates demonstrating good power with prismatic cell (140-150 $W.h.kg^{-1}$) and for the 18650 cells (95 $W.h.kg^{-1}$) energy densities.

4. Hurdles in SIBs

Although there are advancement and use of other coating methods [4]. Generally, tape casting type coating method is used for the application of the electrode to the current collectors. Electrodes are manufactured via a many steps process, starting with an electrode ink production. However, wide space in layers of transition metal and water can more easily intercalate [16, 17]. Thus w.r.t. Li analogs, the oxides of Na layer, specifically oxides of O3-type layer are at higher water absorption risk. Therefore, this parameter causes a hurdle in manufacturing of sturdy ink for coating of the electrode.

In LIBs, if anode 10% capacity is increased for safety reason, the Li dendrites are less likely to form. However, in SIBs, 20% is the first cycle irreversible capacity. After the 1st cycle, cathode loses Na and irreversibly results in solid electrolyte interphase (SEI) layer formation on the anode. However, such a contributing feature was unexpected in the cell. Thus, it is essential to keep a note on the accurate mass equilibrium of cathode-anode and voltage control within a battery.

Diffusion of cation through electrode is slowly restricted by DME (Dimethoxyethane) solvent w.r.t. other solvents in conjunction. Similarly, another co-solvent Dimethyl carbonate (DMC) demonstrates better rate capabilities by a decrease in interface resistance [18].

5.   Next-generation battery research

A concern that the technical limitations of LIBs have arrived is bothering [19]. There is a need for a new innovative model of a battery with sturdy electrode precursor, competitive in cost, feasible for industrial mass production. And also, it is important that this new battery model can reach applications which cannot be accessed by today's technology. Further, it should be better to exist technology in terms of characteristics of energy, shelf life, power, and safety to be a future alternative.

Hence, in research and development, there is a requirement of harmonious attempt to develop new raw material or precursor so that the next generation electrochemical cell construction (battery) fulfills the characteristics or aspects as mentioned above and of course, the continuously rising energy need.

5.1 Se_xS_y-based negative electrode materials (NEMs)

New selenium and selenium-sulfur (Se_xS_y)-based cathode or NEMs precursor for room temperature LIBs, and SIBs is reported [20]. Se_xS_y materials present a wide branch of latest battery electrodes, with higher theoretical capacities (675–1550 mAh g^{-1}) and better conductance at room temperature (rt) cycling process compared to both Se, and S systems. This is attributed to its aspect of Se and S being completely miscible [21] along with numerous accessible solid solutions (e.g., Se_5S, Se_5S_4, Se_5S_2, SeS, Se_3S_5, SeS_7, SeS_2) [22]. SeS_{20} the system with much lesser Se can also be prepared with ease as shown in Table 1.

Table 1. Different materilas for cathode (NEM)

Cathode (NEM)				
Layered O3	Layered P2	Organic Compound	Polyanionic Compounds	Prussian Blue Analogues
$NaCrO_2$	$Na_{0.67}[Ni_{1/3}Mn_{2/2}]O_2$	P(AN-NA)	$Na_4Co_3(PO_4)_2P_2O_7$	$Na_4Fe(CN)_6$
$Na[Ni_{0.25}Co_{0.05}Mn_{0.35}]O_2$	$Na_x[Fe_{1/2}Mn_{1/2}]O_2$	$Na_2C_6O_6$	$Na_4Ni_{0.3}Co_{0.24}Mn_{0.3}(PO_4)_2(P_2O_7)$	$Na_4Fe_2(CN)_6$
O3-$Na[Ni_{1/3}Fe_{1/3}Mn_{1/3}]O_2$	$Na_{0.66}[Co_{0.5}Mn_{0.5}]O_2$	$Na_4C_8H_2O_6$	$NaV_4(P_2O_7)_4(PO_4)$	$KFe_2(CN)_6$
$Na[Fe_{0.5}Co_{0.5}]O_2$	$Na_{0.67}[Ni_{0.15}Fe_{0.2}Mn_{0.65}]O_2$		$Na_2Fe_2(SO_4)_3$	$Na_{1.72}MnFe(CN)_6$
$Na[Ni_{0.6}Co_{0.05}Mn_{0.35}]O_2$	$Na_{2/3}[Mg_{0.28}Mn_{0.72}]O_2$		$Na_3V_2(PO_4)_2F_3$	
$Na[Li_{0.05}(Ni_{0.25}Fe_{0.25}Mn_{0.5})_{0.95}]O_2$			$Na_{1.5}VPO_{4.8}F_{0.7}$	
$NaMnO_2$			$Na_2MnP_2O_7$	
			$Na_3V_2(PO_4)_3$	
			$Na_2Fe_2PO_4F$	
			$NaFePO_4$	
			$NaMnFe_2(PO_4)_3$	

For anodes tunned electrodes, further showed additional potential for this combined system of Se_xS_y. Combining the high capacities of S-rich systems with the high electrical conductivity of Se might be due to the presence of d-electron in Se. This shows excellent electrochemical aspects, along with Li and Na systems. Pair distribution function (PDF) analysis was used for investigating electrode structural mechanisms for Li or Na inclusion or addition.

Both Se system and Se_xS_y system overcome the disadvantage of cycle failure of Li/S systems. Because it can be cycled to high voltages (up to 4.6 V) without fault, this paves the way for a new energy storage system that can facilitate electric automobiles and smart electrical network. Due to its advantages like higher density, voltage efficiency, and shows volumetric energy densities more than even S-based batteries.

As Se is in trace amounts in supplement and personal care products, thus, the toxicity of Se should be analyzed. Se is toxic as S and another element used in common electrode like Ni, Co (Lethal Dose, 50% i.e., LD50: Se ~6.2 g; Ni ~5.0 g; S ~8.4 g and Co ~6.7 g;). Se has a higher price than S, because it is less abundant, and this may affect the commercialization. However, this effect can be reversed by the usage of Na instead of Li or by the usage of combined Se_xS_y systems. In the present propelling force to find novel and better efficiency material. The purpose of this is to seek better electrochemical

Sodium-Ion Batteries: Materials and Applications   Materials Research Forum LLC
Materials Research Foundations **76** (2020) 229-250   https://doi.org/10.21741/9781644900833-10

energy storage, and this research shows a new potential rechargeable battery. Se_xS_y-based electrodes may show higher theoretical capacities than the Se system. Better output and conductance w.r.t. to S system, is seen by the introductory study of SeS_2–C, one mixed chalcogenide system.

In Li/SeS_2–C system even after 30 cycles the discharge capacity is 512 -394 $mAhg^{-1}$, i.e. it is 30% more w.r.t. Li/SeC (with 288 mAh g^{-1} capacity). SeS_2–C can be cycled with Na at rt.

5.2 $Na_3M_2(PO_4)_2F_3$ [$M_{1/4}Ti$, Fe, V] based NEMs

A sequence of compounds $Na_3M_2(PO_4)_2F_3$ [$M_{1/4}Ti$, Fe, V] were produced by a traditional two-step synthesis method, and their electrochemical aspects as the active cathode or NEMs precursor of SIB have also been examined [23]. The best aspects among the sequence of compounds were found in $Na_3M_2(PO_4)_2F_3$ are sturdiness on cycling and capacity approaching 120 $mAhg^{-1}$ vs. Na. And this was due to contraction and expansion in c-axis in co-occurrence with removal and inclusion of Na atoms. This occurs in the (002) aeb plane in the tetragonal structure; this was affirmed by a structural evaluation. This material to possess a tetragonal structure with a space group, P42/ mnm, and were isostructural with b- $Na_3M_2(PO_4)_2F_3$ was shown by X-ray powder diffraction (XRD) measurements. Authors have calculated charge-discharge capacity upon the 40th cycle. These compounds are able to sustain the capacity to 98%. Ex-situ XRD data of $Na_3M_2(PO_4)_2F_3$ electrodes at particular voltage charge-discharge, showed that the expansion of material structure took place in the process of charging up to 4.3 V as shown in Table 1. However on cycling the original size reoccur.

5.3   Inclusion of fluorinated ethylene carbonate (FEC) in the electrolyte

In well recognized organic electrolytes that can cause film formulation in LIBs such as fluorinated ethylene carbonate (FEC), trans-difluoroetyhenecarbonate (DFEC), vinylene carbonate (VC), and ethylene sulfite (ES). While in SIB, it has been observed that the inclusion of FEC in the electrolyte is the most advantageous.

The favorable result or improvement is in hard-carbon and $NaNi_{1/2}Mn_{1/2}O_2$ electrodes in aprotic Na cells in terms of electrochemical Na insertion reversibility. The additive is also capable of higher reversibility with the electrochemical deposition/dissolution of metallic Na. This is because of better passivation and side reaction repression in Na and propylene carbonate solution containing salts of Na. The merging of electrolyte and electrode precursors has to be optimized in the future, to know the best result of this study [24].

5.4   Efficient cycling process by Sb in SIBs

When Sb is taken as NEMs, the result showed that 99 over 160 coulombic efficiency, capacity up to 600 mAh g^{-1} is also maintained at a higher rate cycle [25]. Thus, in SIBs, pure Sb (micrometric) can work excellently as NEM with maximum capacity obtained examined with Li and Na systems. Thus, w.r.t. other compounds this shows much better capacity. In Sb/Li system, the reaction process taking place is of amalgamation and de-amalgamation.

Further, also in concurrence with previous review paper by S.W. Kim et al. [26], the electrochemical mechanism is not as simple and undeviating like Li. In Li, the amalgamation mechanism is as intermediates anticipated in the phase diagram. However, in SIBs, a completely different behavior may be due to more complicated electrochemical reaction mechanism taking place. Yet, the intermediates of SIBs could not be identified in this study. Although it is thought they can be amorphous intermediate phases during the cycling process. An unexpected result of competitive formulation in between hexagonal and the cubic polymorphs of Na_3Sb at end discharge of Sb/Na cell. Na_3Sb are produced only in 1-9 GP pressure. However, Na_3Sb is not stable at atmospheric pressure. Further, the advantageous aspects of Na_3Sb material may be due to the FEC inclusion to the electrolyte, mixed with a suitable electrode of carboxymethyl cellulose (CMC), vapor ground carbon fibers and carbon black.

The advantageous characteristics may also be attributed to the reduction of volume expansion on substitution of Li_3Sb (283.8 Å3) with hexagonal Na_3Sb (237 Å3). The betterment of cycle in the Sb/Na cell as compared to the Sb/Li cell can also be due to the amorphous intermediate because of this amorphous intermediate can reduce the strain by acting as a buffer. Thus, such excellent aspects of Sb/Na, which are absent in Sb/Li can lead to increase of NEM development and research in the Sb/Na system. Thus intermetallic or conversion phases which are examined in the Li system can be seen as future prospects if the Li-ion intermetallic phases can be better in terms of potential than Na/Na^+ or Na system. Thus for future opportunities, all conversion-type precursor may be an engrossed material for the anode in SIBs. This topic should be seen again to obtain excellent electrochemical efficiency.

5.5   SnSb as NEMs

SnSb is reported as potential NEMs for rechargeable battery due to its noteworthy electrochemical execution [27]. Great results such as exceeding coulombic efficiency, reversible capacity, and cycles: 97%, 525 mAh g^{-1}, 125 cycles at a C/2 rate (55 mAg^{-1}) respectively. This was found when CMC binder and an electrolyte containing fluoroethylene carbonate (FEC) was used. For an electrode material vs. Na, this research showed the maximum life cycle ever reported.

On the NEMs of SIBS, many investigations are present. However, many more research is on a positive electrode precursor. The reaction of insertion and amalgamation is influenced by substitution of Li^+ with Na^+. Due to the increase of ionic radius of Na wr.rt. Li, which further on cycling, causes more volume enlargement. Alteration to the traditional precursor of NEMs with Sn- and Sb-based NEMs have demonstrated to be a feasible substitution for hard carbon. Sn- and Sb-based NEMs show various non-fading electrochemical cycle and more than 400 $mAhg^{-1}$ reversible capacity [28, 29]. Na vs. Li system on cycling demonstrates reduction of capacity. Whereas Sb-based NEMs can show about 160 cycles at C/2 (55 mAg^{-1}) capacity sustainability vs. Na system [30].

It has been demonstrated that there is the betterment of cycle life by the inclusion of slight quantity FEC in the electrolyte. And this is also shown by past few studies that this addition leads to a stable SEI layer in Li and Na batteries [30-32]. However, the addition is still in doubt, due to other current research that shows this inclusion or addition of FEC may be disadvantageous [33].

A previous report by L. Xiao et al. showed the intriguing result of high-energy mechanical milling produced SnSb/C nanocomposites [28]. SnSb was created by 60 min short ball milling of Sn and Sb and proper formation of the electrode. By precursory study vs. Li, nanocomposite shows 2-plateau discharge mechanism. The charge-discharge plot or profile of SnSb/C nanocomposite electrode shows two plateaus. The 1st, 2nd plateau is due to the formation of Na_3Sb and metallic Sn via Na with SnSb reaction, slow tin sodiation respectively. This composite showed 1st irreversible cycle capacity of approximately 200 mAh g^{-1}. This is because of continuous gradual fading in the retention of capacity in about 50 cycles. The current report demonstrates excellent cycle results by SnSb.

Ali Darwiche et al. showed that at 50th and 125th, there is increment, decrement, and again increment in cycle ~500 to reach ~550 $mAhg^{-1}$ and 525 $mAhg^{-1}$ respectively [27]. Till 160th cycle, the capacity is increased and preserved, at 650 $mAhg^{-1}$ and 0.02 V plateau is an extension seen the due to approaching of lower potential limit to 0.0 V. The plateau on 1st sodiaton shows three peaks 0.5, 0.3 and approximate 0.02 V strong peak. On 2nd sodiation and following ones, there is an increase in the potential of the 1st plateau at 0.65, while the other at 0.3 and 0.05 is not affected. Conversion-type materials (either vs. Li or Na) mostly showing 1st particular discharge at lesser potential is similar to that seen in 1st plateau potential increment. In 1st conversion, the raw material electrochemical grinding leads to reduction and restructuring of particle size thus leads to a decrease in polarization. A. Darwiche et al. concluded that the cycle life of SnSb electrode is much better than SnSb/C [27, 28]. On numerous cycles at C/2 (111 mA) aspects like stable capacity, a reversible capacity is 500–550 mAhg^{-1}, 700 mAhg^{-1} at C/10 (11 mA/g) respectively is shown by SnSb electrode with 4.9 Na. The reversible capacity is increased due to the increase of 0.02 to 0 V potential window. When we consider the more volume increment due to SnSb with ~4.9 Na (226%) reaction, the excellent cycle life was not expected.

Hence the result which contrasts the expectation is interesting. The unexpected result can be investigated for further complete information of electrochemical mechanism, which is taking place. Till today the SnSb reaction mechanism with Na unlike with Li, cannot be explained by addition reaction mechanism of pure Sn and Sb with Na. This demonstrates that SnSb with Na has a clearly defined different mechanism, makes it the 1st report to do so. Thus, if the coulombic efficiency problem resolves in the future, the Sn and Sb-based intermetallic will be a practical NEMs (Table 1).

6.   Economic perspective of SIBs

Cell price for lithium nickel manganese cobalt oxide (victorious example of Li-ion systems) battery (which has nickel manganese cobalt (NMC) cathode) is 188 USD/kWh, 250 USD/kWh for SIBs, 256 USD/kWh for Lithium-iron-phosphate (LFP) [34]. The final price of SIB is mostly dependent on the cathode and anode components. Thus, these component price fluctuation may lead to a change in storage capacity price/kWh.

The raw materials price has a major effect, such as in SIB anode hard carbon (as anode) cost affect the final or net worth of SIBs. Hard carbon is produced by organic by-product and fossil fuel. In SIBs, Cu anode replacement with Al for current collectors further decreases the price of SIBs. The final price is not much differed by electrolyte substitution of Li with Na. This is because there is a symbolic cost decrement from 18 (Li) to 17.7 USD/L (Na). However, in terms of energy density, SIBs are at lower stage w.r.t. LIBs. If this issue is solved, SIB can be seen as future generation alternative to LIBs.

6.1   Battery Performance and Cost model (BatPaC model)

A modified form of BatPaC model was used to evaluate the final price of individual 18650 round cells. The assessed SIBs were based on a layered oxide cathode: sodium nickel   manganese   magnesium titanate oxide   (NMMT); $Na_{1.1}Ni_{0.3}Mn_{0.5}Mg_{0.05}Ti_{0.05}O_2$) [35] with a hard carbon anode and an electrolyte made of sodium hexafluorophosphate ($NaPF_6$) salt in an organic solvent [36, 37]. Remaining battery component such as collector foils, electrode binders, the separator was same in SIBs and LIBs.

6.2　Cost of cathode

For the NMC, NMMT precursor price of 26.23 USD/kg, 13.96 USD/kg respectively were evaluated. Due to the absence of a market of NMMT, the cost of NMMT cathode was assessed by using the production and precursor price. The evaluation of NMC was done in similar lines for unbiased results and to maintain the validity of price estimation. Similarly, it was done for LFP, and price evaluation concluded at 15.57 USD.

6.3　Cost of anode

Graphite is mostly used anode precursor in LIBs. However, in SIBs Na does not undergo spontaneous intercalation with graphite, resulting in low capacity. Thus, hard carbon is used in SIBs [13, 23]. Sodium ion does not undergo intercalation with graphite due to its more ionic radius. Therefore there is a need for research on adjusting or tuning the interlayer gap and make the most effective use of available graphene-based materials for Na^+ intercalation [38]. On 250 cycle at 200 mA g^{-1} current density, the specific capacity approaches 100 mAhg^{-1} [39]. And in SIBs on ten cycles at a current density of 50 mA g^{-1} caused specific capacity to approach ~300 mAhg^{-1}. This demonstrated the effective support of more interlayer gap in tuned graphene-based anode [40]. Because of no market price availability of hard carbon, the synthetic graphite price was presumed to be 15.38 USD/kg [41,42].

6.4　Cost of electrolyte

The price evaluation of SIBs electrolyte was like LIBs due to the resemblance in the manufacturing of lithium hexafluorophosphate ($LiPF_6$) (17.99 USD/l) and $NaPF_6$(17.74 USD/l). The only difference is in alkali metal in the precursor (Li_2CO_3/Na_2CO_3); thus, the difference in price being minimal.

6.5　Fluctuations or variation in price

In battery production, the alteration in the price of the precursor is the primary concern. Thus, the investigation was carried out by taking the arithmetic mean of the highest and lowest precursor price in the previous decade [42]. In the future, further research of various hard carbon in terms of differing electrochemical aspects will be a favorable study. Energy density can be better by attenuation of anode precursor quantity. And for this purpose, the characterization of higher efficiency hard carbon is needed in the future. The maximum variation in price was seen in Co and Ni in cathode precursor. The Li and

Cu lead to fluctuation in the final cost of the cell. These metals being stable, their price did not alter in the previous decade. However, the LIBs final price is strongly influenced by battery demand. The increase in battery demand leads to a current increment in Li price. In case of SIB because of the presence of expensive Ni in high amount in the cathode; this may lead to an interest in substituted nickel-free SIB cathode.

6.6   Limitation of BatPaC model

The commercialization of SIB still does not occur. The current evaluation of LIBs and SIBs manufacturing was presumed to be in Germany, for fair comparison. Thus, cell cost will be fluctuated by different production rates in each country. Although no precise details of such fluctuation are given by the BatPaC model, as it does not specify all necessary information. Asia forms the maximum part of LIBs production. Decrease of final cell rate to 164.9 USD/kWh (NMC batteries), 219.7 USD/kWh (LFP), and 212.9 USD/kWh (SIB) by presuming human resources costs 5.6 USD/h. Furthermore, decrement in rate can be observed by the mass production of batteries. But this is negligibly noticed because this will lead to changes in all battery characteristics similarly.

Since the price of all types of battery will be affected identically by scale effects, further research on scale effects is thus not investigated in the study by J.F. Peters et al. [34]. However, w.r.t. per single cell, NMC-type cells are the most expensive while SIB being inexpensive. Although storage capacity is not noticed in per cell price for the SIB, per single cell cathode precursor being less in price leads to a decrease in price [34]. They have concluded that LIBs is already at a developed stage. However, SIB is still mature and has more aspects of being explored in the future. And thus, in future further SIBs feature investigation is anticipated, so that it can be competitive to LIBs. Due to the presence of less information and research on SIB, the steady assurance of the results of the final price of SIBs is less [34].

7.   A materialistic outlook of SIBs

Research on electrochemical sodium intercalation in the battery system has been reported since the early 1980s, but SIBs are not yet materialistic. Precursor component of SIB such as anode, cathode, and electrolyte salt are analyzed [4]. Thus SIB technology is flourishing. Many cathodes and anode materials for SIBs can be characterized by groups. For cathodes: layered O3, layered P2, polyanionic compounds, and Prussian blue analogs; for anodes: carbonaceous, alloy, phosphoric, and metal oxide/sulfide [44,45]. For further interpretation and effectiveness of SIBs system anodes and cathodes loading, anode-cathode mass equilibrium, which now is not always recorded, must be filed in the future. The cells with layered oxide cathode usually show higher reversible capacities compared to polyanion cells.

The pre-sodiation is possible or not for large scale commercialization should be investigated. However, as expected, cells with pre-sodiated anode demonstrate more reversible specific capacity.

8.   Challenges of SIBs

8.1   Limitations and materialistic barriers

Limitations like economic value, performance, and safety, environment-friendly, industrial validation, improvement in the material are the barriers to battery technology at commercial scale. These technical hurdles, if solved, will become the fundamental key for possible next-generation rechargeable batteries development and commercialization. Cost mostly depend on the manufacturing of components of a cell-like cathode, anode, and electrolyte salt. The improvement in SIB system is not so much restricted by presently available precursor. However, reduction in the cost of salt like $NaPF_6$ used in SIB electrode is presumed, as commercial need increases. Next generation technology has a broad scope if it tackles the challenges and applications which do not work with existing technology. By preventing cell degradation process like corrosion, the cell can have a better shelf life [2].

The commercialization or materialization of SIB is restricted due to some issues such as air unstability of few materials specifical oxides of O3-type layers. Thus on industrial scale, this issue can lead to difficulty in the mixing and coating processes(which are for the betterment of SIB shelf life) of materials. For SIB, it is not feasible to approach the high energy density of already accepted LIB technology, and this is due to the more atomic weight of Na and more SEP w.r.t. SHE [4].

8.2   Challenges of NEMs

A feasible SIB needs a NEM that produces an adequate amount of volumetric energy density. The investigation shows in SIB the NEM with the formulation of Na alloys. This NEM approaches approximately 50% of volumetric energy than that of Li. This is due to more Na size w.r.t. to Li. Thus, present SIB cannot be comparable to LIB.

Sodiated hard carbon volumetric density will be remarkably lower than for lithiated graphite. Thus, these results pave the way for a new innovative plan for NEM, or positive electrode materials (PEMs) of SIB to show more maintained energy density w.r.t. LIB.

Thus, PEMs advantage may mask the disadvantage of current NEMs. Table 2 shows different materilas for anode (PEM).

Table 2. Different materilas for anode (PEM)

ANODE (PEM)			
Carbonaceous	Metal Alloy	TMO, TMS	Organic
Natural Graphite	Sn	$NaTiO_2$	$Na_2C_{10}H_2O_4$
Non-graphitic Carbon (hard carbon)	Ge	$Li_4Ti_5O_{12}$	Na_2DBQ
S, N-doped carbon, Graphene	Sb	TiO_2	Schiff Bases
	SnSb	$Na_4Ti_5O_{12}$	$Na_2C_8H_4O_4$
	Sn_xP_3	$Na_2Ti_3O_7$	
	P	Fe_3O4, Fe_2O_3	
		$CoS_x, MoS_2, Co_3O_4SnO_x$	
		FeS_x	
		$Cu_xO_7 SnS_x$	
		Z_xS	

Thus, the substitution of precursor or to imitate the present, LIB may not show SIB as a feasible competitive contestant of LIB. Therefore, the promotion of a novel plan in SIB as a future outlook is the need of the hour. To improve the lifetime of cell and moisture sensitive materials, sustain ink stability, further investigation of precursor and electrode sturdiness at diverse humid and temperature level must be performed. Aspects of cell production, i.e., anode-cathode equilibrium and the formulation process regulate the shelf life and cell efficiency.

9. Future opportunities

In precise, SIB has a full scope of development and research. More literature survey needs in the examining, identification of the precursor used. The composite electrodes, construction of the materials, examining procedures, and real characteristics of these materials may be discovered in the future. There is a need to tackle the gap of LIBs system and applications in which LIBs are not applicable. Generally, undeviating move from LIB to SIB may not be a first path. This is due to additional processing is needed in SIB w.r.t. LIB. However, SIB is tagged as "drop-in" technology for LIB is emphasized here.

The insights obtained from LIBs research and its similarity and dissimilarities with SIBs will be useful for future research on SIBs. Challenges remain in the analysis of battery processes that control operation and restrict performance. But the research holds the key for further development of SIBs. Regardless of this, there is scope for future generation SIBs system for bridging the gap where LIBs are not suitable. And other factors like Li reservoirs are geographically restricted is unfavorable for future increase of this technology. However, on the planet, Na is present in seawater and in the form of a mineral. Na is the 6th most available element [4, 13], thus can be seen to be an inexpensive substitution to LIB in the future.

Beyond LIBs, the future foundational investigation remains in solving the problem of battery processes, that regulate working and cause limitations [3]. The energy density of commercially well-utilized LIB has just increased approximately by a factor of 4 in the previous 25 years. Thus, battery development velocity remains a challenge [46]. State-of-the-art commercial LIBs have an energy density of less than 300 Whkg^{-1} [48, 49], which is lesser than the U.S. Department of Energy target of 400 Whkg^{-1} in 2017. The LIBs will be a significant part of commercial batteries used until LIBs are not possible to fulfill the increasing future energy need. Thus, research on novel battery technology is the need of the hour [50].

The SIB electrode, electrolyte, and their many interfacial characteristics are still not studied, which are the cause of its improvement in some aspects w.r.t. LIB. Thus understanding the basic battery process is needed to obtain the genuine prospects of SIB. Any new technology development brings advantages and disadvantages of its own [51]. SIB has advantages like same production lines as LIB and low cost due to the profusion of Na source. However, SIB has disadvantages like low energy density because of more ionic weight. And if we consider other next-generation batteries such as the multivalent ion battery (MIB) of Mg, Al, etc. has a disadvantage of various charge transfer. Thus in MIB, a new hurdle of less ion conductance is formed. And main challenge occurs in all battery types is during the transmission of scientific innovation to materialistic or commercial victory.

Acknowledgment

Dinesh Kumar is also thankful DST, New Delhi for financial support to this work (sanctioned vide project Sanction Order F. No. DST/TM/WTI/WIC/2K17/124(C).

References

[1] M. Sawicki, L.L. Shaw, Advances and challenges of sodium ion batteries as post-lithium-ion batteries, RSC Adv. 5 (2015) 53129–53154. https://doi.org/10.1039/c5ra08321d

[2] C. Ma, Y. Cheng, K. Yin, J. Luo, A. Sharafi, J. Sakamoto, J. Li, K.L. More, N.J. Dudney, M. Chi, Interfacial stability of Li metal-solid electrolyte elucidated via in situ electron microscopy, Nano Lett. 16 (2016) 7030–7036. https://doi.org/10.1021/acs.nanolett.6b03223

[3] A. Longfield, Foundations for the future, Early Years Educ. 13 (2014) 1–8.

[4] S. Roberts, E. Kendrick, The re-emergence of sodium ion batteries: Testing, processing, and manufacturability, Nanotechnol. Sci. Appl. 11 (2018) 23. https://doi.org/10.2147/nsa.s146365

[5] R. Yazami, Y. Ozawa, H. Gabrisch, B. Fultz, Mechanism of electrochemical performance decay in $LiCoO_2$ aged at high voltage, Electrochim. Acta 50 (2004) 385–390. https://doi.org/10.1016/j.electacta.2004.03.048

[6] J. Ye, H. Chen, Q. Wang, P. Huang, J. Sun, S. Lo, Thermal behavior and failure mechanism of lithium-ion cells during overcharge under adiabatic conditions, Appl. Energy 182 (2016) 464–474. https://doi.org/10.1016/j.apenergy.2016.08.124

[7] D. Milan, D. Peal, (12) Patent Application Publication (10) Pub. No .: US 2002/0187020 A1, 1 (2013).

[8] K. Smith, J. Treacher, D. Ledwoch, P. Adamson, E. Kendrick, Novel high energy density sodium layered oxide cathode materials: from material to cells. ECS Transactions 75(22) (2017) 13–24. https://doi.org/10.1149/07522.0013ecst

[9] J.B. Robinson, D.P. Finegan, T.M.M. Heenan, K. Smith, E. Kendrick, D.J.L. Brett, P.R. Shearing, Microstructural analysis of the effects of thermal runaway on li-ion and na-ion battery electrodes, J. Electrochem. Energy Convers. Storage 15 (2017) 011010. https://doi.org/10.1115/1.4038518

[10] Y. You, A. Manthiram, Progress in high-voltage cathode materials for rechargeable sodium-ion batteries, Adv. Energy Mater. 8 (2017) 1–11. https://doi.org/10.1002/aenm.201701785

[11] R.J. Clément, P.G. Bruce, C.P. Grey, Review—Manganese-based P2-type transition metal oxides as sodium-ion battery cathode materials, J. Electrochem. Soc. 162 (2015) A2589–A2604. https://doi.org/10.1149/2.0201514jes

[12] T.A. Pham, K.E. Kweon, A. Samanta, V. Lordi, J.E. Pask, Solvation and dynamics of sodium and potassium in ethylene carbonate from ab initio molecular dynamics simulations, J. Phys. Chem. C. 121 (2017) 21913–21920. https://doi.org/10.1021/acs.jpcc.7b06457

[13] F. Bonaccorso, V. Pellegrini, Graphene and Other 2D crystals for rechargeable batteries, material matters, (2016), Retrieved from https://www.sigmaaldrich.com/technical-documents/articles/material-matters/graphene-and-other-2d-crystals.html

[14] M.D. Farrington, Proposed amendments to UN ST/SG/AC.10/11: Transport of dangerous goods - lithium batteries, J. Power Sources 80 (1999) 278–285. https://doi.org/10.1016/s0378-7753(99)00077-4

[15] M.D. Farrington, Safety of lithium batteries in transportation, J. Power Sources 96 (2001) 260–265. https://doi.org/10.1016/s0378-7753(01)00565-1

[16] K. Takada, M. Itose, K. Fukuda, R. Ma, T. Sasaki, Y. Ebina, X. Yang, Highly Swollen Layered Nickel Oxide with a Trilayer Hydrate Structure, Chem. Mater. 20 (2007) 479–485. https://doi.org/10.1021/cm702981a

[17] D. Buchholz, L.G. Chagas, C. Vaalma, L. Wu, S. Passerini, Water sensitivity of layered P2/P3-$Na_xNi_{0.22}Co_{0.11}Mn_{0.66}O_2$ cathode material, J. Mater. Chem. A. 2 (2014) 13415–13421. https://doi.org/10.1039/c4ta02627f

[18] A. Ponrouch, R. Dedryvère, D. Monti, A.E. Demet, J.M. Ateba Mba, L. Croguennec, C. Masquelier, P. Johansson, M.R. Palacín, Towards high energy density sodium ion batteries through electrolyte optimization, Energy Environ. Sci. 6 (2013) 2361–2369. https://doi.org/10.1039/c3ee41379a

[19] Next-generation battery research (2019). Retrieved from https://www.internationalbatteryseminar.com/battery-research, cited on March 28, 2019

[20] A. Abouimrane, D. Dambournet, K.W. Chapman, P.J. Chupas, W. Weng, K. Amine, A new class of lithium and sodium rechargeable batteries based on selenium and selenium-sulfur as a positive electrode, J. Am. Chem. Soc. 134 (2012) 4505–4508. https://doi.org/10.1021/ja211766q

[21] P. M. Hansen, K. Anderko, Construction of binary alloys, New York 1958; p 1162.

[22] M.F. Kotkata, S.A. Nouh, L. Farkas, M.M. Radwan, Structural studies of glassy and crystalline selenium-sulphur compounds, J. Mater. Sci. 27 (1992) 1785–1794. https://doi.org/10.1007/bf01107205

[23] K. Chihara, A. Kitajou, I.D. Gocheva, S. Okada, J.I. Yamaki, Cathode properties of $Na_3M_2(PO_4)_2F_3$ [M = Ti, Fe, V] for sodium-ion batteries, J. Power Sources 227 (2013) 80–85. https://doi.org/10.1016/j.jpowsour.2012.10.034

[24] S. Komaba, T. Ishikawa, N. Yabuuchi, W. Murata, A. Ito, Y. Ohsawa, Fluorinated ethylene carbonate as electrolyte additive for, ACS Appl. Mater. Interfaces 3 (2011) 4165–4168. https://doi.org/10.1021/am200973k

[25] A. Darwiche, C. Marino, M.T. Sougrati, B. Fraisse, L. Stievano, L. Monconduit, Better cycling performances of bulk Sb in Na-ion batteries compared to Li-ion systems: an unexpected electrochemical mechanism, J. Am. Chem. Soc. 135 (2012) 10179. https://doi.org/10.1021/ja4056195

[26] S.W. Kim, D.H. Seo, X. Ma, G. Ceder, K. Kang, Electrode materials for rechargeable sodium-ion batteries: Potential alternatives to current lithium-ion batteries, Adv. Energy Mater. 2 (2012) 710–721. https://doi.org/10.1002/aenm.201200026

[27] A. Darwiche, M.T. Sougrati, B. Fraisse, L. Stievano, L. Monconduit, Facile synthesis and long cycle life of SnSb as negative electrode material for Na-ion batteries, Electrochem. Commun. 32 (2013) 18–21. https://doi.org/10.1016/j.elecom.2013.03.029

[28] L. Xiao, Y. Cao, J. Xiao, W. Wang, L. Kovarik, Z. Nie, J. Liu, High capacity, reversible alloying reactions in SnSb/C nanocomposites for Na-ion battery applications, Chem. Commun. 48 (2012) 3321–3323. https://doi.org/10.1039/c2cc17129e

[29] M.K. Datta, R. Epur, P. Saha, K. Kadakia, S.K. Park, P.N. Kumta, Tin and graphite based nanocomposites: Potential anode for sodium-ion batteries, J. Power Sources 225 (2013) 316–322. https://doi.org/10.1016/j.jpowsour.2012.10.014

[30] J. Qian, Y. Chen, L. Wu, Y. Cao, X. Ai, H. Yang, High capacity Na-storage and superior cyclability of nanocomposite Sb/C anode for Na-ion batteries, Chem. Commun. 48 (2012) 7070–7072. https://doi.org/10.1039/c2cc32730a

[31] S. Komaba, Y. Matsuura, T. Ishikawa, N. Yabuuchi, W. Murata, S. Kuze, Redox reaction of Sn-polyacrylate electrodes in aprotic Na cell, Electrochem. Commun. 21 (2012) 65–68. https://doi.org/10.1016/j.elecom.2012.05.017

[32] H.A. Wilhelm, C. Marino, A. Darwiche, L. Monconduit, B. Lestriez, Significant electrochemical performance improvement of TiSnSb as anode material for Li-ion batteries with composite electrode formulation and the use of VC and FEC electrolyte

additives, Electrochem. Commun. 24 (2012) 89–92.
https://doi.org/10.1016/j.elecom.2012.08.023

[33] A. Ponrouch, A.R. Goñi, M.R. Palacín, High capacity hard carbon anodes for sodium ion batteries in additive-free electrolyte, Electrochem. Commun. 27 (2013) 85–88. https://doi.org/10.1016/j.elecom.2012.10.038

[34] J.F. Peters, A. Peña Cruz, M. Weil, Exploring the Economic Potential of Sodium-Ion Batteries, Batteries 5 (2019) 10. https://doi.org/10.3390/batteries5010010

[35] J. Barker, R. Heap, Doped nickelate compounds. International Patent Application No. WO2014/009710 A1, 16 January 2014

[36] M.F. Kotkata, S.A. Nouh, L. Farkas, M.M. Radwan, Structural studies of glassy and crystalline selenium-sulfur compounds, J. Mater. Sci. 27 (1992) 1785–1794. https://doi.org/10.1007/bf01107205

[37] J. Peters, D. Buchholz, S. Passerini, M. Weil, Life cycle assessment of sodium-ion batteries, *Energy Environ. Sci.* 9(5), (2016), 1744-1751. https://doi.org/10.1039/c6ee00640j

[38] T. Liu, M. Leskes, W. Yu, A.J. Moore, L. Zhou, P.M. Bayley, G. Kim, C.P. Grey, U. CA Cambardella, USDA Agricultural Research Service, Ames, Formation and Decomposition, Sci. Technol. 350 (2005) 2004.

[39] Y.X. Wang, S.L. Chou, H.K. Liu, S.X. Dou, Reduced graphene oxide with superior cycling stability and rate capability for sodium storage, Carbon 57 (2013) 202–208. https://doi.org/10.1016/j.carbon.2013.01.064

[40] J. Ding, H. Wang, Z. Li, A. Kohandehghan, K. Cui, Z. Xu, carbon nanosheet frameworks derived from peat moss as high-performance NIBs, ACS Nano (2013) 11004–11015. https://doi.org/10.1021/nn404640c

[41] P.A. Nelson, K.G. Gallagher, I.D. Bloom, D.W. Dees, Modeling the performance and cost of lithium-ion batteries for electric-drive vehicles – 2 Ed, (2012). https://doi.org/10.2172/1209682

[42] U.S. Geological Survey, Mineral commodities summaries, Miner. Commod. Summ. 2016. (2016) 205.

[43] F. Renard, 2020 cathode materials cost competition for large scale applications and promising LFP best-in-class performer in term of price per kWh. In Proceedings of the International Conference on Olivines for Rechargeable Batteries, Montreal, QC, Canada, 25–28 May 2014

[44] J.Y. Hwang, S.T. Myung, Y.K. Sun, Sodium-ion batteries: Present and future, Chem. Soc. Rev. 46 (2017) 3529–3614. https://doi.org/10.1039/c6cs00776g

[45] W. Ren, Z. Zhu, Q. An, L. Mai, Emerging prototype sodium-ion full cells with nanostructured electrode materials, Small 13 (2017) 1–31. https://doi.org/10.1002/smll.201604181

[46] G. Crabtree, E. Kócs, L. Trahey, The energy-storage frontier: Lithium-ion batteries and beyond, MRS Bull. 40 (2015) 1067–1078. https://doi.org/10.1557/mrs.2015.259

[47] R.C. Armstrong, C. Wolfram, K.P. de Jong, R. Gross, N.S. Lewis, B. Boardman, A.J. Ragauskas, K. Ehrhardt-Martinez, G. Crabtree, M. V. Ramana, The frontiers of energy, Nat. Energy. 1 (2016) 15020. https://doi.org/10.1038/nenergy.2015.20

[48] SANYO Energy (U.S.A.) Corporation, Panasonic NCR18650B Specifications, (2012) 18650.

[49] Richard Van Noorden. (2014, March 5). The rechargeable revolution: A better battery. Retrieved from https://doi.org/10.1038/507026a

[50] J.W. Choi, D. Aurbach, Promise and reality of post-lithium-ion batteries with high energy densities, Nat. Rev. Mater. 1 (2016) 16013. https://doi.org/10.1038/natrevmats.2016.13

[51] V.L. Chevrier, G. Ceder, Challenges for Na-ion Negative Electrodes, J. Electrochem. Soc. 158 (2011) A1011. https://doi.org/10.1149/1.3607983

Sodium-Ion Batteries: Materials and Applications   Materials Research Forum LLC
Materials Research Foundations 76 (2020) 251-271   https://doi.org/10.21741/9781644900833-11

Chapter 11

Conducting Polymers for Sodium-Ion Batteries

Suzhe Liang[1,2], Yonggao Xia[2,3*], Peter Müller-Buschbaum[1,4*], Ya-Jun Cheng[2,5*]

[1]Lehrstuhl für Funktionelle Materialien, Physik-Department, Technische Universität München, James-Franck-Str. 1, 85748, Garching, Germany

[2]Ningbo Institute of Materials Technology & Engineering, Chinese Academy of Sciences, 1219 Zhongguan West Rd, Zhenhai District, Ningbo, Zhejiang Province, 315201, P.R. China

[3]Center of Materials Science and Optoelectronics Engineering, University of Chinese Academy of Sciences, 19A Yuquan Rd, Shijingshan District, Beijing 100049, P.R. China

[4]Heinz Maier-Leibnitz Zentrum (MLZ), Technische Universität München, Lichtenbergstr. 1, 85748, Garching, Germany

[5]Department of Materials, University of Oxford, Parks Rd, OX1 3PH, Oxford, United Kingdom

*xiayg@nimte.ac.cn, muellerb@ph.tum.de, chengyj@nimte.ac.cn

Abstract

Sodium-ion batteries are regarded as the most promising substitute of lithium-ion batteries in the future, due to the low cost and sustainability. The appropriate electrode material is the key to achieve commercialization of sodium-ion batteries. Conducting polymers can be applied on both cathodes and anodes separately or as a component of the composite, because of the good conductivity and flexibility. In this chapter, the recent progress of conducting polymer applied in SIBs will be reviewed. There will be two main parts focused on cathodes and anodes, respectively.

Keywords

Conducting Polymers, Cathode, Anode, Sodium-Ion Batteries

Contents

1.   Introduction

After over twenty-year development, lithium-ion batteries (LIBs) have not only dominated the portable electronics market but also become the prime candidate of power supply for electric vehicles (EV) and plug-in hybrid electric vehicles (PHEV) [1, 2]. However, the dramatically increasing demand for higher energy density gradually exposes several limits of the first-generation lithium-ion batteries. On the one hand, the lack of appropriate electrode materials leads to difficulties for achieving high energy density. On the other hand, the limitation and high cost of global lithium resources also sound the alarm for the sustainability of Li-ion batteries. Up to now, a considerable number of novel electrode materials have been developed and countless efforts have been done to improve the electrochemical performance of these materials. In order to realize sustainable and low-cost energy storage systems, sodium-ion batteries (SIBs) started to return to the vision of academia in about 2010 [2]. The origin of SIBs can be dated back to the 1960s when high-temperature Na-based batteries started to be studied [3]. In the following decade, high-temperature Na-S [4, 5] and Na metal chloride [6, 7] batteries could be regarded as the predecessors of sodium-ion batteries, which were developed and successfully commercialized in some specific applications. Since 1980, using ions (Li and Na ions) as charge carriers became a research hotspot to realize the normal-temperature electrochemical energy storages [8]. At that time, both LIBs and SIBs were expected to realize commercialization in the near future. The appearance of graphite-based anodes greatly promoted the development of lithium-ion batteries which were finally commercialized in 1992 by Sony [9-11]. In contrast, due to the relative-low energy density and lack of appropriate anode materials, studies of sodium-ion batteries gradually stepped into a frozen period [8]. After a 20-year dormancy, the destiny of sodium-ion batteries finally embraced a turning point in around 2010.

Sodium, an element belongs to the same family of lithium, has similar physical and chemical properties of lithium, which is the best candidate to substitute lithium for battery applications. Compared to lithium, the abundant reserve and low cost of sodium resource make it possible to establish low-cost and scale-up systems for energy storage

[12]. However, due to some natural characteristics of sodium, sodium-ion batteries are inferior to lithium-ion batteries in some aspects, such as the lower energy density derived from the higher equivalent weight of Na. In addition, because of the difference between ion radii, some electrode materials which exhibit good performance in LIBs cannot be applied on SIBs directly [13, 14]. As for sodium-ion batteries, developing appropriate and high-performance electrode materials is also the key point to promote SIBs to the market as soon so possible. Currently, normal electrode materials of SIBs include transition metal oxides, phosphates, fluorides, sulfides, phosphides as cathodes, and carbonaceous materials, semi-metals (Ge, Sb, Sn) as anodes [8, 15, 16]. Moreover, organic electrode materials for SIBs are gaining increasing research interest due to a number of intrinsic advantages, such as lighter weight, lower cost, more sensitive to sodium ions, better plasticity and flexibility than inorganic materials [17-19]. Generally, most of organic electrode materials (small organic molecules and polymers) suffer from their poor conductivity, leading to insufficient electrochemical performance [20, 21]. Fortunately, there is an exception which is the conducting polymers.

Conducting polymers refer to those polymers which can present electric conductivity derived from the conjugated structures in their molecule chains [22-24]. Alan MacDiarmid, Hideki Shirakawa, and Alan J. Heeger discovered the unique characteristic of conducting polymers in 1976 and they were awarded the Noble Prize in Chemistry in 2002 for this [22, 25, 26]. Since the 1980s, due to the unique property, conducting polymers have been applied in various electronic devices, including transistors, sensors, memories, actuators/artificial muscles, supercapacitors, and lithium-ion batteries [27-32]. In recent years, the renaissance of sodium-ion batteries provided opportunities for conducting polymers to explore more applications. Inversely, conducting polymers could be used to construct appropriate electrode materials to achieving high energy density sodium-ion batteries. Conducting polymers can not only be applied as cathodes or anodes directly due to the redox-doping mechanism but also be used to fabricate composite electrodes with conventional materials [20, 21]. As a part of the electrode, conducting polymer is able to enhance the conductivity of the electrode, and alleviate the volume changes of those materials which suffered from this problem[33, 34]. Normally, the conducting polymers applied on sodium-ion batteries include polypyrrole (PPy), polyaniline (PANI), polythiophene (PTH), poly(3,4-ethylene dioxythiophene) (PEDOT), and so on. Molecular structures of these conducting polymers are presented in Fig. 1. At present, conducting polymers have been regarded as an important branch of SIB electrode materials and additives. In this chapter, we will review the progress of the studies about applications of conducting polymers in sodium-ion batteries. Hereinafter, there will be two main parts which focus on the studies of cathodes and anodes, respectively. In each part, different applying types of conducting polymers will be discussed separately.

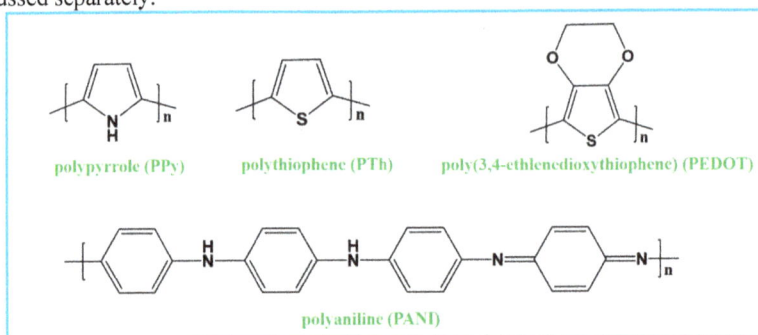

Figure 1.　Molecular structures of typical conducting polymers.

2.　Applications on cathode materials

2.1　Doped and pure conducting polymer cathodes

Due to the redox-doping reaction mechanism, the electrochemical activity of conducting polymers have been demonstrated and they have been widely applied as electrode materials for lithium-ion batteries since about 2000 [35-38]. Up to the 2010s, conducting polymers started to be considered to act as cathodes for sodium-ion batteries. However, when conducting polymers were directly used as cathodes in SIBs, they would react with large electrolyte anions via a p-doping/de-doping mechanism, leading to a poor capacity utilization of the polymer chains [39]. In order to solve this problem, one strategy appeared recently which is doping some ionizable groups in the conducting polymers to change the redox reaction mechanism from conventional p-doping/de-doping processes of large electrolyte anions to the insertion-extraction reactions of small cations (Na ions) [39-47].

Cao and Yang synthesized redox-active $Fe(CN)_6^{4-}$-doped polypyrrole (PPy-FC) which exhibited good Na-storage performance as a cathode for SIBs [41]. The PPy-FC cathode could keep a reversible capacity of 110 mAh g^{-1} after 100 cycles at a rate of 50 mA g^{-1}, corresponding to a capacity retention of 85 %. In addition, it could also deliver a specific capacity of ca. 70 mAh g^{-1} at a high rate of 1600 mA g^{-1}, and recover mostly its initial capacity of 125 mAh g^{-1} when the current density was changed back to 50 mA g^{-1}. Based on the analysis of Fourier-transform infrared spectroscopy (FTIR), the reaction mechanism of PPy-FC cathode at the first discharge can be presented as followed equation (Eq. 1):

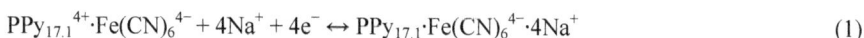

$$PPy_{17.1}^{4+} \cdot Fe(CN)_6^{4-} + 4Na^+ + 4e^- \leftrightarrow PPy_{17.1} \cdot Fe(CN)_6^{4-} \cdot 4Na^+ \tag{1}$$

When the charging continued, the oxidation of the $Fe(CN)_6^{4-}$ dopant and electrochemical redox of PPy chains would happen, as presented by Eq. 2 and Eq. 3:

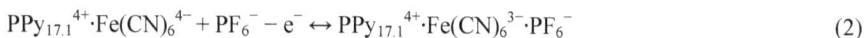

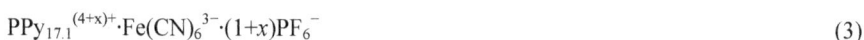

$$PPy_{17.1}^{4+} \cdot Fe(CN)_6^{4-} + PF_6^- - e^- \leftrightarrow PPy_{17.1}^{4+} \cdot Fe(CN)_6^{3-} \cdot PF_6^- \tag{2}$$

$$PPy_{17.1}^{4+} \cdot Fe(CN)_6^{3-} \cdot PF_6^- + xPF_6^- - xe^- \leftrightarrow$$

$$PPy_{17.1}^{(4+x)+} \cdot Fe(CN)_6^{3-} \cdot (1+x)PF_6^- \tag{3}$$

According to these equations, the theory capacity of PPy-FC cathode could be regarded as a combination of Na-insertion/extraction (80 mAh g^{-1}), $Fe(CN)_6^{3-/4-}$ redox (20 mAh g^{-1}), and p-doping reaction (35 mAh g^{-1}), which was close agreement with the experiment value. Besides, the chemical structure and redox mechanism of PPy-FC conducting polymer are displayed by Fig. 2(a).

Polyaniline (PANI) is also a promising electrode material for sodium-ion batteries. In 2015, Jiang and Wang reported their work about sulfonated polyaniline, poly(aniline-co-amino benzene sulfonic sodium) (PANS), as the cathode for sodium-ion batteries [45]. They selected the sulfonate as doping group which has stronger electron-withdrawing ability onto the redox-active polyaniline chains, in order to improve the drawbacks derived from the large doping group (benzene sulfonic sodium) as they reported previously [44]. The PANS electrode was fabricated by a chemical oxidative copolymerization. The chemical structure of PANS and its reaction mechanism during the charge/discharge process are presented in Fig. 2(b). Based on the reversible Na-insertion/extraction reaction, this PANS cathode exhibited a reversible capacity of 133 mAh g^{-1} after 200 cycles (100 mA g^{-1}), corresponding to high capacity retention over 96 %.

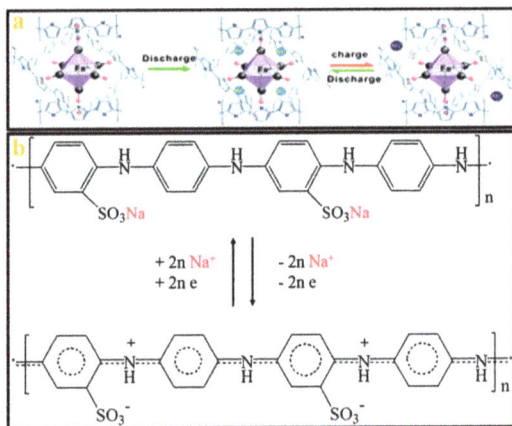

Figure 2.   *a) Chemical structure and redox mechanism of PPy-FC conducting polymer, ref.[41], copyright the Royal Society of Chemistry 2012. b) Molecular structure and reaction mechanism of PANS, ref.[45], copyright the Royal Society of Chemistry 2015.*

Except for modifying active groups on polymer chains, another strategy to enhance the Na-storage performance of CPs is constructing nanostructures in order to facilitate electrolyte penetration and ionic transfer [48-51]. In 2015, Wang and Su synthesized PPy hollow nanospheres using poly(methyl methacrylate) (PMMA) nanospheres as templates, as shown in Fig. 3(a). The average values of the inner and outer diameters of the as-prepared PPy hollow nanospheres were 136.5 nm and 242 nm, respectively. Due to such hollow structure, this PPy cathode exhibited stable cyclability as well as good rate performance. In 2016, Feng and Mai fabricated 2D mesoporous polypyrrole (mPPy) nanosheets with controlled pore sizes via synergic and demonstrated its superior electrochemical performance as SIB cathode [49]. In this work, amphiphilic aliphatic amine (octadecylamine, OTA) and polystyrene-block-poly(ethylene oxide) (PS-*b*-PEO) block copolymers were selected as templates. Because of the self-assembly property, OTA and PS-*b*-PEO firstly formed sandwich-like assemblies by monolayer arrangement of spherical micelles (PS-*b*-PEO) on both surfaces of the 2D bilayers (OTA). Then pyrrole monomers were polymerized on the surface of this 2D templates due to the adsorption of poly(ethylene oxide) domains. After removal the templates, 2D mPPy nanosheets were obtained with pore sizes of 6.8 – 13.6 nm, thickness of 25 – 30 nm, and surface area of 96 m^2 g^{-1}. The pore size of as-prepared mPPy nanosheets could be controlled by varying PS block lengths of PS-*b*-PEO deblocks. As for the electrochemical performance, this PPy cathode maintained its initial capacity of 83 mAh g^{-1} after 120 cycles at a high rate of 300 mA g^{-1}. Moreover, polyaniline hollow nanofibers (Fig. 3(b)) and normal nanofibers were synthesized at most recently, and they exhibited competitive sodium storage performance [50, 51].

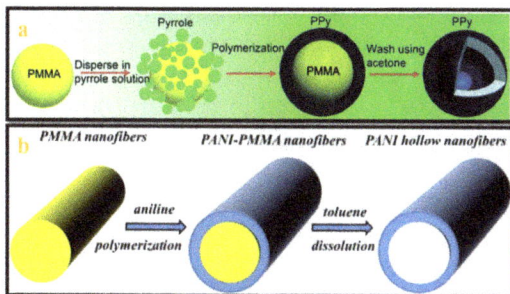

Figure 3.   *a) Schematic diagram of the synthesis of PPy hollow nanospheres, ref.[48], copyright the Royal Society of Chemistry 2015. b) Scheme for the fabrication of the PANI hollow nanofibers, ref.[50], copyright 2019 Elsevier Ltd..*

2.2   Conducting polymer-based composite cathode

Up to now, a verity of inorganic cathode materials have been developed and almost dominated the cathode field of sodium-ion batteries, including transition metal oxides [52], phosphates [53], phosphorus-based polyanion materials [54-56], Prussian blue (PB) and its analogs sodium manganese hexacyanoferrates (NMHFC) [57-59], and so on. Although these materials display high theoretical capacities and potential plateaus, most of them suffer from low conductivity, leading to poor electrochemical performance. Conducting polymers are ideal partners for these conventional cathode materials, which can not only enhance the electric conductivity of the electrode but also decrease the dosage of additional binders and conductive agents. In recent years, inorganic/organic (CPs) composite SIB cathodes have attracted much attention and gained good effects [33, 60-67].

In 2015, Chou *et al.* synthesized ClO_4-doped polypyrrole coated $Na_{1+x}MnFe(CN)_6$ composite (NMHFC@PPy) via a facile self-polymerization approach, which was successfully applied as high-performance cathode for SIBs [60]. In this composite, PPy played multiple important roles, including conductive coating layer to enhance the conductivity, protective layer to reduce the dissolution of Mn in the electrolyte, and

Sodium-Ion Batteries: Materials and Applications   Materials Research Forum LLC
Materials Research Foundations **76** (2020) 251-271   https://doi.org/10.21741/9781644900833-11

active materials to increase the capacity. The electrochemical reaction mechanism of NMHFC@PPy composite is displayed in Fig. 4(a). Due to these advantages above, the NMHFC@PPy composite cathode exhibited improved electrochemical performance than that of bare NMHFC electrode, as shown in Fig. 4(a). This composite cathode could maintain capacity retention of 67 % after 200 cycles (at 0.1 C, 1 C=120 mA g^{-1}), and keep 46 % of initial capacity under 40 C rate. In 2017, Wang *et al.* reported a novel nanostructured composite cathode where a string of nickel hexacyanoferrate (NiCHF) nanocubes coaxially anchored on bipolar conducting polymer (BCP) coated CNTs (CNTs@BCP) [63], as presented in Fig 4(b). The bipolar conducting polymer was derived from intrinsic PANI (p-doping BCP) and Fe(CN)$_6$$^{3-}$-doped PANI (n-doping BCP). The synthesis of CNTs@NiHCF-BCP cathode was realized via an electrochemical route. At first, a CNT modified electrode was obtained by distributing the CNTs (solution) on a stainless-steel sheet and undergoing a drying process. Then, the NiHCF and BCP were coaxially grown on it directly in a three-electrode system. The final hybrid cathode was obtained after rinsing and drying, which could be used in batteries without extra addition of binder and conductive agent. This CNT@NiHCF-BCP cathode exhibited a good cycling stability, which retained a reversible capacity of about 122 mAh g^{-1} after 500 cycles at a rate of 0.5 C. Moreover, it also remained at 97 mAh g^{-1} at a high rate of 15 C, corresponding to 77 % retention of its reversible capacity at 1 C. Such good electrochemical performance of CNT@NiHCF-BCP cathode could be attributed to the synergistic effect of the three active components of CNTs, NiHCF nanocubes, and BCP. In terms of structure, the network built by these three components benefited the electronic and ionic transportations. In addition, all these three components could provide capacity via different reaction mechanisms. The NiHCF can react with sodium ions electrochemically, as described by Eq. 4:

$$NaNiFe(CN)_6 + Na^+ + e^- \leftrightarrow Na_2NiFe(CN)_6 \tag{4}$$

In case of PANI, the electrochemical reactions are realized by p-doping/de-doping of the ClO$_4$$^-$ anions and p-doping/de-doping of the Fe(CN)$_6$$^{3-}$ anions, which presented by Eq. 5 and Eq. 6:

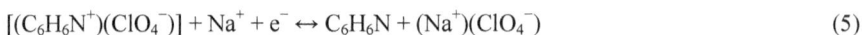

$$[(C_6H_6N^+)(ClO_4^-)] + Na^+ + e^- \leftrightarrow C_6H_6N + (Na^+)(ClO_4^-) \tag{5}$$

$$[(C_6H_6N^+)_3Fe(CN)_6^{3-}] + 4Na^+ + 4e^- \leftrightarrow [(C_6H_6N^+)_3Fe(CN)_6^{4-}(Na^+)_4] \tag{6}$$

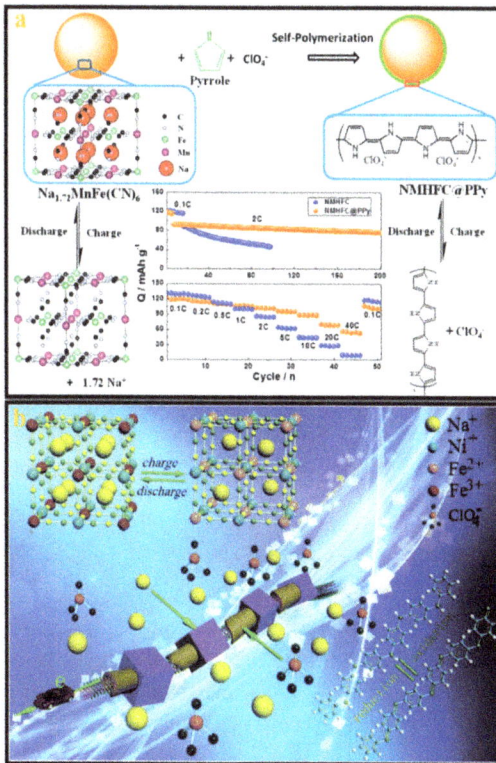

*Figure 4.   a) Schematic illustration of synthesis, sodium storage mechanism and
electrochemical performance of the NMHFC@PPy, ref.[60], copyright 2015 Elsevier
Ltd.. b) Schematic illustration showing paths for sodium-ion diffusion and electron
conduction in the CNT@NiHCF-BCP nanohybrid cathode, ref.[63], copyright the Royal
Society of Chemistry 2017.*

3.   Applications on anode materials

3.1   Doped and pure conducting polymer anodes

Similar to the applications on cathodes, conducting polymers can also be used as anode
materials for sodium-ion batteries based on the redox-doping mechanism [68-71]. In
2016, Liu and Wang applied polypyrrole to act as anode for SIBs for the first time and
compared different electrochemical performance between submicron polypyrrole (S-PPy)

and bulk polypyrrole (B-PPy) [68]. In result, the S-PPy electrode exhibited better Na-storage property which could be ascribed to the enhanced electrical contact between PPy particles due to the smaller size. Besides, the smaller particles size also benefited the penetration of the electrolyte into active material, leading to improved electrochemical performance. Except for the typical conducting polymers, like PPy and PANI, several novel conductive polymers were synthesized recently and their potential applications as SIB anodes have been demonstrated. Armand *et al.* synthesized a family of polySchiff/oligoether terpolymers which exhibited electrochemical activity with Na^+ ions and gained good performance as SIB anodes [70]. In addition, Jiang *et al.* prepared cyclized polyacrylonitrile nanofibers (cPAN-NFs) as anode for SIBs, through a scalable electrospinning technique followed by thermal stabilization [71]. cPAN was a typical poly(*N*-heteroacene) with the substitution of one nitrogen atom in each aromatic ring, which displayed enhanced n-dopable reactivity. The cPAN-NFs anode exhibited a high specific capacity of 527 mAh g^{-1} at 0.05 A g^{-1}. Moreover, it could maintain a reversible capacity of ca. 220 mAh g^{-1} after 3500 cycles at 5 A g^{-1}, corresponding to a capacity retention of 99.4 %.

3.2   Conducting polymer-based composite anode

Although some conducting polymers exhibited stable cyclability as SIB anodes, the specific capacities were still insufficient compared to those inorganic anodes. However, the popular inorganic anodes suffer from serve volume expansion upon sodiation, such as tin, antimony materials and their oxides [2]. Conducting polymers, as ideal buffer matrix, have been widely used to constructing composite anodes for sodium-ion batteries [34, 72-85].

Antimony (Sb) is one of the favorite anode materials for sodium-ion batteries because of its high theoretical capacity of 660 mAh g^{-1} [86]. In 2015, Nam and Kwon combined conducting polymer with Sb-based material, in order to improve its electrochemical performance [34]. Firstly, the PPy nanowire network was prepared through an electrochemical approach. PPy nanowires grew on an oxidizable metal substrate via a cathodic electro-polymerization method, as shown in Fig. 5(a) [87]. Then, Sb and Sb_2O_3 nanocrystals were electrodeposited on the PPy network via Eq. 7 and Eq. 8, forming Sb/Sb_2O_3-PPy composite anode material.

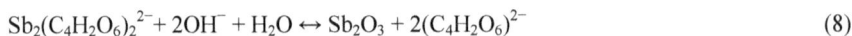

$$Sb_2(C_4H_2O_6)_2{}^{2-} + 4H^+ + 6e^- \leftrightarrow 2Sb + 2(C_4H_2O_6)^{2-} \tag{7}$$

$$Sb_2(C_4H_2O_6)_2{}^{2-} + 2OH^- + H_2O \leftrightarrow Sb_2O_3 + 2(C_4H_2O_6)^{2-} \tag{8}$$

The porous network generated by crosslinked PPy nanowires could not only alleviate volumetric changes of Sb and Sb_2O_3 during cycling processes but also enhance the electric and ionic conductivity of the electrode. As a result, this composite anode presented good stability within 100 cycles (at 60 mA g^{-1}), and retained a reversible capacity of 512 mAh g^{-1} with a retention of 98 %. The existence of PPy network also endowed a good rate performance of this composite electrode, which maintained a capacity of 299 mAh g^{-1} at a high current density of 3300 mA g^{-1}.

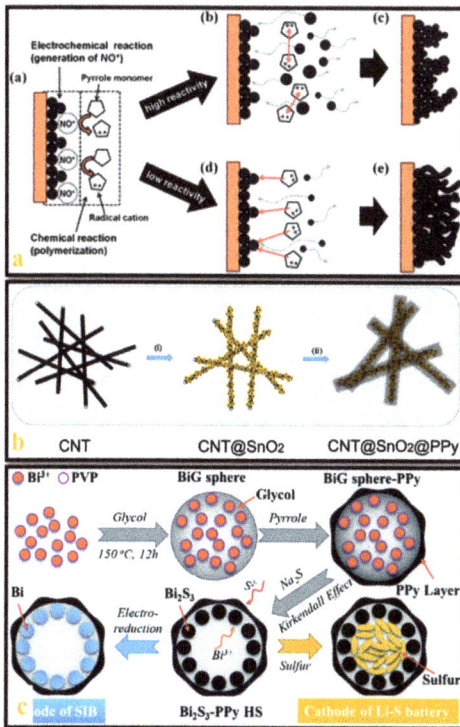

Figure 5.   a) Schematic diagram showing the effects of the reactivity of radical cations on the formation of polypyrrole nanostructures in cathodic electropolymerization, ref.[87], copyright the Royal Society of Chemistry 2013. b) Scheme illustration of synthesis for the CNT@SnO₂@PPy nanocomposites, ref.[82], copyright 2018 Elsevier B.V.. c) Schematic illustration of the preparation of the amorphous Bi₂S₃–polypyrrole hollow spheres (Bi₂S₃–PPy HSs) electrode, ref.[85], copyright the Royal Society of Chemistry 2013.

Tin oxides are another kind of promising anode material for Na-ion batteries, and their applications need effective methods to improve the cycling stability. Wang *et al.* fabricated SnO_2-PPy composite anode by an *in-situ* hydrothermal synthesis, where SnO_2 nanoparticles are homogeneously distributed on the surfaces of the PPy nanotubes [75]. Li *et al.* prepared sandwich-like CNT@SnO_2@PPy nanocomposites, in which SnO_2 nanoparticles were deposited on the surface of CNTs via a liquid-phase deposition (LPD) approach and PPy were further coated on the surface of CNT@SnO_2 hybrids by a chemical-polymerization route (Fig. 5(b)) [82]. Due to the assistance of PPy, the electrochemical performance of this anodes was effectively improved. Besides, Tang *et al.* synthesized amorphous Bi_2S_3 hollow spheres (Bi_2S_3-PPy HSs) which could be used as anode for SIBs after reduction as well as sulfur host of Li-S batteries after sulfuration [85]. As illustrated in Fig. 5(c), the BiG spheres were prepared by a one-pot hydrothermal method, in which Bi nanoparticles were encapsulated in large glycol spheres. Then, PPy was coated on the surface of BiG spheres to form BiG-PPy spheres. Driven by the Kirkendall Effect, the hollow structured Bi_2S_3-PPy spheres were gained. After an electro-reduction process, the as-obtained Bi-PPy hollow sphere could be used as anode for sodium-ion batteries. Due to the hollow structure and PPy shell, this anode exhibited good rate performance with a reversible capacity of 278 mAh g^{-1} at 3000 mA g^{-1} and a long lifespan upon 5000 cycles when cycled at 1000 mA g^{-1}. Because of the outstanding plasticity and conductivity, polypyrrole almost became the first choice for constructing CPs-based anodes for sodium-ion batteries. There were more studies about high-performance SIB anodes assisted by PPy, including graphene oxide-PPy composites[76], $CoFe_2O_4$-PPy nanotubes [77], ZnS@PPy spheres [78], TsONa/PPy/TiO_2/Ti composite [79], Sb_2Se_3@PPy microclips [81], CoP@PPy nanowires/carbon paper composite [83], and flowerlike Sb_2S_3/PPy microspheres.

Conclusions & Outlooks

Since the discovery of conducting polymers in 1976, they have been widely applied in various energy storage devices due to their unique properties. Sodium-ion batteries, as one of the most promising techniques for future energy storages, embraced an explosive development and gained great progress since 2010. Constructing high-performance electrode materials has been regarded as the requirement to promote SIBs to practical applications. Conducting polymers have been demonstrated that they could play significant roles in both cathodes and anodes of SIBs. One the one hand, doped or pure conducting polymers could be directly used as cathodes or anodes due to the redox-doping mechanism. Moreover, assisted with nanostructures, CPs electrodes could boost better electrochemical performance. On the other hand, conducting polymers could effective improve the electrochemical performance of inorganic electrodes via forming composite materials. Particularly, conducting polymers could enhance the electric conductivity of conventional inorganic cathodes, and also act as buffer matrix for those anodes which suffer from the volumetric expansion during the cycling process. Meanwhile, conducting polymers could also act as binder and conductive agent, which is beneficial to increase the energy density of batteries. After several-year development, various conducting polymers have been applied in sodium-ion batteries, including PPy, PANI, PTh, and PEDOT. Among them, PPy attracted most attention especially in the field of composite anodes.

With the increasing demand for low-cost and sustainable energy storage devices, sodium-ion batteries have become the premier candidate for the large-scale energy storage system in the future. At present, one bottleneck of SIB development is the lack of appropriate electrode materials. The emerge of conducting polymers can be considered as a good complement for the electrode material resource library of SIBs. It is convinced that sodium-ion batteries will realize commercial applications in the near future with the assistant of conducting polymers.

Acknowledgment

This research is funded by the National Key R&D Program of China (Grant No. 2016YFB0100100), Natural Science Foundation of China (51702335, 21773279), the CAS-EU S&T cooperation partner program (174433KYSB20150013), and Key Laboratory of Bio-based Polymeric Materials of Zhejiang Province. S.L. acknowledges the China Scholarship Council (CSC). P.M-B acknowledges funding by the International Research Training Group 2022 Alberta/Technical University of Munich International Graduate School for Environmentally Responsible Functional Hybrid Materials (ATUMS).

References

[1] Z. Yang, J. Zhang, M.C. Kintner-Meyer, X. Lu, D. Choi, J.P. Lemmon, J. Liu, Electrochemical energy storage for green grid, Chem. Rev. 111 (2011) 3577-3613. https://doi.org/10.1021/cr100290v

[2] V. Palomares, P. Serras, I. Villaluenga, K.B. Hueso, J. Carretero-González, T. Rojo, Na-ion batteries, recent advances and present challenges to become low cost energy storage systems, Energy Environ. Sci. 5 (2012) 5884-5901. https://doi.org/10.1039/c2ee02781j

[3] Y.F.Y. Yao, J. Kummer, Ion exchange properties of and rates of ionic diffusion in beta-alumina, J. Inorg. Nucl. Chem. 29 (1967) 2453-2475. https://doi.org/10.1016/0022-1902(67)80301-4

[4] J. Sudworth, A. Tiley, Sodium Sulphur Battery, Springer Science & Business Media, New York, 1985, pp. 257-277.

[5] T. Oshima, M. Kajita, A. Okuno, Development of sodium-sulfur batteries, Int. J. Appl. Ceram. Tec. 1 (2004) 269-276.

[6] R. Bones, D. Teagle, S. Brooker, F. Cullen, Development of a Ni, $NiCl_2$ positive electrode for a liquid sodium (ZEBRA) battery cell, J. Electrochem. Soc. 136 (1989) 1274-1277. https://doi.org/10.1002/chin.198937021

[7] C.H. Dustmann, Advances in ZEBRA batteries, J. Power Sources 127 (2004) 85-92.

[8] N. Yabuuchi, K. Kubota, M. Dahbi, S. Komaba, Research development on sodium-ion batteries, Chem. Rev. 114 (2014) 11636-11682. https://doi.org/10.1021/cr500192f

[9] A. Yoshino, K. Sanechika, T. Nakajima, USP4, 668,595, 1985; A, Yoshino, K. Sanechika, T. Nakajima, JP1989293 (1985).

[10] M. Yoshio, R.J. Brodd, A. Kozawa, Lithium-ion batteries, Springer, Switzerland, 2009, pp. 213-233.

[11] A. Yoshino, The Birth of the Lithiumthium-ion batteries. Chem. Int. Ed. 51 (2012) 5798-5800.

[12] S.P. Ong, V.L. Chevrier, G. Hautier, A. Jain, C. Moore, S. Kim, X. Ma, G. Ceder, Voltage, stability and diffusion barrier differences between sodium-ion and lithium-ion intercalation materials, Energy Environ. Sci. 4 (2011) 3680-3688. https://doi.org/10.1039/c1ee01782a

[13] P. Ge, M. Fouletier, Electrochemical intercalation of sodium in graphite, Solid State Ionics 28 (1988) 1172-1175. https://doi.org/10.1016/0167-2738(88)90351-7

[14] M.M. Doeff, Y. Ma, S.J. Visco, L.C. De Jonghe, Electrochemical insertion of sodium into carbon, J. Electrochem. Soc. 140 (1993) L169-L170. https://doi.org/10.1149/1.2221153

[15] M. Lao, Y. Zhang, W. Luo, Q. Yan, W. Sun, S.X. Dou, Alloy-based anode materials toward advanced sodium-ion batteries, Adv. Mater. 29 (2017) 1700622. https://doi.org/10.1002/adma.201700622

[16] Q. Wang, C. Zhao, Y. Lu, Y. Li, Y. Zheng, Y. Qi, X. Rong, L. Jiang, X. Qi, Y. Shao, D. Pan, B. Li, Y.S. Hu, L. Chen, Advanced Nanostructured Anode Materials for

Sodium-Ion Batteries, Small 13 (2017) 1701835.
https://doi.org/10.1002/smll.201701835

[17] Z. Song, H. Zhou, Towards sustainable and versatile energy storage devices: an overview of organic electrode materials, Energy Environ. Sci. 6 (2013) 2280-2301. https://doi.org/10.1039/c3ee40709h

[18] A. Abouimrane, W. Weng, H. Eltayeb, Y. Cui, J. Niklas, O. Poluektov, K. Amine, Sodium insertion in carboxylate based materials and their application in 3.6 V full sodium cells, Energy Environ. Sci. 5 (2012) 9632-9638. https://doi.org/10.1039/c2ee22864e

[19] R. Emanuelsson, M. Sterby, M. Strømme, M. Sjödin, An all-organic proton battery, J. Am. Chem. Soc. 139 (2017) 4828-4834. https://doi.org/10.1021/jacs.7b00159

[20] Q.L. Zhao, A.K. Whittaker, X.S. Zhao, Polymer electrode materials for sodium-ion batteries, materials 11 (2018) 2567-2585. https://doi.org/10.3390/ma11122567

[21] X.Y. Cao, J.B. Liu, L.M. Zhu, L.L. Xie, Polymer electrode materials for high-performance lithium/sodium-ion batteries: A review, Energy Technol. 7 (2019)1800759. https://doi.org/10.1002/ente.201800759

[22] A.J. Heeger, Semiconducting and metallic polymers: the fourth generation of polymeric materials, J. Phys. Chem. B 105 (2001) 8475-8491. https://doi.org/10.1021/jp011611w

[23] J. Heinze, Electrochemistry of conducting polymers, Synthetic Met. 43 (1991) 2805-2823. https://doi.org/10.1016/0379-6779(91)91183-b

[24] X.C. Li, Y.S. Jiao, S.J. Li, The synthesis, properties and application of new conducting polymers, Eur. Polym. J. 27 (1991) 1345-1351.

[25] H. Shirakawa, E.J. Louis, A.G. MacDiarmid, C.K. Chiang, A.J. Heeger, Synthesis of electrically conducting organic polymers: Halogen derivatives of polyacetylene,$(CH)_x$, Chem. Commun. (1977) 578-580. https://doi.org/10.1039/c39770000578

[26] C. Chiang, M. Druy, S. Gau, A. Heeger, E. Louis, A.G. MacDiarmid, Y. Park, H. Shirakawa, Synthesis of highly conducting films of derivatives of polyacetylene,$(CH)x$, J. Am. Chem. Soc. 100 (1978) 1013-1015. https://doi.org/10.1021/ja00471a081

[27] G. Inzelt, Conducting polymers: A new era in electrochemistry, Springer, New York, 2008, pp. 225-263.

[28] L.J. Pan, H. Qiu, C.M. Dou, Y. Li, L. Pu, J.B. Xu, Y. Shi, Conducting polymer nanostructures: Template synthesis and applications in energy storage, Int. J. Mol. Sci. 11 (2010) 2636-2657. https://doi.org/10.3390/ijms11072636

[29] X.F. Lu, W.J. Zhang, C. Wang, T.C. Wen, Y. Wei, One-dimensional conducting polymer nanocomposites: Synthesis, properties and applications, Prog. Polym. Sci. 36 (2011) 671-712. https://doi.org/10.1016/j.progpolymsci.2010.07.010

[30] T.K. Das, S. Prusty, Review on conducting polymers and their applications, Polym-Plast. Technol. 51 (2012) 1487-1500.

[31] Y. Shi, L.L. Peng, G.H. Yu, Nanostructured conducting polymer hydrogels for energy storage applications, Nanoscale 7 (2015) 12796-12806. https://doi.org/10.1039/c5nr03403e

[32] M.H. Naveen, N.G. Gurudatt, Y.B. Shim, Applications of conducting polymer composites to electrochemical sensors: A review, Appl. Mater. Today 9 (2017) 419-433. https://doi.org/10.1016/j.apmt.2017.09.001

[33] H.Y. Kang, Y.C. Liu, M.H. Shang, T.Y. Lu, Y.J. Wang, L.F. Jiao, NaV_3O_8 nanosheet@polypyrrole core-shell composites with good electrochemical performance as cathodes for Na-ion batteries, Nanoscale 7 (2015) 9261-9267. Nanoscale 7 (2015) 9261-9267.

[34] D.H. Nam, K.S. Hong, S.J. Lim, M.J. Kim, H.S. Kwon, High-performance Sb/Sb_2O_3 anode materials using a polypyrrole nanowire network for na-ion batteries, Small 11 (2015) 2885-2892. https://doi.org/10.1002/smll.201500491

[35] P. Novak, K. Muller, K.S.V. Santhanam, O. Haas, Electrochemically active polymers for rechargeable batteries, Chem. Rev. 97 (1997) 207-281. https://doi.org/10.1021/cr941181o

[36] D. Kumar, R. Sharma, Advances in conductive polymers, Eur. Polym. J. 34 (1998) 1053-1060.

[37] M.E. Abdelhamid, A.P. O'Mullane, G.A. Snook, Storing energy in plastics: a review on conducting polymers & their role in electrochemical energy storage, RSC Adv. 5 (2015) 11611-11626. https://doi.org/10.1039/c4ra15947k

[38] M.E. Bhosale, S. Chae, J.M. Kim, J.Y. Choi, Organic small molecules and polymers as an electrode material for rechargeable lithium ion batteries, J. Mater. Chem. A 6 (2018) 19885-19911. https://doi.org/10.1039/c8ta04906h

[39] L. Zhu, Y. Shen, M. Sun, J. Qian, Y. Cao, X. Ai, H. Yang, Self-doped polypyrrole with ionizable sodium sulfonate as a renewable cathode material for sodium ion

batteries, Chem. Commun. 49 (2013) 11370-11372.
https://doi.org/10.1039/c3cc46642f

[40] R. Zhao, L. Zhu, Y. Cao, X. Ai, H.X. Yang, An aniline-nitroaniline copolymer as a high capacity cathode for Na-ion batteries, Electrochem. Commun. 21 (2012) 36-38. https://doi.org/10.1016/j.elecom.2012.05.015

[41] M. Zhou, L. Zhu, Y. Cao, R. Zhao, J. Qian, X. Ai, H. Yang, Fe $(CN)_6^{-4}$-doped polypyrrole: a high-capacity and high-rate cathode material for sodium-ion batteries, RSC Adv. 2 (2012) 5495-5498. https://doi.org/10.1039/c2ra20666h

[42] W. Deng, X. Liang, X. Wu, J. Qian, Y. Cao, X. Ai, J. Feng, H. Yang, A low cost, all-organic Na-ion battery based on polymeric cathode and anode, Sci. Rep. 3 (2013) 2671. https://doi.org/10.1038/srep02671

[43] M. Zhou, Y. Xiong, Y.L. Cao, X.P. Ai, H.X. Yang, Electroactive organic anion-doped polypyrrole as a low cost and renewable cathode for sodium-ion batteries, J. Polym. Sci. Pol. Phys. 51 (2013) 114-118. https://doi.org/10.1002/polb.23184

[44] Y. Shen, D. Yuan, X. Ai, H. Yang, M. Zhou, Poly (diphenylaminesulfonic acid sodium) as a cation-exchanging organic cathode for sodium batteries, Electrochem. Commun. 49 (2014) 5-8. https://doi.org/10.1016/j.elecom.2014.09.016

[45] M. Zhou, W. Li, T. Gu, K. Wang, S. Cheng, K. Jiang, A sulfonated polyaniline with high density and high rate Na-storage performances as a flexible organic cathode for sodium ion batteries, Chem. Commun. 51 (2015) 14354-14356. https://doi.org/10.1039/c5cc05654c

[46] H.Y. Hou, Q.S. Liao, J.X. Duan, S. Liu, Y. Yao, Observation on sodium p-toluenesulfonate-doped polypyrrole cathode for sodium ion battery, Surf. Innov. 6 (2018) 56-62. https://doi.org/10.1680/jsuin.17.00036

[47] Q.S. Liao, H.Y. Hou, X.X. Liu, Y. Yao, Z.P. Dai, C.Y. Yu, D.D. Li, L-lactic acid and sodium p-toluenesulfonate co-doped polypyrrole for high performance cathode in sodium ion battery, J. Phys. Chem. Solids 115 (2018) 233-237. https://doi.org/10.1016/j.jpcs.2017.12.015

[48] D. Su, J. Zhang, S. Dou, G. Wang, Polypyrrole hollow nanospheres: stable cathode materials for sodium-ion batteries, Chem. Commun. 51 (2015) 16092-16095. https://doi.org/10.1039/c5cc04229a

[49] S.H. Liu, F.X. Wang, R.H. Dong, T. Zhang, J. Zhang, X.D. Zhuang, Y.Y. Mai, X.L. Feng, Dual-template synthesis of 2D mesoporous polypyrrole nanosheets with controlled pore size, Adv. Mater. 28 (2016) 8365-8370. https://doi.org/10.1002/adma.201603036

Materials Research Forum LLC
https://doi.org/10.21741/9781644900833-11

[50] H.X. Han, H.Y. Lu, X.Y. Jiang, F.P. Zhong, X.P. Ai, H.X. Yang, Y.L. Cao, Polyaniline hollow nanofibers prepared by controllable sacrifice-template route as high-performance cathode materials for sodium-ion batteries, Electrochim. Acta 301 (2019) 352-358. https://doi.org/10.1016/j.electacta.2019.02.002

[51] J. Manuel, T. Salguero, R.P. Ramasamy, Synthesis and characterization of polyaniline nanofibers as cathode active material for sodium-ion battery, J. Appl. Electrochem. 49 (2019) 529-537. https://doi.org/10.1007/s10800-019-01298-y

[52] R. Berthelot, D. Carlier, C. Delmas, Electrochemical investigation of the P_2–Na_xCoO_2 phase diagram, Nat. Mater. 10 (2011) 74. https://doi.org/10.1038/nmat2920

[53] B. Ellis, W. Makahnouk, Y. Makimura, K. Toghill, L. Nazar, A multifunctional 3.5 V iron-based phosphate cathode for rechargeable batteries, Nat. Mater. 6 (2007) 749. https://doi.org/10.1038/nmat2007

[54] Y. Zhu, Y. Xu, Y. Liu, C. Luo, C. Wang, Comparison of electrochemical performances of olivine $NaFePO_4$ in sodium-ion batteries and olivine $LiFePO_4$ in lithium-ion batteries, Nanoscale 5 (2013) 780-787. https://doi.org/10.1039/c2nr32758a

[55] P. Barpanda, G. Liu, C.D. Ling, M. Tamaru, M. Avdeev, S.-C. Chung, Y. Yamada, A. Yamada, $Na_2FeP_2O_7$: a safe cathode for rechargeable sodium-ion batteries, Chem. Mater. 25 (2013) 3480-3487. https://doi.org/10.1002/chin.201346011

[56] C. Zhu, K. Song, P.A. van Aken, J. Maier, Y. Yu, Carbon-coated $Na_3V_2(PO_4)_3$ embedded in porous carbon matrix: an ultrafast Na-storage cathode with the potential of outperforming Li cathodes, Nano Lett. 14 (2014) 2175-2180. https://doi.org/10.1021/nl500548a

[57] M. Okubo, C.H. Li, D.R. Talham, High rate sodium ion insertion into core–shell nanoparticles of Prussian blue analogues, Chem. Commun. 50 (2014) 1353-1355. https://doi.org/10.1039/c3cc47607c

[58] Y. You, X.L. Wu, Y.X. Yin, Y.G. Guo, High-quality Prussian blue crystals as superior cathode materials for room-temperature sodium-ion batteries, Energy Environ. Sci. 7 (2014) 1643-1647. https://doi.org/10.1039/c3ee44004d

[59] M. Okubo, D. Asakura, Y. Mizuno, J.D. Kim, T. Mizokawa, T. Kudo, I. Honma, Switching Redox-Active Sites by Valence Tautomerism in Prussian Blue Analogues $A_xMn_y[Fe(CN)_6]_n H_2O$ (A: K, Rb): Robust Frameworks for Reversible Li Storage, J. Phys. Chem. Lett. 1 (2010) 2063-2071. https://doi.org/10.1021/jz100708b

[60] W.J. Li, S.L. Chou, J.Z. Wang, J.L. Wang, Q.F. Gu, H.K. Liu, S.X. Dou, Multifunctional conducting polymer coated $Na_{1+x}MnFe(CN)_6$ cathode for sodium-ion

batteries with superior performance via a facile and one-step chemistry approach, Nano Energy 13 (2015) 200-207. https://doi.org/10.1016/j.nanoen.2015.02.019

[61] Y.H. Cao, D. Fang, X.Q. Liu, Z.P. Luo, G.Z. Li, W.L. Xu, M. Jiang, C.X. Xiong, Sodium vanadate nanowires @ polypyrrole with synergetic core-shell structure for enhanced reversible sodium-ion storage, Compos. Sci, Technol. 137 (2016) 130-137. https://doi.org/10.1016/j.compscitech.2016.10.032

[62] F.D. Hu, L. Li, X.L. Jiang, Hierarchical Octahedral $Na_2MnFe(CN)_6$ and $Na_2MnFe(CN)_6$@PPy as cathode materials for sodium-ion batteries, Chinses J. Chem. 35 (2017) 415-419. https://doi.org/10.1002/cjoc.201600713

[63] Z.D. Wang, Y. Liu, Z.J. Wu, G.Q. Guan, D. Zhang, H.Y. Zheng, S.D. Xu, S.B. Liu, X.G. Hao, A string of nickel hexacyanoferrate nanocubes coaxially grown on a CNT@bipolar conducting polymer as a high-performance cathode material for sodium-ion batteries, Nanoscale 9 (2017) 823-831. https://doi.org/10.1039/c6nr08765e

[64] J.X. Zhang, T.C. Yuan, H.Y. Wan, J.F. Qian, X.P. Ai, H.X. Yang, Y.L. Cao, Surface-engineering enhanced sodium storage performance of $Na_3V_2(PO_4)_3$ cathode via in-situ self-decorated conducting polymer route, Sci. China Chem. 60 (2017) 1546-1553. https://doi.org/10.1007/s11426-017-9125-y

[65] H. Lim, J.H. Jung, Y.M. Park, H.N. Lee, H.J. Kim, High-performance aqueous rechargeable sulfate- and sodium-ion battery based on polypyrrole-MWCNT core-shell nanowires and $Na_{0.44}MnO_2$ nanorods, Appl. Surf. Sci. 446 (2018) 131-138. https://doi.org/10.1016/j.apsusc.2018.02.021

[66] D.S. Kim, H. Yoo, M.S. Park, H. Kim, Boosting the sodium storage capability of Prussian blue nanocubes by overlaying PEDOT:PSS layer, J. Alloy. Compd. 791 (2019) 385-390. https://doi.org/10.1016/j.jallcom.2019.03.317

[67] D. Lu, Z.J. Yao, Y. Zhong, X.L. Wang, X.H. Xia, C.D. Gu, J.B. Wu, J.P. Tu, Polypyrrole-Coated Sodium Manganate Hollow Microspheres as a Superior Cathode for Sodium Ion Batteries, ACS Appl. Mater. Interfaces 11 (2019) 15630-15637. https://doi.org/10.1021/acsami.9b02555

[68] X.Y. Chen, L. Liu, Z.C. Yan, Z.F. Huang, Q. Zhou, G.X. Guo, X.Y. Wang, The excellent cycling stability and superior rate capability of polypyrrole as the anode material for rechargeable sodium ion batteries, RSC Adv. 6 (2016) 2345-2351. https://doi.org/10.1039/c5ra22607d

[69] L. Yang, X. Huang, A. Gogoll, M. Stromme, M. Sjodin, Conducting redox polymer based anode materials for high power electrical energy storage, Electrochim. Acta 204 (2016) 270-275. https://doi.org/10.1016/j.electacta.2016.03.163

[70] N. Fernandez, P. Sanchez-Fontecoba, E. Castillo-Martinez, J. Carretero-Gonzalez, T. Rojo, M. Armand, Polymeric redox-active electrodes for sodium-ion batteries, ChemSusChem 11 (2018) 311-319. https://doi.org/10.1002/cssc.201701471

[71] T.T. Gu, M. Zhou, B. Huang, M.Y. Liu, X.L. Xiong, K.L. Wang, S.J. Cheng, K. Jiang, Highly conjugated poly(N-heteroacene) nanofibers for reversible Na storage with ultra-high capacity and a long cycle life, J. Mater. Chem. A 6 (2018) 18592-18598. https://doi.org/10.1039/c8ta06724d

[72] L. Zhu, Y. Niu, Y. Cao, A. Lei, X. Ai, H. Yang, n-Type redox behaviors of polybithiophene and its implications for anodic Li and Na storage materials, Electrochimi. Acta 78 (2012) 27-31. https://doi.org/10.1016/j.electacta.2012.05.152

[73] K. Dai, H. Zhao, Z. Wang, X. Song, V. Battaglia, G. Liu, Toward high specific capacity and high cycling stability of pure tin nanoparticles with conductive polymer binder for sodium ion batteries, J. Power Sources 263 (2014) 276-279. https://doi.org/10.1016/j.jpowsour.2014.04.012

[74] Z.A. Zhang, J. Zhang, X.X. Zhao, F.H. Yang, Core-sheath structured porous carbon nanofiber composite anode material derived from bacterial cellulose/polypyrrole as an anode for sodium-ion batteries, Carbon 95 (2015) 552-559. https://doi.org/10.1016/j.carbon.2015.08.069

[75] B.Y. Ruan, H.P. Guo, Q.N. Liu, D.Q. Shi, S.L. Chou, H.K. Liu, G.H. Chen, J.Z. Wang, 3-D structured SnO_2-polypyrrole nanotubes applied in Na-ion batteries, RSC Adv. 6 (2016) 103124-103131. https://doi.org/10.1039/c6ra21139a

[76] H.W. Wang, Y. Zhang, W.P. Sun, H.T. Tan, J.B. Franklin, Y.Y. Guo, H.S. Fan, M. Ulaganathan, X.L. Wu, Z.Z. Luo, S. Madhavi, Q.Y. Yan, Conversion of uniform graphene oxide/polypyrrole composites into functionalized 3D carbon nanosheet frameworks with superior supercapacitive and sodium-ion storage properties, J. Power Sources 307 (2016) 17-24. https://doi.org/10.1016/j.jpowsour.2015.12.104

[77] Q.M. He, K. Rui, C.H. Chen, J.H. Yang, Z.Y. Wen, Interconnected $CoFe_2O_4$-polypyrrole nanotubes as anode materials for high performance sodium ion batteries, ACS Appl. Mater. Interfaces 9 (2017) 36927-36935. https://doi.org/10.1021/acsami.7b12503

[78] T.Y. Hou, G.J. Tang, X.H. Sun, S. Cai, C.M. Zheng, W.B. Hu, Perchlorate ion doped polypyrrole coated ZnS sphere composites as a sodium-ion battery anode with superior rate capability enhanced by pseudocapacitance, RSC Adv. 7 (2017) 43636-43641. https://doi.org/10.1039/c7ra07901j

[79] Q.S. Liao, H.Y. Hou, J.X. Duan, S. Liu, Y. Yao, Z.P. Dai, C.Y. Yu, D.D. Li, Composite sodium p-toluenesulfonate/polypyrrole/TiO_2 nanotubes/Ti anode for

sodium ion battery, Int. J. Hydrogen Energy 42 (2017) 12414-12419. https://doi.org/10.1016/j.ijhydene.2017.03.116

[80] Y. Wei, Q. Hu, Y.H. Cao, D. Fang, W.L. Xu, M. Jiang, J. Huang, H. Liu, X. Fan, Polypyrrole nanotube arrays on carbonized cotton textile for aqueous sodium battery, Org. Electron. 46 (2017) 211-217. https://doi.org/10.1016/j.orgel.2017.04.008

[81] Y.J. Fang, X.Y. Yu, X.W. Lou, Formation of polypyrrole-coated Sb_2Se_3 microclips with enhanced sodium-storage properties, Angew. Chem.-Int. Edi. 57 (2018) 9859-9863. https://doi.org/10.1002/anie.201805552

[82] J.J. Yuan, Y.C. Hao, X.K. Zhang, X.F. Li, Sandwiched CNT@SnO_2@PPy nanocomposites enhancing sodium storage, Colloid Surface A 555 (2018) 795-801. https://doi.org/10.1016/j.colsurfa.2018.07.023

[83] J. Zhang, K. Zhang, J. Yang, G.H. Lee, J. Shin, V. Wing-hei Lau, Y.M. Kang, Bifunctional conducting polymer coated CoP core-shell nanowires on carbon paper as a free-standing anode for sodium ion batteries, Adv. Energy Mater. 8 (2018). https://doi.org/10.1002/aenm.201800283

[84] T. Zheng, G.D. Li, L.X. Zhao, Y.X. Shen, Flowerlike Sb_2S_3/PPy microspheres used as anode material for high-performance sodium-ion batteries, Eur. J. Inorg. Chem. (2018) 1224-1228. https://doi.org/10.1002/ejic.201701364

[85] B. Long, Z.P. Qiao, J.N. Zhang, S.Q. Zhang, M.S. Balogun, J. Lu, S.Q. Song, Y.X. Tong, Polypyrrole-encapsulated amorphous Bi_2S_3 hollow sphere for long life sodium ion batteries and lithium-sulfur batteries, J. Mater. Chem. A 7 (2019) 11370-11378. https://doi.org/10.1039/c9ta01358j

[86] S.Z. Liang, X.Y. Wang, Y.G. Xia, S.L. Xia, E. Metwalli, B. Qiu, Q. Ji, S.S. Yin, S. Xie, K. Fang, Scalable synthesis of hierarchical antimony/carbon micro-/nanohybrid lithium/sodium-ion battery anodes based on dimethacrylate monomer, Acta Metall. Sin-Engl. 31 (2018) 910-922. https://doi.org/10.1007/s40195-018-0733-5

[87] D.H. Nam, M.J. Kim, S.J. Lim, I.S. Song, H.S. Kwon, Single-step synthesis of polypyrrole nanowires by cathodic electropolymerization, J. Mater. Chem. A 1 (2013) 8061-8068. https://doi.org/10.1039/c3ta11227f

Keyword Index

About the Editors

Dr. Inamuddin is currently working as Assistant Professor in the Chemistry Department, Faculty of Science, King Abdulaziz University, Jeddah, Saudi Arabia. He is a permanent faculty member (Assistant Professor) at the Department of Applied Chemistry, Aligarh Muslim University, Aligarh, India. He obtained Master of Science degree in Organic Chemistry from Chaudhary Charan Singh (CCS) University, Meerut, India, in 2002. He received his Master of Philosophy and Doctor of Philosophy degrees in Applied Chemistry from Aligarh Muslim University (AMU), India, in 2004 and 2007, respectively. He has extensive research experience in multidisciplinary fields of Analytical Chemistry, Materials Chemistry, and Electrochemistry and, more specifically, Renewable Energy and Environment. He has worked on different research projects as project fellow and senior research fellow funded by University Grants Commission (UGC), Government of India, and Council of Scientific and Industrial Research (CSIR), Government of India. He has received Fast Track Young Scientist Award from the Department of Science and Technology, India, to work in the area of bending actuators and artificial muscles. He has completed four major research projects sanctioned by University Grant Commission, Department of Science and Technology, Council of Scientific and Industrial Research, and Council of Science and Technology, India. He has published 162 research articles in international journals of repute and eighteen book chapters in knowledge-based book editions published by renowned international publishers. He has published 80 edited books with Springer (U.K.), Elsevier, Nova Science Publishers, Inc. (U.S.A.), CRC Press Taylor & Francis Asia Pacific, Trans Tech Publications Ltd. (Switzerland), IntechOpen Limited (U.K.), and Materials Research Forum LLC (U.S.A). He is a member of various journals' editorial boards. He is also serving as Associate Editor for journals (Environmental Chemistry Letter, Applied Water Science and Euro-Mediterranean Journal for Environmental Integration, Springer-Nature), Frontiers Section Editor (Current Analytical Chemistry, Bentham Science Publishers), Editorial Board Member (Scientific Reports-Nature), Editor (Eurasian Journal of Analytical Chemistry), and Review Editor (Frontiers in Chemistry, Frontiers, U.K.). He is also guest-editing various special thematic special issues to the journals of Elsevier, Bentham Science Publishers, and John Wiley & Sons, Inc. He has attended as well as chaired sessions in various international and national conferences. He has worked as a Postdoctoral Fellow, leading a research team at the Creative Research Initiative Center for Bio-Artificial Muscle, Hanyang University, South Korea, in the field of renewable energy, especially biofuel cells. He has also worked as a Postdoctoral Fellow at the Center of Research Excellence in Renewable Energy, King Fahd University of Petroleum and Minerals, Saudi Arabia, in the field of polymer electrolyte membrane fuel cells and computational fluid dynamics of polymer electrolyte membrane fuel cells. He is a life member of the Journal of the Indian Chemical Society. His research interest includes ion exchange materials, a sensor for heavy metal ions, biofuel cells, supercapacitors and bending actuators.

Dr. Rajender Boddula is currently working as Chinese Academy of Sciences-President's International Fellowship Initiative (CAS-PIFI) at National Center for Nanoscience and Technology (NCNST, Beijing). His academic honors include University Grants Commission National Fellowship and many merit scholarships, study-abroad fellowships from Australian Endeavour Research fellowship and CAS-PIFI. He has published many scientific articles in international peer-reviewed journals and has authored six book chapters, and also serving as editorial board member and referee for reputed international peer-reviewed journals. He has published edited books with Springer, United Kingdom, Elsevier, CRC Press Taylor & Francis Asia Pacific and Materials Research Forum LLC, U.S.A. His specialized areas of energy conversion and storage, which include nanomaterials, graphene, polymer composites, heterogeneous catalysis, photoelectrocatalytic water splitting, biofuel cell, and supercapacitors.

Dr. Mohd Imran Ahamed received his Ph.D degree on the topic "Synthesis and characterization of inorganic-organic composite heavy metals selective cation-exchangers and their analytical applications", from Aligarh Muslim University, Aligarh, India in 2019. He has published several research and review articles in the journals of international recognition. He has also edited various books which are published by Springer, CRC Press Taylor & Francis Asia Pacific and Materials Research Forum LLC, U.S.A. He has completed his B.Sc. (Hons) Chemistry from Aligarh Muslim University, Aligarh, India, and M.Sc. (Organic Chemistry) from Dr. Bhimrao Ambedkar University, Agra, India. His research work includes ion-exchange chromatography, wastewater treatment, and analysis, bending actuator and electrospinning.

Prof. Abdullah M. Asiri is the Head of the Chemistry Department at King Abdulaziz University since October 2009 and he is the founder and the Director of the Center of Excellence for Advanced Materials Research (CEAMR) since 2010 till date. He is the Professor of Organic Photochemistry. He graduated from King Abdulaziz University (KAU) with B.Sc. in Chemistry in 1990 and a Ph.D. from University of Wales, College of Cardiff, U.K. in 1995. His research interest covers color chemistry, synthesis of novel photochromic and thermochromic systems, synthesis of novel coloring matters and dyeing of textiles, materials chemistry, nanochemistry and nanotechnology, polymers and plastics. Prof. Asiri is the principal supervisors of more than 20 M.Sc. and six Ph.D. theses. He is the main author of ten books of different chemistry disciplines. Prof. Asiri is the Editor-in-Chief of King Abdulaziz University Journal of Science. A major achievement of Prof. Asiri is the discovery of tribochromic compounds, a new class of compounds which change from slightly or colorless to deep colored when subjected to small pressure or when grind. This discovery was introduced to the scientific community as a new terminology published by IUPAC in 2000. This discovery was awarded a patent from European Patent office and from UK patent. Prof. Asiri involved in many committees at the KAU level and on the national level. He took a major role in the advanced materials committee working for KACST to identify the national plan for science and technology in 2007. Prof. Asiri played a major role in advancing the chemistry education and research in KAU. He has been awarded the best researchers from KAU for the past five years. He also awarded the Young Scientist Award from the Saudi Chemical Society in 2009 and also the first prize for the distinction in science from the Saudi Chemical Society in 2012. He also received a recognition certificate from the American Chemical Society (Gulf region Chapter) for the advancement of chemical science in the Kingdome. He received a Scopus certificate for the most publishing scientist in Saudi Arabia in chemistry in 2008. He is also a member of the editorial board of various journals of international repute. He is the Vice- President of Saudi Chemical Society (Western Province Branch). He holds four USA patents, more than one thousand publications in international journals, several book chapters and edited books.